Lecture Notes in Mathematics

A collection of informal reports and seminars
Edited by A. Dold, Heidelberg and B. Eckmann, Zürich

72

The Syntax and Semantics of Infinitary Languages

Edited by Jon Barwise
Yale University, New Haven, Connecticut

1968

Springer-Verlag Berlin · Heidelberg · New York

Library of Congress Catalog Card Number 68-57175 Title No. 3678

This book grew out of a symposium on infinitary logic held at UCLA on December 28-30, 1967. The purpose of the book is to make much of the current work in infinitary logic available to the student as quickly as possible. Many of the papers represent work in progress; it is expected that they will appear elsewhere in final form.

Jon Barwise

Jon Barwise

CONTENTS

1. Jon Barwise, Implicit definability and compactness in infinitary languages 1

2. C. C. Chang, Some remarks on the model theory of infinitary languages 36

3. Erwin Engeler, Remarks on the theory of geometrical constructions 64

4. Harvey Friedman and Ronald Jensen, Note on admissible ordinals 77

5. Carol Karp, An algebraic proof of the Barwise compactness theorem 80

6. H. J. Keisler, Formulas with linearly ordered quantifiers 96

7. R. D. Kopperman and A. R. D. Mathias, Some problems in group theory 131

8. G. Kreisel, Choice of infinitary languages by means of definability criteria; Generalized recursion theory 139

9. David W. Kueker, Definability, automorphisms, and infinitary languages 152

10. Jerome Malitz, The Hanf number for complete $L_{\omega_1,\omega}$ sentences 166

11. A. Preller, Quantified algebras 182

12. W. W. Tait, Normal derivability in classical logic 204

13. Gaisi Takeuti, A determinate logic 237

14. Joseph Weinstein, (ω_1,ω) properties of unions of models 265

IMPLICIT DEFINABILITY AND COMPACTNESS IN INFINITARY LANGUAGES

JON BARWISE

In this paper we continue the investigation begun in [1] of infinitary logic and admissible sets. Our task is an analysis of the compactness results first discovered in [1] for countable languages, in the hope of extending them into the uncountable. In this we have been partially successful, though the subject is far from complete. It is true however, that the compactness phenomenon is nowhere near as singular an occurance as suggested by the negative results of Hanf [5], provided one is more careful in the choice of basic notions.

In §1 we introduce the infinitary languages $\mathcal{L}_A$ which we wish to discuss, and review some of the results known from [1] for the case where A is a countable admissible set. In §2 we continue the investigation begun by Kunen [9] on implicit definability, relating it to compactness. We use the results of §2 in §3 to prove a compactness theorem for uncountable languages. In an appendix we pursue briefly some topics in recursion on sets suggested by results in §§2 and 3.

We are greatly indebted to Kenneth Kunen and Professor Georg Kriesel for interesting conversations on the topic of this paper. The paper was written while the author was an N.S.F. Postdoctoral Fellow.

1. Introduction and review of the countable case.

In this paper we are concerned with certain sublanguages $\mathcal{L}_A$ of the language $\mathcal{L}$ $(= L_{\infty,\omega})$ which allows finite strings of quantifiers and arbitrary conjunction and disjunctions. We consider formulas φ of $\mathcal{L}$ to be sets, and the language $\mathcal{L}_A$ is $\mathcal{L} \cap A$. For example, for any cardinal κ we let $H(\kappa)$ be the set of sets x such that the transitive closure of x, $TC(x)$, has cardinality less than κ. For κ regular $\mathcal{L}_{H(\kappa)}$ is the language usually denoted by $L_{\kappa,\omega}$.

To make this precise we now define the language $\mathcal{L}$. By (x,y) we mean the usual ordered pair $\{\{x\},\{x,y\}\}$. For each set b and natural number n we have:

(a) a constant symbol $(0,b)$, denoted by c_b

(b) a variable $(1,b)$, denoted by v_b

(c) an n-ary relation symbol $(2,b,n)$, denoted by $\underline{R}_{b,n}$, $\underline{R}_b$, or by $\underline{R}$, $\underline{S}$, etc., if no confusion can arise.

(d) an n-ary function symbol $(3,b,n)$, denoted by $\underline{f}_{b,n}$, or simply by $\underline{f}$, $\underline{g}$ etc.

The class of <u>terms</u> of $\mathcal{L}$ is built up as usual from constants, variables and function symbols. For example, if $t_1,\ldots,t_n$ are terms and $\underline{f}$ is an n-ary function symbol, then $(\underline{f},t_1,\ldots,t_n)$ is a term, denoted by $\underline{f}(t_1,\ldots,t_n)$. Similarly, if $t_1,\ldots,t_n$ are terms and $\underline{R}$ is an n-ary relation symbol then $(\underline{R},t_1,\ldots,t_n)$ is an <u>atomic formula</u>, denoted by $\underline{R}(t_1,\ldots,t_n)$.

DEFINITION 1.1. The class $\mathcal{L}$ of finite quantifier formulas is defined inductively as follows:

(a) if φ is an atomic formula, then φ is a formula;

(b) if φ is a formula, then $(4,\varphi)$ is a formula, denoted by $\neg\varphi$;

(c) if φ is a formula and v a variable, then $(5,v,\varphi)$ and $(6,v,\varphi)$ are formulas, denoted by $\forall v\varphi$ and $\exists v\varphi$ respectively;

(d) if Γ is a <u>set</u> of formulas, then $(7,\Gamma)$ and $(8,\Gamma)$ are formulas, denoted by $\bigwedge\Gamma$ and $\bigvee\Gamma$ respectively.

1.2. For any transitive set A, let $\mathcal{L}_A$ be $\mathcal{L}\cap A$.

$\bigwedge\Gamma$ is to be the conjunction of the formulas in Γ, $\bigvee\Gamma$ the disjunction. Thus conjunctions and disjunctions are to be taken over sets, not well-ordered sequences. We often refer to elements of $\mathcal{L}_A$ as A-formulas. The notion of a variable being free or bounded is defined as usual. A <u>sentence</u> of $\mathcal{L}_A$ is a formula with no free variables. The set of such is denoted by $\mathcal{S}_A$.

Satisfaction of formulas is defined in the usual way. If $\mathfrak{A}$ is a structure, f a function which assigns elements of $\mathfrak{A}$ to the free variables of φ, then we abbreviate "f satisfies φ in $\mathfrak{A}$" by $\mathfrak{A}\models\varphi[f]$. If φ has only a finite number of free variables, say $v_1,\ldots,v_n$, we write

$$\mathfrak{A}\models\varphi[a_1,\ldots,a_n]$$

for $\mathfrak{A} \models \varphi[f]$, where $a_i = f(v_i)$. A formula is valid if for all $\mathfrak{A}$ and f, $\mathfrak{A} \models \varphi[f]$. If φ is a sentence, we use $\mathfrak{A} \models \varphi$ as an abbreviation for "φ is true in $\mathfrak{A}$". We assume that the binary relation $\underline{R}_{0,2}$ is always interpreted as the equality relation. We denote $\underline{R}_{0,2}(t_1,t_2)$ by $t_1 \approx t_2$. In the future we shall refer to $\approx$ as the equality symbol, not as a relation symbol.

In order to insure that $\mathfrak{L}_A$ is a sensible language, we must require that A satisfy certain closure conditions. We need the following terminology to express these closure conditions. By the language of set theory, we mean the usual finitary language with ε (membership) the only relation symbol; equality is taken as defined. By an extended language of set theory, we mean the finitary language which allows some relation symbols $\underline{S}_0,\ldots,\underline{S}_{k-1}$ in addition to ε; $\underline{S}_i$ is n_i-ary. We use θ, with or without subscripts, to range over formulas in the (possibly extended) language of set theory. Notice that ε is used as a formal symbol whereas $\in$ is used in our metalanguage.

DEFINITION 1.3.

(a) The $\Delta_0(\underline{S}_1,\ldots,\underline{S}_k)$-formulas of set theory form the smallest collection Y such that

(i) if θ is atomic (i.e., if θ is $x \varepsilon y$ or $\underline{S}_i(x_1,\ldots,x_n)$) then θ and $\neg\theta$ are in Y

(ii) if θ_0 and θ_1 are in Y, then so are $\theta_0 \vee \theta_1$ and $\theta_0 \wedge \theta_1$

(iii) if θ is in Y, then so are $\forall x[x \varepsilon y \rightarrow \theta]$ and $\exists x[x \varepsilon y \wedge \theta]$, (denoted by $\forall x \varepsilon y \theta$ and $\exists x \varepsilon y \theta$ respectively).

(b) The $\Sigma(\underline{S}_1,\ldots,\underline{S}_k)$-formulas of set theory form the smallest collection Y closed under (i), (ii), (iii) and

(iv) if θ is in Y, $\exists x\theta$ is in Y.

(c) The $\Sigma_1(\underline{S}_1,\ldots,\underline{S}_k)$-formulas of set theory form the smallest collection Y containing the $\Delta_0(\underline{S}_1,\ldots,\underline{S}_k)$ formulas and closed under (iv).

We call θ a Δ_0-formula if it is a $\Delta_0(\ )$-formula, i.e., if $k = 0$ in the above. Similarly for Σ and Σ_1-formulas. What we call a Σ-formula was called Σ_1^* by Platek [12]; our notation follows that of Feferman-Kreisel [2].

To see the importance of these classes of formulas we recall the definition of end-extension of Feferman-Kreisel [2]. (The terminology "end extension" is due to Gaifman). Let $\langle A;E,S_1,\ldots,S_k\rangle$ be a relational system with E a binary relation on A. A relational system $\langle A';E',S_1',\ldots,S_k',T_1',\ldots,T_\ell'\rangle$

is an <u>end extension</u> of $\langle A;E,S_1,\ldots,S_k\rangle$ if:

$$\langle A';E',S_1',\ldots,S_k'\rangle \text{ is an extension of } \langle A;E,S_1,\ldots,S_k\rangle$$

and

$$\text{if } aE'b \text{ and } b \in A \text{ then } a \in A .$$

For any set A we let ϵ_A denote the membership relation restricted to A, i.e., the set $\{(x,y) \mid x,y \in A \text{ and } x \in y\}$. Then if $A \subseteq A'$ and A is transitive, then $\langle A';\epsilon_{A'}\rangle$ is an end-extension of $\langle A;\epsilon_A\rangle$ by the above definition.

If $a_1,\ldots,a_n \in A$ satisfy the Σ-formula $\theta(x_1,\ldots,x_n)$ in $\langle A;E\rangle$, and if $\langle A',E'\rangle$ is an end extension of $\langle A;E\rangle$, then $a_1,\ldots,a_n$ satisfy $\theta(x_1,\ldots,x_n)$ in $\langle A',E'\rangle$. One of the main results of Feferman-Kreisel [2] is the converse: if θ is preserved under end extensions, then θ is equivalent to a Σ-formula. We will not need this result, however.

For any formula θ and variable y of set theory, $\theta^{(y)}$ is the Δ_0-formula obtained from θ by relativizing all quantifiers in θ to y. And finally, for any set y, the transitive closure of y, $TC(y)$, is the least transitive set x with $y \subseteq x$. That is

$$TC(y) = y \cup (\bigcup y) \cup (\bigcup\bigcup y) \cup \cdots .$$

If one were interested merely in the syntax of the languages $\mathcal{L}_A$, the most natural development would be to show that all syntatic functions and relations on $\mathcal{L}$ are set primitive recursive (in the sense of Platek [12] or Jensen-Karp [6]) and then require that the set A be closed under all set primitive recursive functions. We, however, are interested in those $\mathcal{L}_A$ which have nice semantic, as well as syntactic, properties. This leads us naturally to study $\mathcal{L}_A$ for admissible sets A. The definition of admissible set is due to Platek [12].

DEFINITION 1.4. A nonempty transitive set A is <u>rudimentary</u> if A satisfies the following:

(a) if $a,b \in A$, then $a \times b$ and $TC(\{a\})$ are in A

(b) (<u>Δ_0-separation</u>) if θ is any Δ_0-formula any y is a variable not free in θ, then the following is universally true in A:

$$\exists y \forall x[x \in y \leftrightarrow x \in w \wedge \theta] .$$

1.5. A is admissible if A is rudimentary and satisfies:

(c) ($\underset{\sim}{\Sigma}$-reflection principle) if θ is a Σ-formula and y is a variable not free in θ, then the following is universally true in A:

$$\theta \to \exists y[y \text{ is transitive } \wedge \theta^{(y)}] .$$

If R is an n-ary relation on A and $\underline{R}$ an n-ary relation symbol then A is R-rudimentary if 1.4(b) holds with Δ_0 replaced by $\Delta_0(\underline{R})$. A is R-admissible if A is R-rudimentary and 1.5(c) holds with Σ replaced by $\Sigma(\underline{R})$.

Let A be a transitive set. We say that a set $X \subseteq A$ is Σ_1 on A if there is a Σ_1-formula which defines X on $\langle A,\in\rangle$. X is $\underset{\sim}{\Sigma}_1$ on A if it is definable by a Σ_1 formula with parameters from A; that is, if there is a Σ_1-formula $\theta(x,y_1,\ldots,y_n)$ and elements $b_1,\ldots,b_n \in A$ such that

$$X = \{a \in A \mid \langle A,\in\rangle \models \theta[a,b_1,\ldots,b_n]\} .$$

(We read "X is Σ_1 in parameters"). The notions Δ_0, $\underset{\sim}{\Delta}_0$, Σ and $\underset{\sim}{\Sigma}$ on A are defined analogously. A set is Δ [resp. Δ_1] on A if both X and $A \sim X$ are Σ [resp. Σ_1] on A. Notice that if A is admissible then every $\underset{\sim}{\Sigma}$ subset of A is already a $\underset{\sim}{\Sigma}_1$ subset, since

$$\theta(a) \leftrightarrow \exists w[w \text{ is transitive } \wedge \theta^{(w)}(a)]$$

holds in A for all Σ formulas θ.

The above terminology makes it clear why we called 1.5(c) the $\underset{\sim}{\Sigma}$-reflection principle rather than the Σ-reflection principle. By the Σ-reflection principle we should mean 1.5(c) restricted to θ which are Σ-sentences of set theory. We shall, however, have no cause to consider the Σ-reflection principle.

If A is rudimentary and $a,b \in A$, then $\{a,b\}$, $a \cup b$ and $a \sim b$ are in A. Thus, every finite subset of A is an element of A. In particular, $H(\omega) \subseteq A$. If A is rudimentary, then $\mathcal{L}_A$ has the following closure properties:

(i) if $\varphi \in \mathcal{L}_A$ then $\neg\varphi \in \mathcal{L}_A$.

(ii) if $\varphi \in \mathcal{L}_A$ and $a \in A$ then $(\forall v_a\, \varphi) \in \mathcal{L}_A$.

(iii) if Γ is a finite subset of $\mathcal{L}_A$ then $\bigwedge\Gamma \in \mathcal{L}_A$.

(iv) if $\Gamma \subseteq \mathcal{L}_A$, $\Gamma \in A$ then $\bigwedge\Gamma \in \mathcal{L}_A$.

Furthermore, if A is rudimentary, and $\Gamma = \{x \approx c_b \mid b \in a\}$ where $a \in A$, then $\Gamma \in A$. Hence for $a \in A$, the sentence

$$\forall x[x \in c_a \leftrightarrow \bigvee_{b \in a} x \approx c_b]$$

is in $\mathcal{L}_A$. The set of such sentences, for $a \in A$, is a Σ_1 (in fact, a Δ_0) subset of A.

The notion of admissible set was introduced by Platek in [12] for the study of recursively regular (i.e. admissible) ordinals. Platek develops a recursion theory on admissible sets by calling a function F (with domain and range subsets of A) A-<u>recursive</u> its graph is $\underset{\sim}{\Sigma}_1$ on A. A set $X \subseteq A$ is A-<u>recursive</u> if its characteristic function is A-recursive, and X is A-<u>recursively enumerable</u> (A-r.e.) if it is the range of an A-recursive function. A set $X \subseteq A$ is A-<u>finite</u> if $X \in A$. It is easy to see that $X \subseteq A$ is A-recursively enumerable just in case X is $\underset{\sim}{\Sigma}_1$ on A and is A-recursive just in case X is $\underset{\sim}{\Delta}_1$ on A. However, we shall use the terms A-recursive, A-recursively enumerable and A-finite only in the case where A is countable; the reasons for this will be explained in §2.

To see the way in which admissible sets arise in the study of infinitary logic, we make the following definition.

DEFINITION 1.6. A rudimentary set A is <u>$\underset{\sim}{\Sigma}_1$-compact</u> if for every $\underset{\sim}{\Sigma}_1$ set Φ of sentences of $\mathcal{L}_A$, either Φ has a model or else there is some $\Phi_0 \subseteq \Phi$ with $\Phi_0 \in A$ such that Φ_0 has no model.

A is Σ_1-compact if the above holds with $\underset{\sim}{\Sigma}_1$ replaced by Σ_1. If A is rudimentary and R is a relation on A, then A is $\underset{\sim}{\Sigma}_1(R)$-compact if the above holds with $\underset{\sim}{\Sigma}_1$ replaced by $\underset{\sim}{\Sigma}_1(R)$. These notions are refinements of the usual notions of compactness. For example it is not difficult to see that an inaccessible cardinal κ is weakly compact if and only if $H(\kappa)$ is $\underset{\sim}{\Sigma}_1(R)$-compact for all relations R on $H(\kappa)$.

THEOREM 1.7. <u>If</u> A <u>is</u> <u>$\underset{\sim}{\Sigma}_1$-compact then</u> A <u>is admissible. More generally, if</u> A <u>is</u> <u>$\underset{\sim}{\Sigma}_1(R)$-compact then</u> A <u>satisfies the</u> $\underset{\sim}{\Sigma}(R)$ <u>reflection principle</u>.

<u>Proof</u>. Suppose that $R \subseteq A^{\ell}$ and A is $\underset{\sim}{\Sigma}_1(R)$-compact. Let $\underline{R}$ be an ℓ-ary relation symbol and $\theta(x_1,\ldots,x_n)$ a $\Sigma(\underline{R})$ formula such that

$$\langle A,\epsilon,R\rangle \models \theta[a_1,\ldots,a_n]$$

for some elements $a_1,\ldots,a_n \in A$. We need to find a transitive set $w \in A$ such that $\theta^{(w)}$ holds in A at $a_1,\ldots,a_n$; that is, such that

$$\langle w,\epsilon_w,R \cap w^{\ell}\rangle \models \theta[a_1,\ldots,a_n] .$$

We can consider θ as a formula of $\mathcal{L}_A$ and let θ' be the sentence

$$\theta(c_{a_1},\ldots,c_{a_n})$$

of $\mathcal{L}_A$. Let Φ_1 be the set of the following sentences:

$$\forall x[x \in c_a \leftrightarrow \bigvee_{b\in a} x \approx c_b]$$

for all $a \in A$. Since A is rudimentary, $\Phi_1 \subseteq \mathcal{L}_A$ and Φ_1 is Σ_1 on A. Let Φ_2 be the set of the following sentences:

$$\underline{R}(c_{a_1},\ldots,c_{a_\ell}) \quad \text{for} \quad (a_1,\ldots,a_\ell) \in R,$$

$$\neg\underline{R}(c_{a_1},\ldots,c_{a_\ell}) \quad \text{for} \quad (a_1,\ldots,a_\ell) \notin R.$$

Φ_2 is $\Sigma_1(R)$ on A. Thus $\Phi = \Phi_1 \cup \Phi_2 \cup \{\neg\theta'\}$ is $\underset{\sim}{\Sigma}_1(R)$ on A. (We need the $\sim$ because of the parameters in θ'). Every model of $\Phi_1 \cup \Phi_2$ is an end extension of $\langle A,\epsilon,R\rangle$. Since Σ formulas are preserved under end extensions Φ can have no model. Let $\Phi_0 \subseteq \Phi$, $\Phi_0 \in A$ be such that Φ_0 has no model, and let $w = TC(\Phi_0)$. Now since $w \in A$ and A is transitive, $w \subseteq A$. Hence $\langle w,\epsilon_w,R \cap w^2\rangle$ is a model of $\Phi_0 \cap (\Phi_1 \cup \Phi_2) = \Phi_0 - \{\neg\theta'\}$. Since Φ_0 has no model, $\neg\theta'$ is not true in $\langle w,\epsilon_w,R \cap w^2\rangle$. Hence θ' holds in this structure, as desired.

REMARKS. 1. If A is compact for all arbitrary sets Φ of $\mathcal{L}_A$ sentences, then A satisfies the $\underset{\sim}{\Sigma}(R)$ reflection principle with respect to all predicates R on A. One easy consequence of this is that $\bar{\bar{a}} < \bar{\bar{A}}$ for all $a \in A$. In particular, the only countable set A for which we could possibly have compactness for arbitrary sets Φ is $A = H(\omega)$, i.e., for the usual finitary first order predicate calculus.

2. The careful reader will have observed that the proof of 1.7 actually establishes a reflection principle much stronger than the $\underset{\sim}{\Sigma}(R)$ reflection principle. See Theorem 2.8.

3. Note that in the proof of 1.7 we did not need $\Phi_0 \in A$, but only that $\Phi_0 \subseteq a$ for some $a \in A$. This observation will be useful when we turn to uncountable sets.

We devote the rest of this section to a review of results about $\mathcal{L}_A$ for admissible sets A which are subsets of $H(\omega_1)$, where ω_1 is the first uncountable cardinal. Thus we are concerned with certain sublanguages of $L_{\omega_1,\omega}$. The proofs of these results appear in [1].

In [11], Lopez-Escobar established the completeness of a Gentyen type system for $L_{\omega_1,\omega}$, and using this obtained the interpolation theorem and Beth's definability theorem for $L_{\omega_1,\omega}$. In [1] we show how to use his completeness theorem to obtain the following results.

THEOREM 1.8. Let A be admissible, $A \subseteq H(\omega_1)$.

(1) (COMPLETENESS). The set of valid sentences of $\mathcal{L}_A$ is Σ_1 on A.

(2) (INTERPOLATION). If $\varphi \to \psi$ is a valid sentence of $\mathcal{L}_A$ then there is a sentence η of $\mathcal{L}_A$ which has constant and relation symbols common to both φ and ψ, and such that $\varphi \to \eta$ and $\eta \to \psi$ are valid.

(3) (DEFINABILITY). Let $\varphi(\underline{R})$ be a sentence of $\mathcal{L}_A$ involving the n-ary relation symbol $\underline{R}$, and let $\varphi(\underline{S})$ be obtained from $\varphi(\underline{R})$ by replacing $\underline{R}$ by a new n-ary relation symbol $\underline{S}$. If

$$\varphi(\underline{R}) \wedge \varphi(\underline{S}) \to \forall x_1,\ldots,x_n\, [\underline{R}(x_1,\ldots,x_n) \leftrightarrow \underline{S}(x_1,\ldots,x_n)]$$

is valid, then there is a formula ψ of $\mathcal{L}_A$ with free variables $x_1,\ldots,x_n$ such that

$$\varphi(\underline{R}) \to \forall x_1,\ldots,x_n\, [\underline{R}(x_1,\ldots,x_n) \leftrightarrow \psi]$$

is valid.

Theorems 1.8(2) and 1.8(3) have interesting applications to the theory of definable subsets of ω and the Baire space ω^ω. We hope to make this the subject of a future paper.

If A is countable, then we have the following converse of 1.7.

COMPACTNESS THEOREM 1.9. Let A be countable. If A is admissible then A is Σ_1-compact.

<u>More generally, if</u> A <u>is</u> R-<u>admissible then</u> A <u>is</u> $\Sigma_1(R)$-<u>compact</u>.

REMARKS 1. This a complete generalization of the compactness theorem for $\mathcal{L}_{H(\omega)}$. For $H(\omega)$ is R-admissible for every predicate R on $H(\omega)$, so 1.9 gives compactness for arbitrary (countable) sets of finite sentences.

2. An interesting special case of 1.9 is where $\langle A,\epsilon\rangle$ is a countable transitive model of ZF. Then A is R-admissible for every definable relation R on A, and hence we have compactness for arbitrary definable sets of $\mathcal{L}_A$ sentences. This remark should have interesting applications in constructing models of set theory.

3. For an elegant application of 1.9 to the theory of admissible ordinals we refer the reader to Friedman and Jensen [3] in this volume.

Combining 1.9 with 1.8(1) we have the following corollary. We say that a sentence φ is a consequence of a set Φ, and write $\Phi \models \varphi$, if every model of Φ is a model of φ.

COROLLARY 1.10. <u>Let</u> A <u>be a countable admissible set. If</u> Φ <u>is a</u> $\underset{\sim}{\Sigma}_1$ <u>set of sentences of</u> $\mathcal{L}_A$ <u>then the set of</u> $\mathcal{L}_A$ <u>sentences which are consequences of</u> Φ <u>is also</u> $\underset{\sim}{\Sigma}_1$ <u>on</u> A. <u>Similarly with</u> $\underset{\sim}{\Sigma}_1$ <u>replaced by</u> Σ_1 <u>throughout</u>.

This result fails for $A = H(\omega_1)$. For an application of 1.10 see Kunen [9], Theorem 4.5.

If an admissible set A is countable then $\mathcal{L}_A$ is countable. Thus for some model theoretic purposes it is to be expected that the $\mathcal{L}_A$ will be more convenient than $L_{\omega_1,\omega}$ which has $2^{\aleph_0}$ formulas. For example, H. J. Keisler has established an analogue of Morely's theorem on categoricity for these languages.

Most of the sentences of $L_{\omega_1,\omega}$ which describe interesting algebraic structures are already sentences of $\mathcal{L}_A$ where A is the smallest admissible set different from $H(\omega)$. And every formula of $L_{\omega_1,\omega}$ is in $\mathcal{L}_A$ for some countable admissible set A. Knowing that a sentence φ of $L_{\omega_1,\omega}$ is actually in $\mathcal{L}_A$ for a particular countable admissible set A often gives one additional information. As an example we consider the question of Hanf numbers.

By the <u>Hanf number</u> of $\mathcal{L}_A$ we mean the least cardinal number κ such that for all sentences φ of $\mathcal{L}_A$, if φ has a model of cardinality κ then φ has models of all infinite cardinalities.

Let $R(\alpha) = \bigcup \{\mathcal{P}(R_\beta) : \beta < \alpha\}$ for each ordinal α, where $\mathcal{P}$ is the power set operation, and let $\beth_\alpha = \overline{\overline{R(\alpha)}}$. This definition of $\beth_\alpha$ differs from the usual one, but only for $\alpha < \omega^2$. We use this definition so that $\beth_\omega = \aleph_0$.

The following theorem is proved in §5 of the author's thesis. It extends the known cases of $\mathcal{L}_{H(\omega)}$ and $\mathcal{L}_{H(\omega_1)}$.

THEOREM 1.11. Let A be an admissible set, $A \subseteq H(\omega_1)$ and let α be the least ordinal not an element of A. The Hanf number of $\mathcal{L}_A$ is $\beth_\alpha$.

What one actually shows is that if $\varphi \in \mathcal{L}_A$ has models of all cardinalities $< \beth_\alpha$ then φ has models of all cardinalities. The method of proof is that used by Morley to obtain the Hauf member for single sentences of ω-logic.

The proof given in [1] has been generalized, independently by K. Kunen and the author, to obtain a description of the Hauf number of $\mathcal{L}_A$ for arbitrary admissible sets A. (For A's of the form $H(\kappa)$, this result was also observed by Morley.) Theorem 1.11 can then be derived from this new result using Compactness Theorem 1.9. We shall not pursue this matter further here.

2. Strict Π_1^1 Predicates and Compactness.

In §1 we assumed that $\underset{\sim}{\Sigma}_1$ and $\underset{\sim}{\Delta}_1$ were reasonable generalizations of r.e. and recursive on an admissible set, or at least on a countable admissible set. For the study of $\mathcal{L}_A$ for uncountable A these generalizations have unpleasant properties (see, e.g., the negative results of Karp [7] or the remark following 1.10). K. Kunen, in [9], has introduced alternate generalizations of r.e. and recursive for an arbitrary admissible set, namely, semi-invariantly implicitly definable (s.i.i.d.) and invariantly implicitly definable (i.i.d.). Using our result 1.10 above, Kunen shows that for countable admissible sets these notions coincide with Σ_1 and Δ_1 respectively. In general, however, they are much wider, and the unpleasant features of Σ_1 and Δ_1 for the study of $\mathcal{L}_A$ disappear when replaced by s.i.i.d. and i.i.d. respectively.

In this section we continue the investigation begun by Kunen in [9]. We obtain an exact characterization of s.i.i.d. predicates (and hence i.i.d. predicates) in terms of quantifier level (Theorem 2.4). This characterization suggests a strengthening of the $\underset{\sim}{\Sigma}$-reflection principle which we call the strict $\underset{\sim}{\Pi}_1^1$ reflection principle. In Theorem 2.6 we prove that this reflection principle is equivalent to $\underset{\sim}{\Sigma}_1$-compactness. The equivalence of the strict $\underset{\sim}{\Pi}_1^1$ reflection principle with $\underset{\sim}{\Sigma}_1$-compactness leads us in §3 to discover new admissible sets which are $\underset{\sim}{\Sigma}_1$-compact (Theorems 3.2

and 3.3).

Let us recall Kunen's definition of s.i.i.d. Let $\mathfrak{A} = \langle A;E,R\rangle$ be a structure for the language of set theory (i.e. $E \subseteq A \times A$), let P be an n-ary relation on A, and let θ be a (finite) sentence with relation symbols ε, $\underline{R}$, $\underline{P}$ corresponding to E, R, P as well as auxiliary relation symbols $\underline{S}_1,\ldots,\underline{S}_m$. Then θ semi-invariantly implicitly (s.i.i.) defines P from R if and only if

(1) there are relations $S_1,\ldots,S_m$ on A such that $(\mathfrak{A},P,S_1,\ldots,S_m) \models \theta$, and

(2) if $\mathfrak{A}' = \langle A',E',R',P',S_1',\ldots,S_m'\rangle$ is an end-extension of $\mathfrak{A}$ and $\mathfrak{A}' \models \theta$ then $P \subseteq P'$.

P is semi-invariantly implicitly definable (s.i.i.d.) from R if there is some θ which s.i.i. defines P from R.

Let $\mathfrak{A} = \langle A,\ldots\rangle$ be a structure and θ some formula with relation symbols $\underline{S}_1,\ldots,\underline{S}_m$ not having interpretations in $\mathfrak{A}$. We write

$$\mathfrak{A} \models \theta[a_1,\ldots,a_n]$$

to mean that for all relations $S_1,\ldots,S_m$ on A,

$$\langle A,\ldots,S_1,\ldots,S_n\rangle \models \theta[a_1,\ldots,a_n] .$$

Thus a predicate P is Π^1_1 on $\mathfrak{A}$ just in case there is some (finite) formula θ such that

$$P(a_1,\ldots,a_n) \text{ if and only if } \mathfrak{A} \models \theta[a_1,\ldots,a_n] .$$

In [9], Kunen shows that $\Sigma_1 \subseteq$ s.i.i.d. $\subseteq \Pi^1_1$. Since he found examples where both extremes were achieved, it looked as though this were the best possible result.

DEFINITION 2.1. Let $\mathfrak{A} = \langle A,E,R\rangle$ be a structure for the language of set theory. A predicate P on $\mathfrak{A}$ is strict Π^1_1 in R if there is some $\Sigma(\underline{R},\underline{S}_1,\ldots,\underline{S}_k)$ formula θ of set theory such that

$$P(a_1,\ldots,a_n) \text{ if and only if } \mathfrak{A} \models \varphi[a_1,\ldots,a_n] .$$

The key fact about strict Π_1^1 predicates, used repeatedly in the following, is expressed by the following simple lemma.

LEMMA 2.2. Let $\mathfrak{A} = \langle A,E,R\rangle$ and let $\mathfrak{A}' = \langle A',E',R'\rangle$ be an end-extension of $\mathfrak{A}$. If θ is a $\Sigma(\underline{R},\underline{S},\ldots,\underline{S})$ formula and $a_1,\ldots,a_n \in A$, then

$$\mathfrak{A} \models \theta[a_1,\ldots,a_n] \text{ implies } \mathfrak{A}' \models \theta[a_1,\ldots,a_n] .$$

The following will also be important for what follows.

LEMMA 2.3. Let A be an admissible set and f any Σ_1 function mapping constants of $\mathcal{L}_A$ into elements of A. Then there is a $\Sigma(\underline{R}_1,\ldots,\underline{R}_k,\underline{S})$ sentence ψ such that for all $R_1,\ldots,R_k,S$ the following are equivalent:

(a) $\langle A,\epsilon,R_1,\ldots,R_k,S\rangle \models \neg\psi$.

(b) S is a (the) satisfaction relation for all quantifier free sentences of $\mathcal{L}_A$ with respect to the structure $\langle A,R_1,\ldots,R_k\rangle$ and assignment f of individuals to constant symbols.

If one accepts s.i.i.d. as a reasonable generalization of r.e., then our next theorem is a generalization of the following observation of Kreisel: the r.e. predicates (on ω) are just those definable in the form $\forall f\, \exists n\, R(\overline{f}(n),x)$ where R is recursive and f ranges over 2^ω. In our terminology, with ω replaced by $H(\omega)$, this says that the strict Π_1^1 predicates on $H(\omega)$ coincide with the Σ_1 predicates.

THEOREM 2.4. Let A be an admissible set. A relation P is s.i.i.d. on A if and only if P is strict Π_1^1.

Proof. Suppose that P is strict Π_1^1 on A, defined by the $\Sigma(\underline{S}_1,\ldots,\underline{S}_k)$ formula $\varphi(x)$. Let $\Phi = \{\forall x(x \in c_a \leftrightarrow \bigvee_{b\in a} x \approx c_b) : a \in A\}$. The following are then equivalent:

$$\begin{aligned}
&P(a)\\
&\langle A,\epsilon\rangle \models \varphi[a]\\
&\mathfrak{A}' \models \varphi[a] \text{ for all end extensions } \mathfrak{A}' \text{ of } \langle A,\epsilon\rangle\\
&\mathfrak{A}' \models \varphi[a] \text{ for all models } \mathfrak{A}' \text{ of } \Phi\\
&\Phi \models \varphi(c_a) .
\end{aligned}$$

But since Φ is Σ_1, the set of consequence of Φ is s.i.i.d by Theorem 3.2 of [9]. Thus P is s.i.i.d.

For the converse, suppose that P_0 is s.i.i. defined by the sentence $\varphi(\underline{P},\underline{Q})$. Put φ into universal Skolem normal form

$$\forall x_1,\ldots,x_k\ \varphi_0(x_1,\ldots,x_k)$$

using relation symbols $\underline{R}_1,\ldots,\underline{R}_m$ so that φ_0 is quantifier-free and

$$\forall x_1 \cdots \forall x_n \varphi_0 \wedge \bigwedge_{i=1}^{m} \forall x_1 \cdots \forall x_{n_i}\ \exists!\, y\ \underline{R}_i(x_1,\ldots,x_{n_i},y)$$

s.i.i. defines P_0. Call this sentence φ_1. We consider equality to be defined in terms of $\in$ in the language of set theory so that $\approx$ does not occur in φ or φ_0. Let φ_2 result from φ_0 by replacing $\in$ by a binary relation symbol $\underline{E}$. Assume for simplicity that φ_0 contains only one ternary Skolem relation $\underline{R}$. Let $\underline{I}$ be a new binary relation symbol used as a congruence relation in the following. We define a set C of constants of $\mathcal{L}_A$ as follows:

(i) for each $a \in A$, $c_{(0,a)} \in C$, denoted by $\overline{a}$

(ii) for each $t_1,t_2 \in C$, $c_{(t_1,t_2)} \in C$, denoted by $t_1 * t_2$.

We use $\overline{a}$ as a name for a and $t_1 * t_2$ as a name for the value of the function (denoted by) $\underline{R}$ applied to the arguments (denoted by) t_1 and t_2. So let Φ be the Δ_0 set of the following quantifier free sentences of $\mathcal{L}_A$:

(1) $\varphi_2(t_1,\ldots,t_k)$ for $t_1,\ldots,t_k \in C$

(2) the appropriate axioms to make $\underline{I}$ a congruence relation with respect to $\underline{E}$, $\underline{P}$, $\underline{Q}$, $\underline{R}$ and the terms in C.

(3) $\underline{R}(t_1,t_2,t_3) \wedge \underline{R}(t_1,t_2,t_4) \to \underline{I}(t_3,t_4)$ for $t_1,\ldots,t_4 \in C$

(4) $\underline{R}(t_1,t_2,t_1 * t_2)$ for $t_1,t_2 \in C$

(5) $\underline{E}(\overline{a},\overline{b})$ for $a,b \in A$, $a \in b$

(6) $\neg\underline{E}(\overline{a},\overline{b})$ for $a,b \in A$, $a \notin b$

(7) $\underline{E}(t_1 * t_2,\overline{a}) \to \bigvee_{b\in a} I(t_1 * t_2,\overline{b})$ for $a \in A$; $t_1,t_2 \in C$.

We claim that for all $a \in A$, the following are equivalent, where C is the least subset of A containing each $(0,a)$ for $a \in A$ and closed under ordered pairs:

2.4.1. $a \in P_0$

2.4.2. if $\langle C,E,P,Q,R,I \rangle$ is a model of Φ then $(0,a) \in P$.

The conclusion of the theorem will follow from this equivalence, together with Lemma 2.3, for

> $a \in P_0$ if and only if for all relations E, I, P, Q, R, S, if S is a satisfaction relation for $\langle C,E,I,P,Q,R \rangle$ which assigns true to all $\varphi \in \Phi$, then $(0,a) \in P$

is a strict Π_1^1 definition of P_0. So we proceed to prove the above assertion. Suppose that 2.4.2 holds for some $a_0 \in A$. We wish to show $a_0 \in P_0$. Let Q_0, R_0 be such that

$$\langle A,\epsilon,P_0,Q_0,R_0 \rangle \models \varphi_1 .$$

Use the function R_0 to define an equivalence relation on C satisfying the following, where a is in A and $(p_1,p_2),(q_1,q_2)$ are in C.

$(p_1,p_2)I(0,a)$	if and only if	$p_1 = 0$ and $p_2 = a$ or
		$p_1 \neq 0$ and there are $b_1,b_2 \in A$ such that
		$p_1I(0,b_1)$, $p_2I(0,b_2)$ and $R(b_1,b_2,a)$
$(0,a)I(p_1,p_2)$	if and only if	$(p_1,p_2)I(0,a)$
$(p_1,p_2)I(q_1,q_2)$	if and only if	there is an $a \in A$ such that
		$(p_1,p_2)I(0,a)$ and $(q_1,q_2)I(0,a)$.

It is easy to see by induction on C that such a relation I does exist. We can then use I to define copies E, P, Q, and R of ϵ, P_0, Q_0, and R_0 respectively on C. For example

$$P = \{x \in C : xI(0,a) \text{ for some } a \in P_0\} .$$

Note that $(0,a) \in P$ just in case $a \in P_0$. This makes $\langle C,E,P,Q,R,I\rangle$ a model of Φ so that $(0,a_0) \in P$ and hence $a_0 \in P_0$. To finish the proof, we need only show that 2.4.1 implies 2.4.2.

So suppose that $a_0 \in P_0$ and $\mathfrak{B} = \langle C,E,I,P,Q,R\rangle$ is a model of Φ. Then $\langle C,E\rangle/I$ is, up to the isomorphism

$$\eta : b \to \frac{(0,b)}{I}$$

an extension of $\mathfrak{A} = \langle A,\in\rangle$ by axioms 2, 5, 6. In fact, due to axiom 7, it is an end extension of $\mathfrak{A}$. Thus, since $\mathfrak{B}/I$ is an end-extension of $\mathfrak{A}$ which is a model of φ_1, P/I must contain P_0, or rather its image under the isomorphism η. Thus $(0,a_0) \in P$, which completes the proof.

The following is a corollary of (the proof of) Theorem 2.4.

COROLLARY 2.5. Let A be admissible and let P be a strict Π_1^1 predicate on A. Then there is a $\Sigma(\underline{S}_1,\ldots,\underline{S}_k)$ formula ψ and predicates $S_1,\ldots,S_k$ on A such that the following are equivalent for all $a_1,\ldots,a_n \in A$:

(a) $P(a_1,\ldots,a_n)$

(b) $\langle A,\in\rangle \models \psi[a_1,\ldots,a_n]$

(c) $\langle A,\in,S_1,\ldots,S_k\rangle \models \psi[a_1,\ldots,a_n]$.

Theorem 2.4 and its Corollary 2.5 have three relativized versions.

THEOREM 2.4A. If A is admissible (not necessarily R-admissible) then a relation on A is s.i.i.d. from R if and only if it is strict Π_1^1 in R.

We define "s.i.i.d. using parameters," $\underset{\sim}{\Pi}_1^1$, and strict $\underset{\sim}{\Pi}_1^1$, by allowing Θ to contain constant symbols for elements of A in the appropriate definitions. We write s.i.i.d. for "s.i.i.d. using parameters" in the following. Thus s.i.i.d. is to s.i.i.d. as $\underset{\sim}{\Sigma}_1$ is to Σ_1.

THEOREM 2.4B. If A is admissible, then a relation on A is s.i.i.d. if and only if it is strict $\underset{\sim}{\Pi}_1^1$.

Just as we are interested in $\underset{\sim}{\Sigma}_1$-compactness, as opposed to Σ_1-compactness, it is s.i.i.d. and strict $\underset{\sim}{\Pi}_1^1$ which interests us. Thus our results are stated in terms of these notions, though most (but not all!) of the unrelativized versions also hold. We make this sacrifice in generality for the sake of exposition. We leave to the reader the task of sorting out the unrelativized results.

The $\underset{\sim}{\Sigma}$-reflection principle insures that the $\underset{\sim}{\Sigma}_1$ predicates on A form a reasonable recursion theory. In view of Theorem 2.4, we would expect the s.i.i.d. predicates to behave especially nicely if A satisfies the following reflection principle.

DEFINITION 2.6. Let A be a transitive set, R some ℓ-place relation on A. A satisfies the strict $\Pi_1^1(R)$ reflection principle if for all relation symbols $\underline{S}_1,\ldots,\underline{S}_k$, all $\Sigma(\underline{R},\underline{S}_1,\ldots,\underline{S}_k)$ formulas $\theta(x_1,\ldots,x_n)$ and elements $a_1,\ldots,a_n \in A$, if

$$\langle A,\in,R\rangle \models \theta[a_1,\ldots,a_n]$$

then there is a transitive set $w \in A$ such that $a_1,\ldots,a_n \in w$ and

$$\langle w,\in,R \cap w^2\rangle \models \theta[a_1,\ldots,a_n] .$$

We say that A satisfies the strict $\underset{\sim}{\Pi}_1^1$ reflection principle if A satisfies the strict $\underset{\sim}{\Pi}_1^1(0)$ reflection principle. Notice that the $\underset{\sim}{\Sigma}(R)$ reflection principle is the special case of the strict $\underset{\sim}{\Pi}_1^1(R)$ reflection principle where the $\underline{S}_1,\ldots,\underline{S}_k$ do not actually occur in θ. In particular, if a rudimentary set A satisfies the strict $\underset{\sim}{\Pi}_1^1$ reflection principle, then A is admissible. The strict $\underset{\sim}{\Pi}_1^1$ reflection principle is in general much stronger than the $\underset{\sim}{\Sigma}$ reflection principle. For example, we shall see in the appendix that $H(\kappa^+)$ never satisfies the strict $\underset{\sim}{\Pi}_1^1$ reflection principle, even though it satisfies the $\underset{\sim}{\Sigma}(R)$ reflection principle for every R.

We shall also see in the appendix that the strict $\underset{\sim}{\Pi}_1^1$ reflection principle does indeed insure that the strict $\underset{\sim}{\Pi}_1^1$ predicates give rise to a reasonable recursion theory. We concentrate here on the connection with infinitary logic.

THEOREM 2.7. Let A be a rudimentary set. Then A is $\underset{\sim}{\Sigma}_1$-compact if and only if A satisfies the strict $\underset{\sim}{\Pi}_1^1$ reflection principle.

Proof. That $\underset{\sim}{\Sigma}_1$-compactness implies strict $\underset{\sim}{\Pi}_1^1$ reflection is the real content of Theorem 1.7. To prove the converse assume that A satisfies the strict $\underset{\sim}{\Pi}_1^1$-reflection principle. It follows from remarks above that A is admissible. We say that a set Φ of sentences violates compactness if every $\Phi_0 \subseteq \Phi$ with $\Phi_0 \in A$ has a model but Φ has no model. So we wish to show that there is no

$\underset{\sim}{\Sigma}_1$ set Φ which violates compactness. Suppose that we could prove the following:

2.7.1. If there is a $\underset{\sim}{\Sigma}_1$ set which violates compactness, then there is a $\underset{\sim}{\Sigma}_1$ of quantifier free sentences involving only constants and one binary relation symbol, no function symbols or $\approx$, which violates compactness.

The theorem will follow from 2.7.1 as follows. Let Φ be a $\underset{\sim}{\Sigma}_1$ set of sentences of $\mathcal{L}_A$, quantifier free, involving only constant symbols and the relation symbol $\underline{R}$. Suppose that Φ has no model. Then:

> $\langle A,\epsilon\rangle \models$ if $\underline{S}$ is a satisfaction relation for $\langle A,R\rangle$ and the assignment of a to c_a for all quantifier free sentences of $\mathcal{L}_A$, then $\exists\, \varphi \in \Phi$ such that $\underline{S}$ assigns false to φ.

It follows from Lemma 2.2 that the part following the $\models$ is a $\Sigma(\underline{R},\underline{S})$ formula, say θ. (θ has the same parameters which occur in the $\underset{\sim}{\Sigma}_1$ definition θ_1 of Φ.) Let w be a transitive element of A such that $\langle w,\epsilon\rangle \models \theta$. Then $\Phi_0 = \{\varphi \in w : \langle w,\epsilon\rangle \models \theta_1[\varphi]\}$ is an element of A by the $\underset{\sim}{\Delta}_0$-separation principle, $\Phi_0 \subseteq \Phi$, and Φ_0 has no model of the form $\langle w,R\rangle$ where the constants c_a are interpreted by a. But then Φ_0 has no model.

To complete the proof we need only prove 2.7.1. We can eliminate function symbols and equality by the usual methods. We next go to a set of universal sentences involving Skolem relations and then take a certain set of substitutions of these universal sentences, much as in the proof of 2.4. We leave it to the reader to check the details. To go from a $\underset{\sim}{\Sigma}_1$ set Φ of quantifier free sentences involving many relation symbols to one involving only one binary relation symbol, replace each $\underline{R}_{a,n}(c_{a_1},\dots,c_{a_n})$ in a sentence of Φ by $\underline{R}(c_{(a,n)},c_{(a_1,\dots,a_n)})$, where $\underline{R}$ is a fixed binary relation symbol. This clearly gives the desired set, and completes the proof.

The natural question to ask is whether $\underset{\sim}{\Sigma}_1$-compactness can be replaced by s.i.i.d.-compactness in 2.8; where by s.i.i.d. compactness we mean that if an s.i.i.d. set Φ has no model then some subset $\Phi_0 \subseteq \Phi$ with $\Phi_0 \in A$ has no model. The following result shows that it can for those sets A with strong separation properties.

THEOREM 2.8. Let A be admissible and satisfy the strict $\underset{\sim}{\Pi}^1_1$ reflection principle. If Φ is an s.i.i.d set of sentences of $\mathcal{L}_A$ which has no model, then there is a set $a \in A$ such that $\Phi \cap a$ has no model.

Proof. Let Φ be an s.i.i.d set of sentences of $\mathcal{L}_A$ with no model. Using Theorem 2.4B and Corollary 2.5B we can choose predicates $T_1,\ldots,T_k$ on A, a $\Sigma(\underline{T}_1,\ldots,\underline{T}_k)$ formula $\theta(x,y_1,\ldots,y_n)$, and elements $b_1,\ldots,b_n \in A$ such that the following are equivalent:

$$a \in \Phi$$
$$\langle A,\epsilon\rangle \models \theta[a,b_1,\ldots,b_n]$$
$$\langle A,\epsilon,T_1,\ldots,T_k\rangle \models \theta[a,b_1,\ldots,b_n] .$$

Assume for simplicity that the only relation symbol occurring in Φ is $\underline{R}$. Let Ψ be the $\underset{\sim}{\Sigma}_1$ set of the following sentences of $\mathcal{L}_A$:

(1) $\forall x[x \in c_a \leftrightarrow \bigvee_{b\in a} x \approx c_b]$ for all $a \in A$

(2) $\underline{S}$ is a satisfaction relation for formulas of $\mathcal{L}_A$ with at most a finite free variables with respect to the structure $\langle A,R\rangle$.

(3) $\theta(c_a,c_{b_1},\ldots,c_{b_n}) \to \underline{S}$ assigns true to c_a for all $a \in A$.

Now Ψ has no models since a model of Ψ would "contain" a model for Φ. So by 2.8 there is some $\Psi_0 \subseteq \Psi$ with $\Psi_0 \in A$ such that Ψ_0 has no model. Let $a = TC(\Psi_1)$. We claim that $\Phi \cap a$ has no model. If $\Phi \cap a$ has a model it has one of the form $\langle A,R\rangle$ by a Lowenheim-Skolem argument. Let S be the satisfaction relation for $\langle A,R\rangle$. We claim that $\mathfrak{A} = \langle A,\epsilon,T_1,\ldots,T_k,R,S\rangle$ is a model for Ψ_0. The only sentences we need to check are (3). But if $\mathfrak{A} \models \theta[\varphi,b_1,\ldots,b_n]$ then $\varphi \in \Phi$. Hence, if the sentence

$$\theta(c_\varphi,c_{b_1},\ldots,c_{b_n}) \to S \text{ asigns true to } c_\varphi$$

is in Ψ_0 then $\varphi \in \Phi \cap a$. Then $\langle A,R\rangle \models \varphi$ and hence S does assign true to φ. Thus the assumption that $\Phi \cap a$ has a model implies that Ψ_0 has a model, contradicting the choice of Ψ_0.

Theorem 2.7 can also be used to good advantage to tell us where to look, or where not to look, for $\underset{\sim}{\Sigma}_1$-compactness. (To obtain the definition of i.i.d., replace "$P \subseteq P'$" by "$P = P' \cap A^n$" in the definition of s.i.i.d. given above.) We use $\mathcal{P}$ to denote both the power set operation and and its graph.

COROLLARY 2.9. Let A be $\underset{\sim}{\Sigma}_1$-compact.

(1) If R is i.i.d. on A then A satisfies the strict $\underset{\sim}{\Pi}^1_1(R)$ reflection principle. Hence, if A also satisfies $\Delta_0(R)$ separation, then A is $\underset{\sim}{\Sigma}_1(R)$-compact and R-admissible.

(2) If A is closed under $\mathcal{P}$, then A is $\underset{\sim}{\Sigma}_1(\mathcal{P})$-compact and every s.i.i.d. predicate is $\underset{\sim}{\Sigma}_1(\mathcal{P})$.

Proof. (1) By 2.8, together with the fact that every $\underset{\sim}{\Sigma}_1(R)$ relation is s.i.i.d., we see that if Φ is an inconsistent $\underset{\sim}{\Sigma}_1(R)$ set of sentences of $\mathcal{L}_A$, then there is an $a \in A$ such that $\Phi \cap a$ is inconsistent. By Remark 3 following Theorem 1.7, this sufficies for the proof of the strict $\underset{\sim}{\Pi}^1_1$ reflection principle.

(2) If A is closed under $\mathcal{P}$, then $\mathcal{P}$ is i.i.d. on A, and every subset of an element of A is an element of A. Thus, the first part of (2) follows from (1). The second part follows from 2.4B, the strict $\underset{\sim}{\Pi}^1_1$ reflection principle, and the fact that for $w \in A$, the predicate

$$\langle w,\epsilon\rangle \models \theta[\vec{x}]$$

is $\Sigma_1(\mathcal{P})$ if θ is a $\Sigma(\underline{S}_1,\ldots,\underline{S}_k)$ formula.

REMARKS. 1. Let κ be an inaccessible. In [9], Kunen calls κ $\underset{\sim}{\Sigma}_1$-compact if every inconsistent $\underset{\sim}{\Sigma}_1(\mathcal{P})$ set of sentences of $\mathcal{L}_{H(K)}$ has an inconsistent subset of cardinality $< \kappa$. Thus, his $\underset{\sim}{\Sigma}_1$-compactness is our $\underset{\sim}{\Sigma}_1(\mathcal{P})$-compactness. Corollary 2.10(2) shows that the two notions of $\underset{\sim}{\Sigma}_1$-compactness coincide in the cases where Kunen's is defined.

2. The second statement of 2.9(2) is a generalization of Theorem 4.10(2) in Kunen [9], though the proof is completely different. We note here that Σ_1 and s.i.i.d. should be replaced by $\underset{\sim}{\Sigma}_1$ and s.i.i.d. respectively in Kunen's 4.10(1).

§3. Some Uncountable $\underset{\sim}{\Sigma}_1$-Compact Sets.

In this section we use results of §2 to prove the existence of uncountable $\underset{\sim}{\Sigma}_1$-compact sets of various kinds. Our main result, Theorem 3.3, gives a sufficient condition that a set A closed under $\underline{\mathcal{P}}$ be $\underset{\sim}{\Sigma}_1$-compact, and hence $\underset{\sim}{\Sigma}_1(\underline{\mathcal{P}})$-compact. Before turning to this result we give a highly non-constructive proof of the existence of $\underset{\sim}{\Sigma}_1$-compact sets of various kinds.

LEMMA 3.1. Let F be a function from ordinals to sets such that the following hold:

(a) for all ordinals α, $F(\alpha)$ is a transitive set and $F(\alpha) \in F(\alpha + 1)$,

(b) if λ is a limit ordinal then $F(\lambda) = \bigcup \{F(\gamma) : \gamma < \lambda\}$,

(c) for every ordinal α there is a β such that $\alpha \in F(\beta)$.

Then there are arbitrarily large limit ordinals α such that $F(\alpha)$ satisfies the strict $\underset{\sim}{\Pi}_1^1$ reflection principle.

Proof. For any sets A and B, $B \subseteq A$, we say that A satisfies the strict $\Pi_1^{1,B}$ reflection principle if for all $\underline{S}_1, \ldots, \underline{S}_k$ and $\Sigma(\underline{S}_1, \ldots, \underline{S}_k)$ formulas $\theta(x_1, \ldots, x_n)$ and $b_1, \ldots, b_n \in B$, if

$$\langle A, \in \rangle \models \theta[b_1, \ldots, b_n]$$

then there is a transitive set $w \in A$ with $b_1, \ldots, b_n \in w$ such that

$$\langle w, \in \rangle \models \theta[b_1, \ldots, b_n] .$$

Thus A satisfies the strict $\underset{\sim}{\Pi}_1^1$ reflection principle if and only if A satisfies the strict $\Pi_1^{1,A}$ reflection principle. We first show that for every α there is a β_0 such that for all $\beta \geq \beta_0$, $F(\beta)$ satisfies the strict $\Pi_1^{1,F(\alpha)}$ reflection principle. For each k, each $\underline{S}_1, \ldots, \underline{S}_k$, each $\Sigma(\underline{S}_1, \ldots, \underline{S}_k)$ formula $\theta(x_1, \ldots, x_n)$ and each $b_1, \ldots, b_n \in F(\alpha)$ let $\gamma(\theta, b_1, \ldots, b_n)$ be the least ordinal γ such that

$$\langle F(\gamma), \in \rangle \models \theta[b_1, \ldots, b_n]$$

if such exists, otherwise let $\gamma(\theta, b_1, \ldots, b_n) = 0$. By the axiom of replacement there is an ordinal β_0 greater than all of these $\gamma(\theta, b_1, \ldots, b_n)$. This ordinal β_0 clearly has the desired property.

We now define a function f from ordinals to ordinals by:

$f(\alpha)$ = the least β_0 such that $\beta_0 > f(\gamma)$ for all $\gamma < \alpha$ and such that $F(\beta)$ satisfies the strict $\Pi_1^{1,F(\alpha)}$ reflection principle for all $\beta \geq \beta_0$.

We claim that f is normal, i.e.,

(i) $\alpha < \beta$ implies $f(\alpha) < f(\beta)$

(ii) if λ is a limit ordinal then $f(\lambda) = \sup\{f(\gamma) : \gamma < \lambda\}$.

We prove (ii) since (i) is immediate. It is clearly sufficient to show that if λ is a limit ordinal and $\beta > f(\gamma)$ for all $\gamma < \lambda$ then $F(\beta)$ satisfies the strict $\Pi_1^{1,F(\lambda)}$ reflection principle. But by (b) every instance of the strict $\Pi_1^{1,F(\lambda)}$ reflection principle is an instance of the strict $\Pi_1^{1,F(\gamma)}$ reflection principle for some $\gamma < \lambda$, and $F(\beta)$ does satisfy each of these reflection principles. Since the function f is normal, it has arbitrarily large limit fixed points. But if $f(\alpha) = \alpha$ then $F(\alpha)$ satisfies the strict $\underset{\sim}{\Pi}_1^1$ reflection principle.

We remark that in all applications, the class F will be definable and so could be eliminated in favor of its definition.

THEOREM 3.2.

(1) <u>There are arbitrarily large cardinals</u> κ <u>such that</u> $H(\kappa)$ <u>is</u> $\underset{\sim}{\Sigma}_1$-<u>compact</u>.

(2) <u>There are arbitrarily large ordinals</u> α <u>such that</u> $R(\alpha)$ <u>is</u> $\underset{\sim}{\Sigma}_1$-<u>compact</u>.

(3) <u>There are arbitrarily large ordinals</u> α <u>such that</u> L_α <u>is</u> $\underset{\sim}{\Sigma}_1$-<u>compact</u>.

<u>Proof</u>. For (1) let $F(\alpha) = H(\aleph_\alpha)$. For (2) let $F(\alpha) = R(\alpha)$. In (3), L_α is the set of sets constructible before ordinal α in the sense of Gödel [4]. Let $F(\alpha) = L_\alpha$. Each of the results then follows from 3.1 together with Theorem 2.7.

REMARKS. 1. Suppose that F satisfies the hypothesis of 3.1 and f is some normal function. Then the composition $F \circ f$ again satisfies the hypothesis of 3.1. Using this we can put more and more conditions on the κ's and α's asserted to exist in 3.2. For example. The α's in (2) and (3) can be chosen to be cardinals. We may in fact find arbitrarily large κ such that all of $H(\kappa)$, $R(\kappa)$ and L_κ are $\underset{\sim}{\Sigma}_1$-compact. Another improvement results in taking $F(\alpha) = H(\beth_\alpha)$. For if α is a limit ordinal then $H(\beth_\alpha)$ is closed under ρ so we can apply Corollary 2.10(2).

2. Most of the κ given by the proof of 3.2(1) are singular. For singular κ, $\mathcal{L}_{H(\kappa)}$ is not what is usually called $L_{\kappa,\omega}$. Rather, it is the union of the $L_{\kappa',\omega}$ for $\kappa' < \kappa$. Thus the most interesting languages are skipped in the definition of $L_{\kappa,\omega}$.

3. The unrelativized version of 3.1 is rather startling. It states there is a fixed

transitive set A_0 such that if A is an admissible set with $A_0 \subseteq A$ then A is Σ_1-compact. If κ is the first inaccessible, however, $H(\kappa)$ is not Σ_1-compact. Let $\kappa \geq \bar{\bar{A}}_0$. Then $H(\kappa^+)$ is Σ_1-compact (since $A_0 \subseteq H(\kappa^+)$), but is not $\underset{\sim}{\Sigma}_1$-compact by Proposition 2.7 and Lemma A9 of the appendix.

We now turn to the main result of the section. We wish to find a sufficient condition for a set A which is closed under $\mathcal{P}$ to be $\underset{\sim}{\Sigma}_1$-compact. By Corollary 2.10(2) we know that we must assume that A is $\mathcal{P}$-admissible. This condition is not sufficient for Σ_1-compactness, let alone for $\underset{\sim}{\Sigma}_1$-compactness.

We say that a set A has cofinality ω, and write $\mathrm{cf}(A) = \omega$, if $A = \bigcup_{n<\omega} A_n$, where each $A_n \in A$. If A is rudimentary and $\mathrm{cf}(A) = \omega$, then $A = \bigcup_{n<\omega} A_n$ where $A_n \in A_{n+1}$ and A_n is transitive, for each $n < \omega$. The compactness result is the following:

THEOREM 3.3. <u>Let</u> A <u>be closed under</u> $\mathcal{P}$, $\mathcal{P}$-<u>admissible, and have cofinality</u> ω. <u>Then</u> A <u>is</u> $\underset{\sim}{\Sigma}_1$-<u>compact</u>.

REMARKS. 1. This result was obtained independently by C. Karp using algebraic methods.

2. The relativized result 3.3A provides another complete generalization of usual compactness theorem for $\mathcal{L}_{H(\omega)}$ $(= L_{\omega,\omega})$, since $H(\omega)$ is closed under $\mathcal{P}$, is admissible with respect to $\mathcal{P}$ and R for all relations R on $H(\omega)$, and has cofinality ω. Our original proof of this result was by means of a Gentzen type formal system with the full distributivity law. The proof which we present here, by means of the strict $\underset{\sim}{\Pi}^1_1$ reflection principle, seems more enlightening with regard both to compactness, and the occasional collapse of s.i.i.d. to $\underset{\sim}{\Sigma}_1$ or $\underset{\sim}{\Sigma}_1(\mathcal{P})$.

Before we prove the above result, we need a "quantifier pushing" lemma about strict Π^1_1 predicates. To simplify the statement of the lemma we introduce the following terminology.
Let $\theta_1(x_1,\ldots,x_n)$ be a $\Sigma(\underline{R}_1,\ldots,\underline{R}_k,\underline{S}_1,\ldots,\underline{S}_k)$ and $\theta_2(x_1,\ldots,x_n)$ be a $\Sigma(\underline{R}_1,\ldots,\underline{R}_k,\underline{T}_1,\ldots,\underline{T}_m)$ formula, where it is not assumed that all of the $\underline{S}_i$ are distinct from all of the $\underline{T}_j$. We say that θ_1 is strict Π^1_1 equivalent to θ_2 relative to $\underline{R}_1,\ldots,\underline{R}_k$, and write

$$\theta_1 \overset{R_1,\ldots,R_k}{\Longleftrightarrow} \theta_2$$

if for every structure $\mathfrak{A} = \langle A,E,R_1,\ldots,R_k\rangle$ and all $a_1,\ldots,a_n \in A$ we have

$$\mathfrak{A} \models \theta_1[a_1,\ldots,a_n] \text{ if and only if } \mathfrak{A} \models \theta_2[a_1,\ldots,a_n] .$$

LEMMA 3.4. Let θ be a $\Sigma(\underline{R}_1,\underline{S}_1,\ldots,\underline{S}_k)$ formula. Then there are some relation symbols $\underline{T}_1,\ldots,\underline{T}_m$ and a $\Sigma_1(\underline{R},\underline{T}_1,\ldots,\underline{T}_m)$ formula θ' such that $\theta \Longleftrightarrow^R \theta'$.

Proof. See Corollary A2 of the appendix.

We now come to the generalization of the Brower-König infinity lemma (as stated on page 187 of Shoenfield [13]) which we mentioned earlier. Let $\underline{P}$ be a binary relation symbol whose interpretation is always the graph of P.

LEMMA 3.5. Let A be closed under P and P-admissible. Suppose further that $A = \bigcup_{i<\omega} A_i$ where $A_i \in A_{i+1}$ and A_i is transitive for each i. Let $\varphi(x_1,\ldots,x_k,y_1,\ldots,y_\ell)$ be a $\Delta_0(\underline{P},\underline{S}_1,\ldots,\underline{S}_m)$ formula and let $b_1,\ldots,b_\ell \in A_{i_0}$. The following are then equivalent:

(i) $\exists S_1 \subseteq A^{n_1} \cdots \exists S_n \subseteq A^{n_m}$ such that $\langle A,\in,P,S_1,\ldots,S_m\rangle \models \forall x_1,\ldots,x_k\ \varphi[b_1,\ldots,b_\ell]$

(ii) $\forall i \geq i_0\ \exists S_1 \subseteq (A_i)^{n_1} \cdots \exists S_m \subseteq (A_i)^{n_m}$ such that

$$\langle A_i,\in,\ P\cap A_i^2,\ S_1,\ldots,S_m\rangle \models \forall x_1,\ldots,x_k\ \varphi[b_1,\ldots,b_2]\ .$$

Proof. (i) $\Rightarrow$ (ii) is immediate since we can take $S_1 \cap (A_i)^{n_1},\ldots,S_m \cap (A_i)^{n_m}$ for the desired relations on A_i.

We first prove the special case of (ii) $\Rightarrow$ (i) where $k = \ell = m = n_1 = 1$, and then indicate the changes required for the general case. So suppose that $\varphi(x,y)$ is a $\Delta_0(\underline{P},\underline{S})$ formula, $b \in A_{i_0}$, and for all $i \geq i_0$ there is a subset S of A_i such that

$$\langle A_i,\in,P\cap(A_i)^2,S\rangle \models \forall x\ \varphi[b]\ .$$

We wish to find a relation $S \subseteq A$ such that

$$\langle A,\in,P,S\rangle \models \forall x\ \varphi[b]\ .$$

We can assume that $i_0 = 0$ since we can always ignore the first i_0 A_i's. For each i let

$$\mathcal{S}_i = \left\{ S \in P(A_i) \mid \langle A_i,\in,P\cap A_i^2,S\rangle \models \forall x\ \varphi[b]\right\}.$$

If $S \in \mathcal{S}_i$ and $S' \in \mathcal{S}_j$ where $i \leq j$, we say S' extends S, and write $S \prec S'$, if $S = S' \cap A_i$. Note that each $\mathcal{S}_i$ is an element of A, and that if $i < j$ and $S \in \mathcal{S}_j$ then $(S \cap A_i) \in \mathcal{S}_i$. We wish to define a sequence $S_0, S_1, \ldots$ with the following properties, for each i,

(1)
$$S_i \prec S_{i+1}$$

(2)
$$\langle A_i, \in, \mathcal{P} \cap A_i^2, S_i \rangle \models \forall x\, \varphi[b], \text{ i.e. } S_i \in \mathcal{S}_i \ .$$

Then $S = \bigcup_{i<\omega} S_i$ will be the desired subset of A. The sequence will be defined by induction on i satisfying (1), (2) and

(3) for all $j > i$ there is an S' in $\mathcal{S}_j$ extending S_i .

For S_0 take any $S \in \mathcal{S}_0$ satisfying (3) with $i = 0$. That there is such an S_0 follows from an argument similar to the following argument for the induction step. So suppose S_i is defined. We wish to show that there is an $S_{i+1} \in \mathcal{S}_{i+1}$ extending S_i which itself has extensions in each $\mathcal{S}_j$ for $j > i + 1$. Assume that no such $S_{i+1} \in \mathcal{S}_{i+1}$ exists. Then for each S in $\mathcal{S}_{i+1}$ which extends S_i there is a j such that S has no extensions in $\mathcal{S}_j$. Hence

$$\forall S \in \mathcal{S}_{i+1}[S_i \prec S \rightarrow \exists w,w'[\mathrm{Tran}(w) \wedge A_{i+1} \subseteq w \wedge w' = \mathcal{P}(w)$$
$$\wedge \ \forall S' \in w'[S' \cap A_{i+1} = S \rightarrow \langle w, \in, \mathcal{P} \cap w^2, S' \rangle \models \exists x \neg \varphi[b]]]] \ .$$

By the $\Sigma(\mathcal{P})$-reflection principle there is a transitive set $W \in A$ such that the above holds in W, and hence in any A_n, $n > i + 1$, such that $W \subseteq A_n$. But then we claim that S_i has no extensions in A_n, contrary to the inductive assumption. For if $S_i \prec S$ for some $S \in \mathcal{S}_n$, then $S \cap A_{i+1} \in \mathcal{S}_{i+1}$. So there is a transitive set $w \in A_n$ such that

$$\langle w, \in, \mathcal{P} \cap w^2, S \cap w \rangle \models \exists x \neg \varphi[b] \ .$$

But then

$$\langle A_n, \in, \mathcal{P} \cap A_n^2, S \rangle \models \exists x \neg \varphi[b] \ ,$$

contradicting $S \in \mathcal{S}_n$.

The general case of (ii) ⇒ (i) is proved in the same way except that the notation becomes more complicated. For example,

$$\mathcal{S}_i = \{(S_1,\dots,S_m) \in P(A_i^{n_1}) \times \cdots \times P(A_i^{n_m}) \mid$$

$$\langle A_i,\in,P \cap A_i^2,S_1,\dots,S_m\rangle \models \forall x_1,\dots,x_k\ \varphi[b_1,\dots,b_\ell]\}.$$

One sequence $(S_1',\dots,S_m') \in \mathcal{S}_j$ extends an $(S_1,\dots,S_m) \in \mathcal{S}_i$ if $i \le j$ and $S_1 = S_1' \cap A_i^{n_1},\dots,$ and $S_m = S_m' \cap A_i^{n_m}$. The proof now proceeds as before.

COROLLARY 3.6. <u>For every $\Sigma(\underline{P},\underline{S}_1,\dots,\underline{S}_k)$ formula</u> θ <u>there is a $\Sigma_1(\underline{P})$ formula</u> $\hat\theta$ <u>such that</u>

(1) <u>if</u> w <u>is transitive then</u>

$$\langle w,\in_w,\mathcal{P} \cap w^2\rangle \models \hat\theta\,[b_1,\dots,b_2] \ \underline{\text{implies}}\ \langle w,\in_w,P \cap w^2\rangle \models \theta[b_1,\dots,b_2]$$

(2) <u>if</u> A <u>is closed under</u> $\mathcal{P}$, $\mathcal{P}$<u>-admissible and</u> $cf(A) = \omega$ <u>then</u>

$$\langle A,\ ,P\rangle \models \theta[b_1,\dots,b_\ell] \ \underline{\text{if and only if}}\ \langle A,\ ,P\rangle \models \hat\theta[b_1,\dots,b_2].$$

<u>Proof</u>. By Lemma 3.4 there is a $\Sigma_1(\underline{P},\underline{T}_1,\dots,\underline{T}_2)$ formula θ' such that $\theta \overset{\mathcal{P}}{\Longleftrightarrow} \theta'$. Thus we may assume that θ is $\Sigma_1(\underline{P},S_1,\dots,S_k)$. Let $w,s_1,\dots,s_k$ be new variables, and for any formula φ let $\varphi^{w,s_1,\dots,s_k}$ be obtained from φ by relativizing all quantifiers in φ to w and replacing $\underline{S}_i(x_1,\dots,x_{n_i})$ by $\exists z \in s_i[z = (x_1,\dots,x_{n_i})]$. Thus $\varphi^{w,s_1,\dots,s_k}$ is a $\Delta_0(\underline{P})$ formula with free variables $w,s_1,\dots,s_k$ and those of φ. The formula $\hat\theta$ which we want is:

$$\exists w \exists w_1,\dots,w_k\Big[\text{Trans}(w) \wedge \bigwedge_{i<k} w_i = \mathcal{P}(w^{n_1}) \wedge \forall s_1 \in w_1,\dots, \forall s_k \in w_k\ \varphi^{w_1,s_1,\dots,s_k}\Big].$$

Condition (1) is satisfied by Lemma 2.1. One half of (2) follows from (1), the other half follows from Lemma 3.5.

3.7. The proof of Theorem 3.3 is now immediate. Let A satisfy the hypothesis of the theorem. We wish to show that A satisfies the strict $\underset{\sim}{\Pi}_1^1$ reflection principle. So let $\theta(x_1,\dots,x_n)$ be some $\Sigma(\underline{P},\underline{S}_1,\ldots,\underline{S}_k)$ formula such that for some $b_1,\ldots,b_\ell \in A$

$$\langle A,\epsilon,\mathcal{P}\rangle \models \theta[b_1,\ldots,b_\ell] .$$

Then by Corollary 3.6(2)

$$\langle A,\epsilon,\mathcal{P}\rangle \models \hat{\theta}[b_1,\ldots,b_\ell] .$$

Hence by the $\underset{\sim}{\Sigma}(\mathcal{P})$ reflection principle there is a $w \in A$ such that $b_1,\ldots,b_\ell \in w$ and

$$\langle w,\epsilon,\mathcal{P} \cap w^2\rangle \models \hat{\theta}[b_1,\ldots,b_\ell] .$$

Then by Corollary 3.6(1)

$$\langle w,\epsilon,\mathcal{P} \cap w^2\rangle \models \theta[b_1,\ldots,b_\ell] .$$

Note also that 3.6(2) implies that every strict $\underset{\sim}{\Pi}_1^1$ relation on A is $\underset{\sim}{\Sigma}(\mathcal{P})$. Of course we already know this from Corollary 2.9(2).

We can apply Theorem 3.3 to get another proof of Theorem 3.2(2). Let σ_α be the αth ordinal σ such that $R(\sigma)$ is an elementary subsystem of the universal V with respect to all $\Sigma_1(\underline{P})$ formulas. By Theorem 36 of Levy's monograph [10] there are arbitrarily large such σ. Note that $\sigma_\lambda = \sup_{\gamma<\lambda} \sigma_\gamma$ for limit ordinals λ. Hence if $cf(\alpha) = \omega$ then $cf(R(\sigma_\alpha)) = \omega$. Theorem 3.3 implies that if $cf(\alpha) = \omega$ then $R(\sigma_\alpha)$ is $\underset{\sim}{\Sigma}_1$-compact.

We remark in closing that if A is closed under $\mathcal{P}$ and $\underset{\sim}{\Sigma}_1$ compact, then one can prove some compactness theorems even for the case where the formulas involve infinite strings of quantifiers.

APPENDIX: Strongly Admissible Sets

In this appendix we pursue briefly the study of strict $\underset{\sim}{\Pi}_1^1$ predicates and the recursion theory to which they give rise. In particular, we justify the claim made in §2 that this recursion theory is well behaved if the underlying set A satisfies the strict $\underset{\sim}{\Pi}_1^1$ reflection principle. For example, the strict $\underset{\sim}{\Pi}_1^1$ reflection principle is needed to show that definition by recursion over ϵ does not lead out of the class of recursive functions, in our sense. The interest in strict

$\underset{\sim}{\Pi}_1^1$ predicates as a recursion theory stems from Theorem 2.4B, where we saw that they correspond to the s.i.i.d predicates in the sense of Kunen [9].

The most natural setting for the study of strict $\underset{\sim}{\Pi}_1^1$ predicates is the language of second order set theory. That is, we now allow quantification over relation symbols in the formulas of set theory. We do not allow function symbols so we do not allow quantification over them; neither do we allow quantifiers to bind $\in$. We continue to use $\theta, \theta', \theta_1, \ldots$ for formulas of first order set theory; we use $\Theta, \Theta', \Theta_1, \ldots$ for formulas of second order set theory. If we wish to emphasize that a relation symbol $\underline{R}$ occurs free in Θ we write $\Theta(\underline{R})$. If we wish to indicate that a relation symbol $\underline{R}$ is n-ary we write $\underline{R}^n$. Structures are still of the form $\langle A, E, \ldots \rangle$ where A is a set and $E \subseteq A \times A$. Typically A is a transitive set and E is $\in \cap A^2$. The relation quantifiers are interpreted in the classical sense. For example,

$$\mathfrak{A} \models \forall \underline{R}^n \, \Theta(\underline{R}^n)$$

means that for every $R \subseteq A^n$,

$$(\mathfrak{A}, R) \models \Theta(\underline{R}^n) .$$

We say that a formula $\Theta(\underline{R}_1, \ldots, \underline{R}_n)$, with no other free relation symbols, is valid if it holds in all structure $\langle A, E, R_1, \ldots, R_n \rangle$ for all assignments to the free individual variables. We say that Θ_1 and Θ_2 are logically equivalent if $(\Theta_1 \leftrightarrow \Theta_2)$ is valid.

The second order formulas with which we actually work are particularly simple ones. We say that a formula Θ is strict Π_1^1, or $s - \Pi_1^1$, if it is of the form

$$\forall \underline{S}_1 \cdots \forall \underline{S}_k \, \theta$$

where θ is a $\Sigma_1(\underline{S}_1, \ldots, \underline{S}_k)$ formula. (Note that this does not correspond to our earlier definition since we require θ to be Σ_1 rather than just Σ. We shall see in A2 that the two are equivalent.) We say that the Θ above is $s - \Pi_1^1(\underline{R}_1, \ldots, \underline{R}_\ell)$ if θ is $\Sigma_1(\underline{R}_1, \ldots, \underline{R}_\ell, \underline{S}_1, \ldots, \underline{S}_k)$.

LEMMA A1. Let Y be the set of formulas Θ such that for some $\underline{R}_1, \ldots, \underline{R}_k$ and some $s - \Pi_1^1(\underline{R}_1, \ldots, \underline{R}_k)$ formula Θ', Θ is logically equivalent to Θ'.

(1) <u>If</u> Θ <u>is atomic then</u> Θ <u>and</u> $\neg\Theta$ <u>are in</u> Y.

(2) <u>If</u> Θ_1 <u>and</u> Θ_2 <u>are in</u> Y <u>then</u> $\Theta_1 \wedge \Theta_2$ <u>and</u> $\Theta_1 \vee \Theta_2$ <u>are in</u> Y.

(3) <u>If</u> Θ <u>is in</u> Y <u>then</u> $\exists x\Theta$ <u>and</u> $\exists x \in y\Theta$ <u>are in</u> Y.

(4) <u>If</u> Θ <u>is in</u> Y <u>then</u> $\forall x \in y\Theta$ <u>is in</u> Y.

(5) <u>If</u> $\Theta(\underline{R})$ <u>is in</u> Y <u>then</u> $\forall \underline{R}\Theta(\underline{R})$ <u>is in</u> Y.

<u>Proof</u>. The proofs of (1) and (5) are trivial, that of (2) is straightforward. The proof of (3) is handled by the usual manipulation

$$\forall x \exists \underline{T}^n(\cdots) \text{ becomes } \exists \underline{T}^{n+1} \forall x(---)$$

where (---) is obtained from $(\cdots)$ by replacing $\underline{T}^n(y_1,\ldots,y_n)$ by $\underline{T}^{n+1}(x,y_1,\ldots,y_n)$. The only interesting case is (4). So suppose that Θ is equivalent to the strict $\Pi_1^1(\underline{R})$ formula

$$\forall \underline{S}_1 \cdots \forall \underline{S}_k \exists x_1,\ldots,\exists x_\ell \theta$$

where θ is $\Delta_0(\underline{R}_1,\underline{S}_1,\ldots,\underline{S}_k)$. Let $\underline{U}$ be a unary relation symbol. We wish to find a $\Sigma_1(\underline{S}_1,\ldots,\underline{S}_k,\underline{U})$ formula θ' such that $\forall x \in y\Theta$ is logically equivalent to

$$\forall \underline{S}_1 \cdots \forall \underline{S}_k \forall \underline{U}\theta' .$$

It will suffice to find a $\Sigma_1(\underline{R},\underline{S}_1,\ldots,\underline{S}_k,\underline{U})$ formula θ' such that $\forall x \in y \exists x_1 \cdots \exists x_n\theta$ is logically equivalent to $\forall \underline{U}\theta'$ and then universally quantify over $\underline{S}_1,\ldots,\underline{S}_k$. Let θ' be the following formula:

$$\exists x_1 \cdots \exists x_n \exists u \exists u'\Big\{[\underline{U}(u) \wedge \underline{U}(u') \wedge u \neq u'] \\ \vee [\exists x \in y\, \underline{U}(x) \wedge \forall x \in y(\underline{U}(x) \rightarrow \theta(x,y,x_1,\ldots,x_n))]\Big\} .$$

Since θ is $\Delta_0(R,S_1,\ldots,S_k)$, θ' is $\Sigma_1(R,S_1,\ldots,S_k,U)$. To see that $\forall x \in y \exists x_1 \cdots \exists x_n\theta$ is equivalent to $\forall \underline{U}\theta'$, let $\mathfrak{A} = \langle A,E,R,S_1,\ldots,S_k\rangle$ be any structure, let $a \in A$ and let $\vec{b}$ be a sequence of elements of A. The following are equivalent:

$\mathfrak{A} \models \forall x \in a\, \exists x_1 \ldots \exists x_n\, \theta[\vec{b}]$

$\mathfrak{A} \models \exists x_1 \ldots \exists x_n\, \theta[x,\vec{b}]$ for all $x \in a$

$(\mathfrak{A},U) \models$ (if $\underline{U}$ is $\{x\}$ for some $x \in a$ then $\exists x_1 \ldots \exists x_n\, \theta[x,\vec{b}]$)

for all $U \subseteq A$.

$\mathfrak{A} \models \forall \underline{U} \theta'$.

This completes the proof of the lemma.

COROLLARY A2. *If* θ *is a* $\Sigma(R,S_1,\ldots,S_k)$*-formula, then* $\forall \underline{S}_1 \ldots \forall \underline{S}_k \theta$ *is logically equivalent to some* s - $\Pi^1_1(R)$ *formula.*

This corollary shows that the notion of strict Π^1_1 used here coincides with that used in §§2 and 3.

An alternate approach would be to use (1)-(5) of A1 as an inductive definition of the class of strict Π^1_1 formulas, and then show that

$$\forall S_1 \ldots \forall_k \exists x_1 \ldots \exists x_\ell \theta$$

where θ is a Δ_0-formula, is a normal form.

Let A be a transitive set with $X \subseteq A$. We say that X is strict $\underset{\sim}{\Pi}^1_1$ on A if X can be defined by a strict Π^1_1 formula with parameters from A. X is strict $\underset{\sim}{\Sigma}^1_1$ if $A \sim X$ is strict $\underset{\sim}{\Pi}^1_1$ on A, and X is strict $\underset{\sim}{\Delta}^1_1$ on A if it is both strict $\underset{\sim}{\Pi}^1_1$ and strict $\underset{\sim}{\Sigma}^1_1$ on A. Similarly for relations R on A.

DEFINITION A3. A nonempty transitive set A is *strongly admissible* if A is closed under TC and satisfies:

(a) (*strict* $\underset{\sim}{\Delta}^1_0$ *separation*) If a is strict Δ^1_1 on w, where w is some transitive element of A, then $a \in A$.

(b) (*strict* $\underset{\sim}{\Pi}^1_1$ *reflection*) If $(A,\epsilon) \models \theta[b_1,\ldots,b_n]$ where θ is strict Π^1_1, then there is a transitive element w of A such that $b_1,\ldots,b_n \in w$ and

$$\langle w,\in\rangle \models \Theta[b_1,\ldots,b_n] .$$

LEMMA A4. Every strongly admissible set is admissible.

Proof. The $\underset{\sim}{\Delta}_0$ separation axiom and $\underset{\sim}{\Sigma}$ reflection principle are the special cases of the strict $\underset{\sim}{\Delta}_0^1$ separation principle and strict $\underset{\sim}{\Pi}_1^1$ reflection principle where no second order quantifiers appear.

We shall see later that the strict $\underset{\sim}{\Pi}_1^1$ reflection principle is in general much stronger than the $\underset{\sim}{\Sigma}$ reflection principle. Nevertheless, for the countable case one does have the following:

THEOREM A5. Every countable admissible set is strongly admissible.

Proof. Let A be a countable admissible set. By Theorem 1.9 A is $\underset{\sim}{\Sigma}_1$-compact, so by Theorem 2.7 , A satisfies the strict $\underset{\sim}{\Pi}_1^1$ reflection principle. One can use Theorem 1.8 (1) to show that every strict $\underset{\sim}{\Delta}_1^1$ subset of a transitive element w of A is $\underset{\sim}{\Delta}_1$ on A and hence an element of A.

It follows from Theorem 4.5 of Kunen [9] together with 2.4 that on a countable admissible set, strict $\underset{\sim}{\Pi}_1^1 = \underset{\sim}{\Sigma}_1$ and hence strict $\underset{\sim}{\Delta}_1^1 = \underset{\sim}{\Delta}_1$. This is quite easy to see directly using Theorem 1.10.

We now state some simple consequences of a set A being strongly admissible. The proofs of A6-A8 are straightforward to anyone familiar with the second half of Platek [12]. Since this work is rather inaccessible, we present brief sketches of the proofs. Unless otherwise indicated, A is always a strongly admissible set.

A6. STRICT $\underset{\sim}{\Delta}_1^1$ SEPARATION. Let X be strict $\underset{\sim}{\Delta}_1^1$ on A. If $X \subseteq a$ for some $a \in A$, then $X \in A$.

Proof. Suppose that

$$x \in X \quad \text{if and only if} \quad \Theta_1[x,b]$$

$$x \notin X \quad \text{if and only if} \quad \Theta_2[x,c] .$$

Then $\forall x \in a[\Theta_1 \vee \Theta_2]$ holds in A. Since it is (equivalent to) a $s - \Pi_1^1$ formula $\Theta(a,b,c)$ there is a transitive set $w \in A$ such that $a,b,c \in w$ and $\Theta(a,b,c)$ holds in w. But then X is strict $\underset{\sim}{\Delta}_1^1$ on w, so $X \in A$ by strict $\underset{\sim}{\Delta}_0^1$ separation.

A7. STRICT $\underset{\sim}{\Pi}_1^1$ REPLACEMENT. Let Θ be a $s - \Pi_1^1$ formula. The universal closure of the following is true in A:

$$\forall x \in y \, \exists ! z \, \Theta(x,z) \rightarrow \exists w \forall z (z \in w \leftrightarrow \exists x \in y \, \Theta(x,z)) .$$

Proof. Suppose that $\forall x \in a \, \exists ! z \, \Theta(x,z,b)$ holds in A. Using $s - \underset{\sim}{\Pi}_1^1$ reflection there is a transitive $w \in A$ such that $a,b \in w$ and $\forall x \in a \, \exists z \in w \, \Theta(x,z,b)$ holds in A. But then

$$\{z \in w \mid \exists x \in a \;\; \Theta(x,z,b)\} = \{z \in w \mid \neg [\, \exists x \in a \, \exists z' \in w(\Theta(x,z') \wedge z \neq z')\}$$

is $s - \underset{\sim}{\Delta}_1^1$ on A. So by A6 it is an element of A.

Let A be strongly admissible. We call a function f, with domain and range subsets of A, K-recursive on A if its graph is $s - \underset{\sim}{\Pi}_1^1$ on A. A set $X \subseteq A$ is K-recursive on A if its characteristic function is K-recursive on A, equivalently, if X is $s - \underset{\sim}{\Delta}_1^1$ on A, or i.i.d on A. Similarly, we call X K-recursively ennumerable on A (K-r.e.) if X is the domain of a K-recursive function; equivalently, if X is $s - \underset{\sim}{\Pi}_1^1$ or s.i.i.d. on A. (The K stands for Kreisel and/or Kunen.)

LEMMA A8. Let A be strongly admissible, α be the least ordinal not in A, and let $g, g_1, \ldots, g_n$ be K-recursive on A. If f is defined from $g, g_1, \ldots, g_n$ by means of S1-S6 below, then A is closed under f and $f \upharpoonright A$ is K-recursive on A.

S1. Definition by cases:

$$f(x_1,x_2,x_3) \simeq \begin{cases} g_1(x_1) & \text{if} \quad x_2 = x_3 \\ g_2(x_1) & \text{otherwise} \end{cases}$$

S2. Replacement:

$$f(x) = \begin{cases} \{g(z) : z \in x\} & \text{if} \quad x \subseteq \text{domain } (g) \\ \text{undefined} & \text{otherwise} \end{cases}$$

S3. Union

$$f(x) \simeq \bigcup g(x)$$

S4. Substitution

$$f(x_1,\ldots,x_n) \simeq g(x_1,\ldots,x_n,g_1(x_1,\ldots,x_n),\ldots,g_k(x_1,\ldots,x_n))$$

S5. Recursion

$$f(x,y) = g(x,y, \{(z,f(x,z)) : z \in y\})$$

S6. ${}_{\alpha}\mathrm{MIN}(g)$

$f(x) \simeq$ the least ordinal β such that $g(x,\beta) \simeq 0$ if such exists, i.e.

$$f(x) \simeq \begin{cases} \beta \text{ if } g(x,\beta) \simeq 0 \text{ and } \forall \gamma < \beta[g(x,\gamma) \text{ is defined and } \neq 0] \\ \text{undefined otherwise} \end{cases}$$

Proof. S1, S3, S4, and S6 are all straightforward. S2 follows simply from A7. The proof of S5 is tedious but not difficult. It uses the $s - \underset{\sim}{\Pi}_1^1$ reflection principle in an essential way. For a published proof of a similar result for $\underset{\sim}{\Sigma}_1$ functions, see T5, page 380, of Karp [7].

The above lemma is really a special case of the following. One can show, under the hypothesis of the lemma, that if f is α-set recursive in $g_1,\ldots,g_n$ in the sense of Platek [12], then A is closed under f and $f \upharpoonright A$ is K-recursive. The fact that one needs the $s - \underset{\sim}{\Pi}_1^1$ reflection principle to prove A8 is an indication that the i.i.d. and s.i.i.d. sets give a truly satisfactory recursion theory only on strongly admissible sets.

We now state some results which show how strong the $s - \underset{\sim}{\Pi}_1^1$ reflection principle is, and what kinds of functions are K-recursive but not $\underset{\sim}{\Sigma}_1$. We use $\bar{\bar{a}}$ for the cardinality of a and κ^+ for the least cardinal $> \kappa$. A set A is super-transitive if every subset of an element of A is an element of A.

LEMMA A9. <u>Let</u> A <u>be strongly admissible,</u> $a \in A$. <u>If</u> $\bar{\bar{a}} < \bar{\bar{A}}$ <u>then there is a</u> $b \in A$ <u>such that</u> $\bar{\bar{a}} < \bar{\bar{b}}$. <u>In particular,</u> $H(\kappa^+)$ <u>is never strongly admissible</u>.

<u>Proof</u>. Let $\underline{S}$ be a binary relation symbol, and let $\Theta(x)$ say that S is not a function with domain $\subseteq x$ and range all of A. That is, let $\Theta(x)$ be the following $s - \Pi^1_1$ formula.

$$\forall \underline{S}[\exists y,z(\underline{S}(y,z) \wedge y \notin z) \vee \exists y,z',z(\underline{S}(y,z) \wedge \underline{S}(y,z') \wedge z \neq z')$$
$$\vee \exists z \forall y \in x(\neg \underline{S}(y,z))] .$$

Then for any transitive set w with $a \in w$

$$\langle w,\epsilon\rangle \models \Theta[a] \text{ if and only if } \bar{\bar{a}} < \bar{\bar{w}} .$$

So Θ holds in A at a, hence in some transitive $w \in A$. Then $\bar{\bar{a}} < \bar{\bar{w}}$.

LEMMA A10. <u>Let</u> A <u>be strongly admissible. The relation</u> $\bar{\bar{a}} < \bar{\bar{b}}$ <u>is</u> K-<u>r.e. on</u> A. <u>If</u> A <u>is super transitive then this relation is</u> K-<u>recursive</u>.

<u>Proof</u>. Let $\Theta(x,x')$ be obtained from the formula Θ in the proof of A9 by replacing the final $\exists z$ by $\exists z \in x'$. Then

$$\langle A,\epsilon\rangle \models \Theta[a,b] \text{ if and only if } \bar{\bar{a}} < \bar{\bar{b}} .$$

The second statement is trivial since if A is super-transitive, the relation $\bar{\bar{a}} \leq \bar{\bar{b}}$ is Σ_1 on A.

COROLLARY A11. <u>Let</u> $H(\kappa)$ <u>be strongly admissible. Then</u> $\kappa = \aleph_\kappa$.

<u>Proof</u>. We know that κ is not a successor cardinal by A9. Suppose that $\kappa = \aleph_\lambda$ for some limit ordinal $\lambda < \kappa$. Define a function f by $f(\alpha)$ = the least $\beta < \kappa$ such that $\overline{\overline{f(\gamma)}} < \bar{\bar{\beta}}$ for all $\gamma < \alpha$, if α is an ordinal; = 0 otherwise. Since $H(\kappa)$ is supertransitive, the relation $\bar{\bar{a}} < \bar{\bar{b}}$ is K-recursive so f is a K-recursive function, defined for all $\alpha < \lambda$.

$$\kappa = \bigcup \{f(\alpha) : \alpha < \lambda\}$$

is an element of $H(\kappa)$, by S2 and S3. This is a contradiction.

REMARK. One could go on to prove stronger and stronger results in the usual manner. For example, let $f(\alpha)$ be the αth κ such that $\kappa = \aleph_\kappa$. If $H(\kappa)$ is strongly admissible then $f(\kappa) = \kappa$. Similarly, if κ is inaccessible and $H(\kappa)$ is strongly admissible, then κ is the κth inaccessible. Thus, we have a recursion theoretic approach to the results of Hauf [5], since $H(\kappa)$ is strongly admissible with respect to some relation R if and only if $H(\kappa)$ is $\underset{\sim}{\Sigma}_1(R)$-compact.

We conclude with some remarks on the notion of finiteness which is suggested by Kunen's notions **i.i.d.** and **s.i.i.d.** In the notations used in the definition of s.i.i.d. in §2, we say that θ *absolutely implicitly* (a.i.) *defines* P from R if and only if (1) and (2) with "$P \subseteq P'$" replaced by "$P = P'$". P is absolutely implicitly definable (a.i.d.) from R if there is some θ which a.i. defines P from R. Similarly, **a.i.d.** means a.i.d. using parameters from A. It is reasonable to suppose that if **i.i.d.** and **s.i.i.d.** are reasonable generalization of recursive and r.e. on a set A, then **a.i.d.** should be the corresponding generalization of finite.

THEOREM A12. *Let* A *be a strongly admissible set. A subset* U *of* A *is* **a.i.d.** *if and only if* U *is an element of* A.

Proof. If $U \in A$, then U is clearly **a.i.d.** For the converse, let U be a.i. defined by the formula $\theta(x_1,\dots,x_n)$ and parameters $b_1,\dots,b_n \in A$. Since every **a.i.d.** set is **i.i.d.**, U is $s - \underset{\sim}{\Delta}{}^1_1$ on A. If $U \subseteq a$ for some $a \in A$, then $U \in A$ by A6. Let $\bar{a}$ be the constant symbol $c_{(0,a)}$ for each $a \in A$, and let c be c_0, and let $\underline{U}$ be the relation symbol used in θ to denote U. Then let Φ be the $\underset{\sim}{\Sigma}_1$ set of the following sentences of $\mathcal{L}_A$:

$$\theta(\bar{b}_1,\dots,\bar{b}_n)$$

$$\underline{U}(c)$$

$$\forall x[x \in \bar{a} \leftrightarrow \bigvee_{b \in a} x \approx \bar{b}] \qquad \text{for} \quad a \in A$$

$$\neg (c \in \bar{a}) \qquad \text{for} \quad a \in A\,.$$

Since θ a.i. defines U from $b_1,\dots,b_n$, Φ has no model. So, since every strongly admissible set is $\underset{\sim}{\Sigma}$-compact, there is some $\Phi_0 \in A$ such that Φ_0 has no model. Then every $x \in U$ must be an element of $TC(\Phi_0)$, for otherwise we could get a model of Φ_0. In other words, $U \subseteq TC(\Phi_0) \in A$, which concludes the proof.

REFERENCES

[1] Barwise, J., Infinitary Logic and Admissible Sets, to appear. See also thesis by same title, Stanford University, 1967.

[2] Feferman, S. and G. Kreisel, Persistent and invariant formulas relative to theories of higher type, Bull. Amer. Math. Soc., 72 (1966), 480-485.

[3] Friedman, H. and R. Jensen, Note on Admissible Ordinals, this volume.

[4] Gödel, K., The consistency of the axiom of choice and of the generalized continuum hypothesis, Proc. Nat. Acad. of Sci., 24 (1938), 556-557.

[5] Hanf, W. Incompactness in languages with infinitely long expressions, Fund. Math., LIII (1964), 309-324.

[6] Jensen, R., and C. Karp, Primitive recursive set functions, to appear.

[7] Karp, C. Nonaxiomatizing results for infinitary systems, J. Symbolic Logic, 32 (1967), 367-384.

[8] Kreisel, G., Choice of infinitary languages by means of definability criteria; generalized recursion theory, this volume.

[9] Kunen, K., Implicit definability and infinitary languages, J. Symbolic Logic, to appear.

[10] Lévy, A., A hierarchy of formulas in set theory, Memoirs of the Amer. Math. Soc., No. 57 (1965).

[11] Lopez-Escobar, E. G. K., An interpolation theorem for denumerably long formulas, Fund. Math., LVII (1965), 254-272.

[12] Platek, R., Foundations of recursion theory, Doctoral Dissertation and Supplement, Stanford, 1966.

[13] Shoenfield, J. Mathematical Logic, Addison-Wesley, 1967.

UNIVERSITY OF CALIFORNIA, LOS ANGELES

SOME REMARKS ON THE MODEL THEORY OF INFINITARY LANGUAGES

C. C. CHANG

This paper is a more or less expository account of certain parts of the model theory of infinitary languages. Its purpose is two-fold: first, to collect together for the first time various known and original results in a single paper, and second, to expose some of the interesting techniques that are used to prove these results. The three parts of the paper deal with the connected topics of (§1) the characterization of a model up to $\infty\,\kappa$-equivalence and isomorphism, (§2) the Hanf number of $L_{\kappa^+\omega}$ and Morley numbers, and (§3) methods of finding models generated from $\kappa\omega$-indiscernibles. Most proofs that are given are quite complete, though not always spelled out in great detail. References and credits are given in the text. We believe that the techniques introduced in various places in the paper will find further use in this field of research.

1. $\lambda\kappa$-equivalence, $\infty\kappa$-equivalence, and isomorphism.

The starting point of our discussion is an ordinary first-order language L with identity and a certain number of finitary predicate and function symbols. Here we have the familiar notions of terms, atomic formulas, formulas, sentences, etc. In order to discuss the extensions $L_{\lambda\kappa}$ of L, we first specify that there shall be a supply of variables v_ξ one for each ordinal ξ. Next we assume that the connectives $\bigvee$ (disjunction) and $\bigwedge$ (conjunction) are capable of being applied to arbitrary sets or sequences of formulas. Similarly, we assume that the quantifiers $\exists$ (there exists) and $\forall$ (for all) can be applied to arbitrary sequences of variables. The set of formulas of $L_{\lambda\kappa}$ is denoted by $F_{\lambda\kappa}$ and is defined as follows: $F_{\lambda\kappa}$ is the smallest set Φ of formulas such that

(i) Φ contains all atomic formulas in the variables v_ξ, $\xi < \kappa$;

Research partially supported by NSF Grant 5200.

(ii) Φ is closed under $\neg$;

(iii) Φ is closed under $\bigvee$ and $\bigwedge$ of fewer than λ formulas, provided the results of $\bigvee$ and $\bigwedge$ contain fewer than κ free variables;

(iv) Φ is closed under universal $\forall$ and existential $\exists$ quantifications over sequences of variables from $\{v_\xi : \xi < \kappa\}$ of length less that κ.

In case $\kappa = \lambda = \omega$, we see that $F_{\omega\omega}$ is the usual set of finitary formulas of L. Henceforth we shall assume in §1 that $\lambda \geq \kappa$, λ, κ are infinite cardinals, and λ is regular. It is obvious that $F_{\lambda\kappa}$ is a monotonically increasing function in both λ and κ. For any fixed κ, we let $F_{\infty\kappa} = \bigcup_\lambda F_{\lambda\kappa}$. As examples of other possible types of formulas of $L_{\lambda\kappa}$ we shall define $E_{\lambda\kappa}$, the set of existential formulas of $L_{\lambda\kappa}$, and $P_{\lambda\kappa}$, the set of positive formulas of $L_{\lambda\kappa}$, as follows: $E_{\lambda\kappa}$ is the smallest set Φ of formulas such that Φ contains all atomic and negations of atomic formulas in the variables v_ξ, $\xi < \kappa$, and which is closed under (iii) and the application of $\exists$ as in (iv). Similarly, $P_{\lambda\kappa}$ is the smallest set Φ of formulas such that (i), (iii), and (iv) hold. The sets $E_{\omega\omega}$ and $P_{\omega\omega}$ give rise to the ordinary first-order existential and positive formulas. We let $E_{\infty\kappa} = \bigcup_\lambda E_{\lambda\kappa}$ and $P_{\infty\kappa} = \bigcup_\lambda P_{\lambda\kappa}$. We shall refer to $F_{\infty\kappa}$, $E_{\infty\kappa}$, $P_{\infty\kappa}$ as the formulas, existential formulas, positive formulas of the language $L_{\infty\kappa}$. For a thorough and systematic study of $L_{\lambda\kappa}$ we refer the reader to Karp [8].

Models for L are denoted by $\mathfrak{A} = \langle A, \ldots\rangle$, $\mathfrak{B} = \langle B, \ldots\rangle$, etc. By convention, α shall always denote the cardinal of $\mathfrak{A}$, $\alpha = |A|$, and similarly β shall denote the cardinal of $\mathfrak{B}$, $\beta = |B|$. We assume that the reader is familiar with the model-theoretical notions of $\mathfrak{A}$ being a submodel of $\mathfrak{B}$, $\mathfrak{A} \subset \mathfrak{B}$, $\mathfrak{A}$ being an extension of $\mathfrak{B}$, $\mathfrak{A} \supset \mathfrak{B}$, $\mathfrak{A}$ is isomorphic to $\mathfrak{B}$, $\mathfrak{A} \cong \mathfrak{B}$, and $\mathfrak{A}$ is homomorphic to $\mathfrak{B}$, $\mathfrak{A} \simeq \mathfrak{B}$. We also assume the notion of satisfaction for all languages $L_{\lambda\kappa}$, so that it is clear what we mean by $\mathfrak{A} \equiv_{\lambda\kappa} \mathfrak{B}$ ($\mathfrak{A}$ and $\mathfrak{B}$ are $\lambda\kappa$-equivalent) and $\mathfrak{A} \prec_{\lambda\kappa} \mathfrak{B}$ ($\mathfrak{A}$ is an $\lambda\kappa$-elementary submodel of $\mathfrak{B}$). We write $\mathfrak{A} \equiv_{\infty\kappa} \mathfrak{B}$ ($\mathfrak{A}$ and $\mathfrak{B}$ are $\infty\kappa$-equivalent) if $\mathfrak{A} \equiv_{\lambda\kappa} \mathfrak{B}$ for all λ; similarly, we write $\mathfrak{A} \prec_{\infty\kappa} \mathfrak{B}$ ($\mathfrak{A}$ is an $\infty\kappa$-elementary submodel of $\mathfrak{B}$) if $\mathfrak{A} \prec_{\lambda\kappa} \mathfrak{B}$ for all λ. In general, if G is any collection of formulas of $F_{\infty\kappa}$, we write $\mathfrak{A}(G)\mathfrak{B}$ to mean that

every sentence in G holding in $\mathfrak{A}$ holds in $\mathfrak{B}$.

(Note that $\mathfrak{A}(G)\mathfrak{B}$ is not necessarily a symmetric relation.) Thus $\mathfrak{A}(E_{\lambda\kappa})\mathfrak{B}$, $\mathfrak{A}(P_{\lambda\kappa})\mathfrak{B}$, $\mathfrak{A}(E_{\infty\kappa})\mathfrak{B}$, $\mathfrak{A}(P_{\infty\kappa})\mathfrak{B}$ are all meaningful statements about the models $\mathfrak{A}$ and $\mathfrak{B}$. The following facts are easy to verify:

$$\text{If } \mathfrak{A} \cong \mathfrak{B}, \text{ then } \mathfrak{A} \equiv_{\infty\kappa} \mathfrak{B}.$$

$$\text{If } \mathfrak{A} \subset \mathfrak{B}, \text{ then } \mathfrak{A}(E_{\infty\kappa})\mathfrak{B}.$$

$$\text{If } \mathfrak{A} \simeq \mathfrak{B}, \text{ then } \mathfrak{A}(P_{\infty\kappa})\mathfrak{B}.$$

We are now ready to consider the following simple question: Given two models $\mathfrak{A}$ and $\mathfrak{B}$ of powers α and β, respectively, what is the least λ such that $\mathfrak{A} \equiv_{\lambda\kappa} \mathfrak{B}$ implies $\mathfrak{A} \equiv_{\infty\kappa} \mathfrak{B}$? We shall see below that λ can be calculated very easily from α, β, and κ. So λ is independent of $\mathfrak{A}$, $\mathfrak{B}$, and (the number of symbols) of L. (We may, of course, ask the same question with $F_{\lambda\kappa}$ replaced by $E_{\lambda\kappa}$, or $P_{\lambda\kappa}$, and $F_{\infty\kappa}$ replaced by $E_{\infty\kappa}$, or $P_{\infty\kappa}$. Again λ exists and is independent of the number of symbols of L.) Readers who are already familiar with the result of Scott [17], will know that if L is countable and $\alpha = \beta = \kappa = \omega$, then $\lambda = \omega_1$. It turns out that this is still true even if L is not countable. However, one then can not obtain a single sentence of $L_{\omega_1\omega}$ which characterizes $\mathfrak{A}$ up to countable isomorphic models. The situation (and indeed the proof) is somewhat like the case of finite models for $L_{\omega\omega}$. If $\mathfrak{A}$ is finite then of course a single sentence of $L_{\omega\omega}$ characterizes $\mathfrak{A}$ up to isomorphism, provided L has only a finite number of symbols. On the other hand if L has an infinite number of symbols then no single sentence of $L_{\omega\omega}$ will do the job. However, it is still true that no matter how many symbols are in L, equivalence implies isomorphism for finite models. In general, $\infty\kappa$-equivalence does not imply isomorphism; it will if the cofinality of $\kappa = \omega$. The following sequence of propositions will lead us gradually to Scott's theorem and generalizations thereof.

We need some further notation. If γ is a cardinal, we let

$$\gamma^{*^\kappa} = \sum_{\mu<\kappa} [\gamma^\mu]^+$$

where μ ranges over cardinals less than κ and γ^μ is the cardinal exponentiation of γ to the μth power, and $[\gamma^\mu]^+$ denotes the cardinal successor of γ^μ. If ν is an ordinal and $a \in {}^\nu A$, then $(\mathfrak{A},a)$ denotes the model obtained from $\mathfrak{A}$ by adjoining the sequence of constants a_ξ, $\xi < \nu$. The assertion $(\mathfrak{A},a) \equiv_{\lambda\kappa} (\mathfrak{B},b)$ for $a \in {}^\nu A$, $b \in {}^\nu B$, $\nu < \kappa$, can be made in the language $(L')_{\lambda\kappa}$ where L' contains all the symbols of L plus an individual constant symbol c_ξ for each $\xi < \nu$. It also has the following interpretation: for every formula $\varphi \in F_{\lambda\kappa}$ with at most the variables v_ξ, $\xi < \nu$, free if $\mathfrak{A} \models \varphi[a]$ then $\mathfrak{B} \models \varphi[b]$. In general the meaning of expressions like $(\mathfrak{A},a)(E_{\lambda\kappa})(\mathfrak{B},b)$,

$(\mathfrak{A},a)(P_{\lambda\kappa})(\mathfrak{B},b)$ should be clear. For each ordinal ξ we let $F_{\xi\kappa}$ $(E_{\xi\kappa},P_{\xi\kappa})$ be the collection of all formulas of $F_{\infty\kappa}$ $(E_{\infty\kappa},P_{\infty\kappa})$ with quantifier rank at most ξ. Evidently, each formula of $F_{\xi\kappa}$ $(E_{\xi\kappa},P_{\xi\kappa})$ has fewer than κ free variables, and we have

$$F_{\infty\kappa} = \bigcup_{\xi} F_{\xi\kappa}, \quad E_{\infty\kappa} = \bigcup_{\xi} E_{\xi\kappa}, \quad P_{\infty\kappa} = \bigcup_{\xi} P_{\xi\kappa}.$$

Note that the meaning of $F_{\lambda\kappa}$ is different depending on whether we regard λ as a regular cardinal or as an ordinal. We hope no confusion will arise through this double usage.

PROPOSITION 1. (i) Suppose that $\mathfrak{A}(E_{\lambda\kappa})\mathfrak{B}$ and $\lambda \geq \beta^{*\kappa}$. Then

(1) for every $\mu < \kappa$ and every $a \in {}^{\mu}A$, there is a $b \in {}^{\mu}B$ such that $(\mathfrak{A},a)(E_{\lambda\kappa})(\mathfrak{B},b)$.

(ii) Suppose that $\mathfrak{A}(P_{\lambda\kappa})\mathfrak{B}$ and $\lambda \geq \alpha^{*\kappa} + \beta^{*\kappa}$. Then

(2) for every $\mu < \kappa$ and every $a \in {}^{\mu}A$, there is a $b \in {}^{\mu}B$ such that $(\mathfrak{A},a)(P_{\lambda\kappa})(\mathfrak{B},b)$, and
for every $\mu < \kappa$ and every $b \in {}^{\mu}B$, there is an $a \in {}^{\mu}A$ such that $(\mathfrak{A},a)(P_{\lambda\kappa})(\mathfrak{B},b)$.

Proof. (i) Let $\mu < \kappa$ and $a \in {}^{\mu}A$. Suppose there is no $b \in {}^{\mu}B$ so that (1) holds. Then for each $b \in {}^{\mu}B$, there is a formula $\varphi_b(x) \in E_{\lambda\kappa}$ such that $\mathfrak{A} \models \varphi_b[a]$ but not $\mathfrak{B} \models \varphi_b[b]$. The sentence $\varphi = \exists x \bigwedge_{b \in {}^{\mu}B} \varphi_b(x)$ is in $E_{\lambda\kappa}$ and $\mathfrak{A} \models \varphi$. So $\mathfrak{B} \models \varphi$. But this leads to the contradiction that $\mathfrak{B} \models \varphi_b[b]$ for some $b \in {}^{\mu}B$.

(ii) The first part of (2) goes exactly as in (1), we only need to observe that $P_{\lambda\kappa}$ is also closed under $\exists$ and $\bigwedge$. For the second part of (2) we agrue as follows. Let $\mu < \kappa$ and $b \in {}^{\mu}B$. If no $a \in {}^{\mu}A$ satisfies the conclusion then for each $a \in {}^{\mu}A$ there is a formula $\varphi_a(x) \in P_{\lambda\kappa}$ such that $\mathfrak{A} \models \varphi_a[a]$ and not $\mathfrak{B} \models \varphi_a[b]$. Let $\varphi = \forall x \bigvee_{a \in {}^{\mu}A} \varphi_a(x)$. Clearly $\varphi \in P_{\lambda\kappa}$ and $\mathfrak{A} \models \varphi$. So $\mathfrak{B} \models \varphi$, but, again this leads to the contradiction that $\mathfrak{B} \models \varphi_a[b]$ for some $a \in {}^{\mu}A$.

An immediate consequence of Proposition 1 is:

PROPOSITION 2. (i) Suppose that $\mathfrak{A}(E_{\lambda\kappa})\mathfrak{B}$ and $\lambda \geq \beta^{*\kappa}$. Then $\mathfrak{A}(E_{\infty\kappa})\mathfrak{B}$.

(ii) Suppose that $\mathfrak{A}(P_{\lambda\kappa})\mathfrak{B}$ and $\lambda \geq \alpha^{*\kappa} + \beta^{*\kappa}$. Then $\mathfrak{A}(P_{\infty\kappa})\mathfrak{B}$.

<u>Proof</u>. We prove (i) and the proof of (ii) will be evident. We show the following by induction on the ordinal ξ:

(3) $\left\{ \text{for all models } \mathfrak{A}, \mathfrak{B} \text{ with } \lambda \geq \beta^{*\kappa}, \text{ if } \mathfrak{A}(E_{\lambda\kappa})\mathfrak{B} \text{ then } \mathfrak{A}(E_{\xi\kappa})\mathfrak{B}. \right.$

Here $E_{\xi\kappa}$ is the set of existential formulas of $E_{\infty\kappa}$ with quantifier rank at most ξ. Note that every sentence of a limit quantifier rank is a Boolean combination of sentences with smaller ranks. So we only need to show that if (3) holds for ξ, then it holds for $\xi + 1$. Let $\mathfrak{A}, \mathfrak{B}$ be given so that $\lambda \geq \beta^{*\kappa}$. Let $\varphi = \exists x \psi(x)$ where ψ is of rank ξ, and suppose that $a \in {}^{\mu}A$ and $\mathfrak{A} \models \psi[a]$. By (1), there is a $b \in {}^{\mu}B$ so that $(\mathfrak{A},a)(E_{\lambda\kappa})(\mathfrak{B},b)$. Since the power of $(\mathfrak{B},b)$ is still only β, the induction hypothesis on (3) applies, so $(\mathfrak{A},a)(E_{\xi\kappa})(\mathfrak{B},b)$. This shows that $\mathfrak{B} \models \psi[b]$. So $\mathfrak{B} \models \varphi$ and the induction is complete. For (ii) we show by induction on ξ that

(4) $\left\{ \text{for all models } \mathfrak{A}, \mathfrak{B} \text{ with } \lambda \geq \alpha^{*\kappa} + \beta^{*\kappa}, \text{ if } \mathfrak{A}(P_{\lambda\kappa})\mathfrak{B} \text{ then } \mathfrak{A}(P_{\xi\kappa})\mathfrak{B}. \right.$

Here the argument varies as whether $\varphi = \exists x \psi$ or $\varphi = \forall x \psi$. In the first case use part one of (2); in the second case use part two of (2).

The basic idea for the proof of Proposition 2 in the case of $L_{\omega\omega}$ goes back to Ehrenfeucht and Fraïssé, see the discussion and results in Karp [9]. The presentation given here boils everything down to the essentials.

Another consequence of Proposition 1 is the following. Given two sequences $a, \bar{a} \in {}^{\kappa}A$, we say that $\bar{a}$ <u>extends</u> a if the range of $\bar{a}$ includes the range of a.

PROPOSITION 3. (i) Suppose that $\mathfrak{A}(E_{\lambda\kappa})\mathfrak{B}$, $\lambda \geq \beta^{*\kappa}$, $a \in {}^{\kappa}A$, and $cf(\kappa) = \omega$. Then there is a $b \in {}^{\kappa}B$ such that $(\mathfrak{A},a)(E_{\omega\omega})(\mathfrak{B},b)$.

(ii) Suppose that $\mathfrak{A}(P_{\lambda\kappa})\mathfrak{B}$, $\lambda \geq \alpha^{*\kappa} + \beta^{*\kappa}$, $a \in {}^{\kappa}A$, $b \in {}^{\kappa}B$, and $cf(\kappa) = \omega$. Then there are extensions $\bar{a} \in {}^{\kappa}A$ of a and $\bar{b} \in {}^{\kappa}B$ of b such that $(\mathfrak{A},\bar{a})(P_{\omega\omega})(\mathfrak{B},\bar{b})$.

Proof. Suppose $\kappa = \Sigma_{n\in\omega} \kappa_n$ where κ_n is an increasing sequence of cardinals less than κ. To prove (i) we suppose that the given sequence $a \in {}^{\kappa}A$ is a concatenation of subsequences a_n, $n \in \omega$, each of length κ_n. By Proposition 1 and a simple induction on n we can prove that there are sequences $b_n \in {}^{\kappa_n}B$ such that

$$(\mathfrak{A}, a_0, a_1, \ldots, a_n)(E_{\lambda\kappa})(\mathfrak{B}, b_0, b_1, \ldots, b_n).$$

Let b be the concatenation of the sequences b_n so that $b \in {}^{\kappa}B$. By the above we (can only) conclude that $(\mathfrak{A}, a)(E_{\omega\omega})(\mathfrak{B}, b)$. To prove (ii) we represent a and b in terms of $a_n \in {}^{\kappa_n}A$ and $b_n \in {}^{\kappa_n}B$, then by a familiar back and forth argument, we get extensions $\bar{a}$ and $\bar{b}$ of a and b so that $(\mathfrak{A}, \bar{a})(P_{\omega\omega})(\mathfrak{B}, \bar{b})$. We should point out that it is impossible to improve (in general) the conclusion to involve $E_{\lambda\kappa}$ and $P_{\lambda\kappa}$, this is because processes involving $E_{\lambda\kappa}$ or $P_{\lambda\kappa}$ are never closed under countable limits.

An easy corollary to Proposition 3 is

PROPOSITION 4. Suppose that $\alpha = \beta = \kappa$, $cf(\kappa) = \omega$, and $\lambda \geq \kappa^{*\kappa}$.

(i) If $\mathfrak{A}(E_{\lambda\kappa})\mathfrak{B}$, then $\mathfrak{A}$ is isomorphically embeddable into $\mathfrak{B}$.

(ii) If $\mathfrak{A}(P_{\lambda\kappa})\mathfrak{B}$, then $\mathfrak{A} \simeq \mathfrak{B}$.

Proof. For (i) we start with $a \in {}^{\kappa}A$ such that the range of a is A. We find $b \in {}^{\kappa}B$ as in Proposition 3. The mapping of a_ξ into b_ξ for $\xi < \kappa$ defines an isomorphic embedding of A into B, as the set $E_{\omega\omega}$ contains all atomic and negations of atomic formulas. For (ii), we start with two sequences $a \in {}^{\kappa}A$ and $b \in {}^{\kappa}B$ such that the range of a is A and the range of b is B. Obviously the extensions $\bar{a}$ and $\bar{b}$ of a and b will have ranges respectively A and B. Now the mapping of $\bar{a}_\xi$ into $\bar{b}_\xi$ for $\xi < \kappa$ defines a homomorphism of A onto B, as the set $P_{\omega\omega}$ contains all atomic formulas.

PROPOSITION 5. In the previous Propositions 1 - 4 we can replace $P_{\lambda\kappa}$ everywhere by $F_{\lambda\kappa}$ so that $\mathfrak{A}(P_{\lambda\kappa})\mathfrak{B}$ becomes $\mathfrak{A} \models_{\lambda\kappa} \mathfrak{B}$, $\mathfrak{A}(P_{\infty\kappa})\mathfrak{B}$ becomes $\mathfrak{A} \equiv_{\infty\kappa} \mathfrak{B}$, and $\mathfrak{A} \simeq \mathfrak{B}$ becomes $\mathfrak{A} \cong \mathfrak{B}$. In particular we have:

(i) If $\lambda \geq \alpha^{*\kappa} + \beta^{*\kappa}$, then $\mathfrak{A} \equiv_{\lambda\kappa} \mathfrak{B}$ if and only if $\mathfrak{A} \equiv_{\infty\kappa} \mathfrak{B}$.

(ii) If $\alpha = \beta = \kappa$, $cf(\kappa) = \omega$, and $\lambda \geq \kappa^{*\kappa}$, then the following are equivalent: $\mathfrak{A} \equiv_{\lambda\kappa} \mathfrak{B}$,

$\mathfrak{A} \equiv_{\infty\kappa} \mathfrak{B}$, $\mathfrak{A} \cong \mathfrak{B}$.

Proof. The proofs of (i) and (ii) depend only on an examination of the earlier proofs.

Note that in 5(ii) above, if $\alpha = \beta = \kappa = \omega$, then we may take $\lambda = \omega_1$, so that we have obtained a version of Scott's theorem with no restriction on the number of symbols in L. The reason that we have not given a parallel development for the notions involving $F_{\lambda\kappa}$ and $\equiv_{\lambda\kappa}$ in Propositions 1 - 4 is that (several people noticed that) 5(i) can be improved by only specifying that $\lambda \geq \alpha^{*\kappa}$. In other words it is sufficient if $\lambda \geq \min(\alpha^{*\kappa}, \beta^{*\kappa})$. The following proof of Proposition 6 was pointed out to me by David Kueker. I was only acquainted with an improvement of 5(i) in case the language L has a small number of symbols. (See Proposition 7 and remarks following it.)

PROPOSITION 6. If $\mathfrak{A} \equiv_{\lambda\kappa} \mathfrak{B}$ and $\lambda \geq \alpha^{*\kappa}$, then $\mathfrak{A} \equiv_{\infty\kappa} \mathfrak{B}$.

Proof. Note that we shall have to make use of the negation connective, as well as the connective $\rightarrow$ which can be defined from $\neg$ and $\vee$. We first make sure that condition (2) of Proposition 1 holds with $P_{\lambda\kappa}$ replaced by $F_{\lambda\kappa}$. It is clear that the second part of (2) will hold, by the argument given there. So, it is sufficient to see that the first part holds. Let $\mu < \kappa$ and $a \in {}^{\mu}A$. For each $b \in {}^{\mu}A$ such that $(\mathfrak{A}, a)$ is not $\lambda\kappa$-equivalent to $(\mathfrak{A}, b)$, find a formula $\varphi_b \in F_{\lambda\kappa}$ such that $\mathfrak{A} \models \varphi_b[a]$ and not $\mathfrak{A} \models \varphi_b[b]$. Let φ be the conjunction of all such φ_b, one for each such $b \in {}^{\mu}A$; φ is still a formula of $F_{\lambda\kappa}$. Note that for any formula $\psi \in F_{\lambda\kappa}$ with at most v_ξ, $\xi < \mu$, free, if $\mathfrak{A} \models \psi[a]$, then we have $\mathfrak{A} \models \forall x (\varphi(x) \rightarrow \psi(x))$. This is because every sequence $b \in {}^{\mu}A$ which satisfies φ, must be $\lambda\kappa$-equivalent with the sequence a. Clearly also $\mathfrak{A} \models \exists x \varphi(x)$. From this and the hypothesis we have:

$$\begin{cases} \text{there is a } b \in {}^{\mu}B \text{ such that } \mathfrak{B} \models \varphi[b], \text{ and furthermore} \\ \mathfrak{B} \models \forall x (\varphi(x) \rightarrow \psi(x)) \text{ for each } \psi \text{ such that } \mathfrak{A} \models \psi[a]. \end{cases}$$

This proves that $(\mathfrak{A}, a) \equiv_{\lambda\kappa} (\mathfrak{B}, b)$. So condition (2) of Proposition 1 is verified. We now proceed as in the proof of Proposition 2 to get the conclusion that $\mathfrak{A} \equiv_{\infty\kappa} \mathfrak{B}$.

So far, in none of the propositions have we used the fact that the number of symbols in L may be small when compared with the cardinal λ. Using an idea originally due to Scott [17] we see that under certain circumstances a single Scott sentence of $L_{\lambda\kappa}$ will serve to characterize a model $\mathfrak{A}$ up to $\infty\kappa$-equivalence, and whence, in case $\mathrm{cf}(\kappa) = \omega$ and $\alpha = \kappa$, up to isomorphism for models of power κ. We let $\alpha^{\underline{\kappa}} = \Sigma_{\mu<\kappa}\alpha^{\mu}$. Note that $\alpha^{\underline{\kappa}} \leq \alpha^{*\kappa} \leq [\alpha^{\underline{\kappa}}]^{+}$ and that $[\alpha^{\underline{\kappa}}]^{+}$ is always the least regular $\lambda \geq \alpha^{*\kappa}$.

PROPOSITION 7. Let ρ be the cardinal of the number of symbols in L, and let $\lambda = (\alpha^{\underline{\kappa}} + \rho)^{+}$. Then for every model $\mathfrak{A}$ of power α, there is a single sentence $\varphi_{\mathfrak{A}}$ of $L_{\lambda\kappa}$ such that for all models $\mathfrak{B}$,

$$\mathfrak{B} \models \varphi_{\mathfrak{A}} \text{ if and only if } \mathfrak{A} \equiv_{\infty\kappa} \mathfrak{B}.$$

Proof. Let $\mathfrak{A}$ be a model of power α, and let a, b range over all sequences in the set ${}^{\underline{\kappa}}A = \bigcup_{\mu<\kappa} {}^{\mu}A$. We see that $|{}^{\underline{\kappa}}A| = \alpha^{\underline{\kappa}} < \lambda$. For each $a \in {}^{\underline{\kappa}}A$ and ordinal $\xi < \lambda$ we define a formula φ_a^{ξ} in $F_{\lambda\kappa}$ by induction on ξ as follows: (If a is a μ-termed sequence then φ_a^{ξ} will be a formula with exactly the free variables $x = \langle v_{\xi}\rangle_{\xi<\mu}$. The symbol $\emptyset$ is used to denote the empty sequence.)

$\varphi_a^{0}(x)$ is the conjunction of all atomic and negation of atomic formulas satisfied by a in the model $\mathfrak{A}$ under the interpretation of a_{ξ} for v_{ξ}. If $a = \emptyset$, φ_a^{0} shall be any true proposition. Suppose $\varphi_a^{\eta}(x)$ has been defined for all $a \in {}^{\underline{\kappa}}A$ and all $\eta < \xi$. If ξ is a limit ordinal, we let

$$\varphi_a^{\xi}(x) = \bigwedge_{\eta<\xi} \varphi_a^{\eta}(x).$$

If $\xi = \eta + 1$, we let

$$\varphi_a^{\xi}(x) = \varphi_a^{\eta}(x) \wedge \bigwedge_{b\in {}^{\underline{\kappa}}A} \left[\exists y\, \varphi^{\eta}_{a\frown b}(x\frown y)\right] \wedge \bigwedge_{\mu<\kappa} \left[\forall z \bigvee_{b\in {}^{\mu}A} \varphi^{\eta}_{a\frown b}(x\frown z)\right].$$

In the above formula it is understood that (i) y is a sequence of variables of the same length as the given sequence b, (ii) $a\frown b$ and $x\frown y$ are the natural concatenations of the sequences a and b and x and y, (iii) z is a sequence of variables of length μ and $x\frown z$ is again the natural concatenation of x and z. Each of the following properties has a simple proof based on an induction on ξ. We omit the details.

(1) φ_a^{ξ} is a formula of $F_{\lambda\kappa}$ with quantifier rank ξ;

$\varphi_{\emptyset}^{\xi}$ is a sentence of $F_{\lambda\kappa}$ with rank ξ;

(2) $\mathfrak{A} \models \varphi_a^{\xi}[a]$ for all $a \in {}^{\underline{\kappa}}A$;

(3) if $\mu < \kappa$ and $a, b \in {}^{\mu}A$, then the following are equivalent:

$$\mathfrak{A} \models \varphi_a^\xi[b], \quad (\mathfrak{A},a) \equiv_{\xi\kappa} (\mathfrak{A},b), \quad \mathfrak{A} \models \varphi_b^\xi[a].$$

Practically the same inductive proofs for (1) - (3) will also establish the following properties for an arbitrary model $\mathfrak{B}$:

(4) if $\mu < \kappa$, $a \in {}^{\mu}A$, $b \in {}^{\mu}B$, then for all $\eta \leq \xi$

$$\mathfrak{B} \models \varphi_a^\xi[b] \rightarrow \varphi_a^\eta[b];$$

(5) if $\mu < \kappa$, $a \in {}^{\mu}A$, $b \in {}^{\mu}B$, then the following are equivalent:

$$\mathfrak{B} \models \varphi_a^\xi[b], \quad (\mathfrak{A},a) \equiv_{\xi\kappa} (\mathfrak{B},b).$$

Note that in (5) it does not make sense to say that $\mathfrak{A} \models \varphi_b^\xi[a]$ as the formulas φ_a^ξ are only defined for sequences $a \in {}^{\kappa\!\smile}\!A$. However, there is a symmetic form of (5) just like (3) if we bothered to go through the same inductive definition of ψ_b^ξ for all $b \in {}^{\kappa\!\smile}\!B$. To obtain the sentence $\varphi_{\mathfrak{A}}$ we proceed as follows: Given $a, b \in {}^{\mu}A$, let $f(a,b)$ be the least ordinal $\xi < \lambda$ such that $\mathfrak{A} \models \neg\, \varphi_a^\xi[b]$, if such exists; otherwise set $f(a,b) = 0$. Note that by (3), $f(a,b) = f(b,a)$. Let ξ_0 be an ordinal less than λ which is greater than the sup of $\{f(a,b) : a,b \in {}^{\mu}A,\ \mu < \kappa\}$. By the definition of $f(a,b)$ we have

$$\text{for all } \mu < \kappa, \text{ all } a, b \in {}^{\mu}A, \quad \mathfrak{A} \models \varphi_a^{\xi_0}[b] \rightarrow \varphi_a^{\xi_0+1}[b].$$

Now let

$$\varphi_{\mathfrak{A}} = \varphi_{\emptyset}^{\xi_0} \wedge \bigwedge_{\mu<\kappa} \; \bigwedge_{a\in{}^{\mu}A} \; \forall x \left(\varphi_a^{\xi_0}(x) \rightarrow \varphi_a^{\xi_0+1}(x) \right).$$

In the above formula, x is a μ-termed sequence of variables. It is quite clear that $\varphi_{\mathfrak{A}} \in F_{\lambda\kappa}$, in fact the quantifier rank of $\varphi_{\mathfrak{A}}$ is $\xi_0 + 2$. It is also clear that if $\mathfrak{B} \equiv_{\infty\kappa} \mathfrak{A}$ then $\mathfrak{B} \models \varphi_{\mathfrak{A}}$. We now show

$$\text{if } \mathfrak{B} \models \varphi_{\mathfrak{A}} \text{ then } \mathfrak{A} \equiv_{\infty\kappa} \mathfrak{B}.$$

As in the proof of Proposition 6, we first verify that condition (2) of Proposition 1 holds in the following special form:

(6) Let $a \in {}^{\mu}A$, $b \in {}^{\mu}B$ be such that $(\mathfrak{A},a) \equiv_{\xi_0\kappa} (\mathfrak{B},b)$, then

$$\begin{cases} \text{for all } \nu < \kappa, \text{ all } a' \in {}^{\nu}A, \text{ there is a } b' \in {}^{\nu}B \text{ such that } (\mathfrak{A}, a \frown a') \equiv_{\xi_0\kappa} (\mathfrak{B}, b \frown b'), \\ \text{for all } \nu < \kappa, \text{ all } b' \in {}^{\nu}B, \text{ there is an } a' \in {}^{\nu}A \text{ such that } (\mathfrak{A}, a \frown a') \equiv_{\xi_0\kappa} (\mathfrak{B}, b \frown b'). \end{cases}$$

Let $\mathfrak{B} \models \varphi_{\mathfrak{A}}$, and let μ, a, b as in (6). By (5) we have $\mathfrak{B} \models \varphi_a^{\xi_0}[b]$. Using $\varphi_{\mathfrak{A}}$ we see that

$$\mathfrak{B} \models \varphi_a^{\xi_0+1}[b].$$

If we now decipher $\varphi_a^{\xi_0+1}$ from the definition we see that

$$\begin{cases} \text{for all } \nu < \kappa, \text{ all } a' \in {}^{\nu}A, \text{ there is a } b' \in {}^{\nu}B \text{ such that } \mathfrak{B} \models \varphi^{\xi_0}_{a \frown a'} [b \frown b'], \\ \text{for all } \nu < \kappa, \text{ all } b' \in {}^{\nu}B, \text{ there is an } a' \in {}^{\nu}A \text{ such that } \mathfrak{B} \models \varphi^{\xi_0}_{a \frown a'} [b \frown b']. \end{cases}$$

If we translate the conclusions above by (5) we have the desired conclusions of (6).

Once (6) is established, then a simple induction like the one given in Proposition 2, but only for the particular models $(\mathfrak{A}, a)$ and $(\mathfrak{B}, b)$, will prove that $\mathfrak{A} \equiv_{\infty\kappa} \mathfrak{B}$.

If we put in some specific values into Proposition 7, say $\alpha = \kappa$ and $\rho \leq \kappa^{\underline{\kappa}}$, then every model $\mathfrak{A}$ of power κ can be characterized up to $\infty\kappa$-equivalence by a single sentence of $L_{\lambda\kappa}$ with $\lambda = [\kappa^{\underline{\kappa}}]^+$. If in addition we assume that $cf(\kappa) = \omega$, then by Proposition 5 every model $\mathfrak{A}$ of power κ can be characterized up to isomorphism in the power κ by a single sentence of $L_{\lambda\kappa}$, $\lambda = [\kappa^{\underline{\kappa}}]^+$. If $\kappa = \omega$, then $\lambda = [\omega^{\underline{\omega}}]^+ = \omega_1$, so we have Scott's result.

We conclude this section with some remarks which will put some limitations on possible generalizations of these propositions. First of all, there is an (unpublish) example due to M. Morley of two models $\mathfrak{A}$ and $\mathfrak{B}$ of power ω_1 such that

$$\mathfrak{A} \equiv_{\infty\omega_1} \mathfrak{B} \quad \text{but} \quad \mathfrak{A} \not\cong \mathfrak{B}.$$

The models $\mathfrak{A}$, $\mathfrak{B}$ each has only one binary relation. This example of Morley's extends easily to any regular $\kappa > \omega$, that is: there are two models $\mathfrak{A}$ and $\mathfrak{B}$ of power κ such that $\mathfrak{A} \equiv_{\infty\kappa} \mathfrak{B}$ but $\mathfrak{A} \not\cong \mathfrak{B}$. The situation for a singular κ whose cofinality is greater than ω is not known. Secondly, one might ask if Propositions 1 - 4 can be improved in the same way Proposition 5 is improved by Propositions 6 and 7. It turns out that one can not hope to show that every countable model $\mathfrak{A}$ (assuming L is also countable) has an existential sentence $\varphi_{\mathfrak{A}} \in E_{\omega_1\omega}$ such that for every model $\mathfrak{B}$,

$\mathfrak{A}$ is isomorphically embeddable into $\mathfrak{B}$ if and only if $\mathfrak{B} \models \varphi_{\mathfrak{A}}$. A counterexample can be found in Barwise [1] to the effect that every ordering relation S stands in the relation $(E_{\omega_1\omega})$ to the ordering on ω_1, i.e. $S(E_{\omega_1\omega})\omega_1$. If we now take S to be the ordering on the negative integers, $S = \omega^*$, then $\omega^*(E_{\omega_1\omega})\omega_1$. So if such a characterizing sentence $\varphi_{\omega^*} \in E_{\omega_1\omega}$ exists, then ω^* is isomorphically embeddable in ω_1, an obvious contradiction. We believe that similar counterexamples will show that the λ in Propositions 1(ii), 2(ii), and 3(ii) can not be reduced to $\lambda \geq \min(\alpha^{*\kappa}, \beta^{*\kappa})$, although we have no simple counterexamples at hand. Exactly the same problem, that is finding a counterexample, for the set of universal formulas of $L_{\omega_1\omega}$ is left as an open question in Barwise [1]. Just as in the case of first-order languages, we may combine notions involving E and P or find other interesting classes of formulas of $F_{\infty\kappa}$ which may give rise to results of the type represented here.

Finally, we should point out that the results in this section are just some simple beginnings of the model theory of the languages $L_{\lambda\kappa}$. The reader should certainly also study the papers of Kueker [10] and Malitz [13] in this same volume for other similar results.

HISTORICAL REMARK. As far as I know, I was the first one to notice that Scott's theorem originally stated for countable languages, countable models and $L_{\omega_1\omega}$ extends to cardinals κ with cofinality ω, with or without restriction on the number of symbols in L. As far as the other results concerning $F_{\infty\kappa}$, $E_{\infty\kappa}$, and $P_{\infty\kappa}$ are concerned, some of them have already been noticed by other people, at least in the case of countable languages and $\kappa = \omega$. (See for instance, Barwise [1] and Lopez-Escobar [11].)

2. <u>Hanf number of $L_{\kappa^+\omega}$ and Morley numbers.</u>

In this section we assume that κ is a fixed infinite cardinal and that the language $L_{\kappa^+\omega}$ has at most κ symbols, so that $|F_{\omega\omega}| \leq \kappa$. A type Σ (of $L_{\omega\omega}$) in the variables $x_1,\ldots,x_n$ is a set of formulas $\sigma(x_1,\ldots,x_n)$ of $F_{\omega\omega}$ with at most the variables $x_1,\ldots,x_n$ free. Notice that a type can consist of a single sentence of $L_{\omega\omega}$. Since $|F_{\omega\omega}| \leq \kappa$, the total number of types is at most 2^κ. Since the discussion for types in a finite number of free variables or in a single free variable is entirely analogous, we shall only use types in the single variable x in the following definitions. A type Σ is realized in a model $\mathfrak{A}$ for L if there is an element $a \in A$ such that $\mathfrak{A} \models \sigma[a]$ for each $\sigma \in \Sigma$; in this case we say that a realizes Σ in $\mathfrak{A}$ and we write $\mathfrak{A} \models \Sigma[a]$. If Σ is not realized in $\mathfrak{A}$ then it is omitted in $\mathfrak{A}$. A set $\mathfrak{S}$ of types is omitted in $\mathfrak{A}$ if each $\Sigma \in \mathfrak{S}$ is omitted in $\mathfrak{A}$.

The Hanf number h_κ of $L_{\kappa^+\omega}$ is defined as the least cardinal λ such that every sentence of $L_{\kappa^+\omega}$ having a model of power λ has arbitrarily large models. The Morley number m_κ is defined as the least cardinal λ such that every set $\mathfrak{S}$ of at most κ types which is omitted by a model of power λ is omitted by arbitrarily large models. A second Morley number n_κ is defined as the least cardinal λ such that every set $\mathfrak{S}$ of types which is omitted by a model of power λ is omitted by arbitrarily large models.

Hanf [6] showed that the number h_κ exists (in fact, Hanf number for any language which is a set exists). Morley [12] first proved that

(A) $$m_\omega = \beth_{\omega_1} \quad \text{and} \quad \beth_{\kappa^+} \leq m_\kappa < \beth_{(2^\kappa)^+} .$$

Then Lopez-Escobar showed that (see [14] p. 273, [19] p. 88)

(B) $$m_\omega = h_\omega, \quad \text{so that} \quad h_\omega = \beth_{\omega_1} .$$

Later using the GCH, Helling [7] proved that

(C) $$\text{if } cf(\kappa) = \omega, \text{ then } m_\kappa = \beth_{\kappa^+} .$$

At about the same time I noticed that (see Vaught [19] p. 88)

(D) $$m_\kappa = h_\kappa \text{ for all } \kappa,$$

and (see [2])

(E) $$m_\kappa \leq n_\kappa \leq \beth_{(2^\kappa)^+} .$$

More recently, Morley and Morley [16] showed that, assuming $V = L$,

(F) $$\text{if } cf(\kappa) > \omega, \text{ then } \beth_{\kappa^+} < m_\kappa .$$

Silver has pointed out that the assumption $V = L$ can be eliminated in the proof of (F); instead he used some non-obvious facts about absoluteness and relative constructibility. We shall give below a very simple proof of (F) without any use of $V = L$ or the heavy machinery of relative constructibility.

This section consists of the proofs of (D), (E), and (F). Essentially each result is proved by a simple observation. It is difficult to assess exactly how useful these simple observations are, but at least in the case of (F) we know that the observation was overlooked.

Proof of (D). We first show the easy direction $m_\kappa \leq h_\kappa$. Let $\mathfrak{S} = \{\Sigma_\xi : \xi < \kappa\}$ be a set of types and suppose that $\mathfrak{A}$ is a model of power h_κ which omits each Σ_ξ. For each $\xi < \kappa$, let $\varphi_\xi = \neg \exists x \bigwedge_{\sigma \in \Sigma_\xi} \sigma$. Let $\varphi = \bigwedge_{\xi < \kappa} \varphi_\xi$. Then $\varphi \in L_{\kappa^+ \omega}$ and for all models $\mathfrak{B}$,

$$\mathfrak{B} \models \varphi \text{ if and only if } \mathfrak{B} \text{ omits each } \Sigma_\xi .$$

From this we see that $\mathfrak{A} \models \varphi$. By the definition of h_κ, there are arbitrarily large models $\mathfrak{B}$ of φ, whence there are arbitrarily large models omitting $\mathfrak{S}$.

In the other direction, let φ be a sentence of $L_{\kappa^+ \omega}$. Note that φ has at most κ subformulas in $F_{\kappa^+ \omega}$, each such subformula having at most a finite number of free variables. For each subformula σ of φ with say n free variables we introduce a new n-placed predicate symbol P_σ. We make the convention that 0-placed predicate symbols P_σ are taken as propositional constants. Let the new language $L' = L \cup \{P_\sigma : \sigma \text{ a subformula of } \varphi\}$. Consider the following set Γ of sentences of L' (in what follows x is always a finite sequence of variables appropriate to the formula):

(i) $\forall x\, (\sigma(x) \leftrightarrow P_\sigma(x))$, for every atomic subformula σ;

(ii) $\forall x\, (P_\tau(x) \leftrightarrow \neg P_\sigma(x))$, for every subformula $\tau = \neg \sigma$;

(iii) $\forall x\, (P_\tau(x) \leftrightarrow \forall y\, P_\sigma(x \frown y))$, for every subformula $\tau = \forall y\, \sigma$;

(iv) $\forall x\, (P_\tau(x) \leftrightarrow \exists y\, P_\sigma(x \frown y))$, for every subformula $\tau = \exists y\, \sigma$;

(v) $\forall x\, (P_\tau(x) \leftrightarrow \bigwedge_{\xi < \kappa} P_{\sigma_\xi}(x))$, for every subformula $\tau = \bigwedge_{\xi < \kappa} \sigma_\xi$;

(vi) $\forall x\, (P_\tau(x) \leftrightarrow \bigvee_{\xi < \kappa} P_{\sigma_\xi}(x))$, for every subformula $\tau = \bigvee_{\xi < \kappa} \sigma_\xi$;

(vii) $P_\varphi \leftrightarrow \forall x\, (x = x)$.

Observe the following three simple facts:

(1) The language L' still has at most κ symbols.

(2) There is a set $\mathfrak{S} = \{\Sigma_\xi : \xi < \kappa\}$ of types of $L'_{\omega\omega}$ such that given any model $\mathfrak{B}$ for L',

$$\mathfrak{B} \models \Gamma \text{ if and only if } \mathfrak{B} \text{ omits } \mathfrak{S}.$$

(3) Given any model $\mathfrak{A}$ for L,

$$\mathfrak{A} \models \varphi \text{ if and only if } \mathfrak{A} \text{ has an expansion } \mathfrak{A}' \text{ which}$$
$$\text{is a model for } L' \text{ and } \mathfrak{A}' \models \Gamma.$$

(An expansion $\mathfrak{A}'$ of $\mathfrak{A}$ is obtained from the model $\mathfrak{A}$ by adding a relation R_σ for each symbol P_σ. Note also that if $\mathfrak{A}$ has an expansion $\mathfrak{A}'$ which satisfies Γ, the expansion is necessarily unique.) Now let $\mathfrak{A}$ be a model of φ of power m_κ. Then an expansion $\mathfrak{A}'$ of $\mathfrak{A}$ is a model of Γ, whence $\mathfrak{A}'$ omits $\mathfrak{S}$. The model $\mathfrak{A}'$ is still of power m_κ, whence there are arbitrarily large models $\mathfrak{B}$ (for L') which omits $\mathfrak{S}$. If we now take the L-reduct of the models $\mathfrak{B}$, we will obtain arbitrarily large models of φ.

HISTORICAL REMARK. The result (D), as far as I know, was first noticed by me during the year 1963-64, when both Vaught and Helling were also at UCLA. I believe that Vaught checked the priority question with a few other people before he inserted the remark in his paper [19, p. 88]. Apparently, later Lopez-Escobar rediscovered the result and put it in [12]. Since the paper is somewhat expository in nature, I have decided to write down a proof of it.

<u>Proof of (E)</u>. First of all, it is obvious that $m_\kappa \leq n_\kappa$. To prove that $n_\kappa \leq \beth_{(2^\kappa)^+}$, let us recall briefly Morley's proof that $m_\kappa \leq \beth_{(2^\kappa)^+}$. It is essentially a combination of the Ehrenfeucht-Mostowski method [5] and generalizations of the Ramsey theorem by Erdős-Rado, see [14] for details and further references. Suppose that $\mathfrak{A}$ is a model of power $\beth_{(2^\kappa)^+}$ which omits $\mathfrak{S} = \{\Sigma_\xi : \xi < \kappa\}$. (Actually, a sequence of models $\mathfrak{A}_\xi$ each of power at least $\beth_\xi$, $\xi < (2^\kappa)^+$, is sufficient.) We may suppose that L has all the Skolem functions. Suppose that

(1) all n-placed Skolem functions of L are arranged in a sequence:

$$f_0, \ldots, f_\eta, \ldots, \eta < \kappa,$$

and suppose also that

(2) each type Σ_ξ, $\xi < \kappa$, to be omitted is also arranged in a sequence:

$$\sigma_0^\xi, \ldots, \sigma_\eta^\xi, \ldots, \eta < \kappa.$$

Morley's main idea is to show that there are a function $g \colon \kappa \times \kappa \to \kappa$ and a sequence of ordered subsets X_ξ of A:

$$X_0, \ldots, X_\xi, \ldots, \xi < (2^\kappa)^+ \quad \text{each } X_\xi \text{ of power at least } \beth_\xi,$$

such that

(3) for all $\xi, \eta < \kappa$, all $\zeta < (2^\kappa)^+$, and all $x_1 < \cdots < x_n$ from X_ζ, we have

$$A \models \neg\, \sigma^\eta_{g(\eta,\xi)}\, [f_\xi(x_1, \ldots, x_n)].$$

The reason that $\beth_{(2^\kappa)^+}$ is needed in the proof is because there are 2^κ possible functions $g : \kappa \times \kappa \to \kappa$. The rest of Morley's proof follows easily from an inductive argument on n based on (3). To show that $n_\kappa \leq \beth_{(2^\kappa)^+}$ some new twist must be added to the argument above since in general an arbitrarily set $\mathfrak{S}$ of types can be of power 2^κ, whence a quick computation shows that number of functions $g : 2^\kappa \times \kappa \to \kappa$ is 2^{2^κ}, so that the obvious bound for n_κ is $\beth_{(2^{2^\kappa})^+}$ rather than $\beth_{(2^\kappa)^+}$. (In fact, this bound of $\beth_{(2^{2^\kappa})^+}$ was asserted by Morley in a letter to the author in 1963.)

It turns out that only a very minor change is required. Let us first list all the complete types Δ realized in $\mathfrak{A}$. (A complete type Δ is a type having the property that for every formula $\sigma(x)$ of $L_{\omega\omega}$ either $\sigma \in \Delta$ or $\neg\sigma \in \Delta$.) There are at most 2^κ such types:

(4) $$\Delta_0, \ldots, \Delta_\xi, \ldots, \xi < 2^\kappa.$$

Referring again to the listings (1), (2), and (4), we now find a function $g : \kappa \to 2^\kappa$ and a sequence of ordered subsets X_ξ of A:

$$X_0,\ldots,X_\xi,\ldots,\xi < (2^\kappa)^+ \quad \text{each } X_\xi \text{ of power at least } \beth_\xi,$$

such that

(5) for all $\xi < \kappa$, all $\zeta < (2^\kappa)^+$, and all $x_1 < \cdots < x_n$ from X_ζ, we have

$$\mathfrak{A} \models \Delta_{g(\xi)}\ [f_\xi(x_1,\ldots,x_n)].$$

By an identical induction on n based on (5), we can find arbitrarily large models $\mathfrak{B}$ which realizes only these complete types (in fact, at most κ of them) from among Δ_ξ, $\xi < 2^\kappa$. Now it is clear that any such model will automatically omit all types omitted by $\mathfrak{A}$. So $n_\kappa \leq \beth_{(2^\kappa)^+}$.

REMARKS. (a) A simple cardinality argument will show the strict inequality in (A), namely, $m_\kappa < \beth_{(2^\kappa)^+}$. From this and (E) we see that

$$\text{either } m_\kappa < n_\kappa \text{ or } n_\kappa < \beth_{(2^\kappa)^+}.$$

However, we do not know which one of the two inequalities (or possibly both) holds. Even in the case $\kappa = \omega$ it is not known whether $m_\omega < n_\omega$ or $n_\omega < \beth_{(2^\omega)^+}$. The point seems to be whether one can really say much more with n_κ rather than with m_κ when altogether the language L has at most κ symbols.

(b) We might make some comments here on the Hanf number of more complicated infinitary languages, in particular, $L_{\omega_1\omega_1}$ (L countable). Let us say that a sentence φ of $L_{\omega_1\omega_1}$ is universal if it starts out with a countable sequence of universally quantified variables followed by a matrix which is quantifier-free and which is a Boolean combination of atomic formulas. (Note that this is the obvious counterpart to the formulas in $E_{\omega_1\omega_1}$ introduced in §1.) In his thesis Silver [18] showed that the Hanf number of universal sentences of $L_{\omega_1\omega_1}$ is less than the first cardinal ρ (if such exist) satisfying the partition property $\rho \to (\omega_1)^{<\omega}$. It is easy to see from the techniques introduced in the proofs of (A) and (D) that the Hanf number of sentences of $L_{\omega_1\omega_1}$ which starts with a countable universal quantifier followed by a Boolean combination of formulas in $L_{\omega_1\omega}$ is also less than the first such partition cardinal if it exists. Once this is understood it is then immediate that the Hanf number of sentences of $L_{\kappa^+\kappa^+}$ which starts with a κ-termed universal quantifier followed by a Boolean combination of formula of $L_{\kappa^+\omega}$ is less than the first cardinal ρ (if such exist) with the partition property $\rho \to (\kappa^+)^{<\omega}$. Recently Kunen showed (unpublished, private communication) that one can not hope to prove that the Hanf number of $L_{\omega_1\omega_1}$ is smaller than the first measurable cardinal if such exist. Some results concerning the poor man's Hanf number of $L_{\kappa\kappa}$ can be found in [4].

Proof of (F). Assume that $cf(\kappa) > \omega$. We suppose that the canonical well-ordering relation $<$ on κ coincides with the $\in$-relation on κ. Similarly, every ordinal $\alpha \in \kappa$ is also a subset of κ. Let A be any set of binary relations r such that r simply orders κ. We use $r \restriction \alpha$ to denote the simple ordering induced on $\alpha \in \kappa$ by r. If r happens to be a well-ordering on κ, not necessarily the canonical well-ordering of κ, then clearly each $r \restriction \alpha$ for $\alpha \in \kappa$ is a well-ordering of α. Since κ is a cardinal, it follows that each $r \restriction \alpha$ is isomorphic to some $\beta \in \kappa$ with β carrying the well-ordering given by $\in$. We shall show that for $r \in A$,

(1) r well-orders κ

if and only if

(2) for every $\alpha \in \kappa$, there is a $\beta \in \kappa$ such that $r \restriction \alpha$ is isomorphic to β.

We have already shown that (1) implies (2). For the converse we argue by contradiction. Suppose r does not well-order κ. Then there will exist an infinite r-descending sequence

$$\alpha_0 >_r \alpha_1 >_r \cdots >_r \alpha_n >_r \cdots$$

of ordinals $\alpha_n \in \kappa$. Since κ is not cofinal with ω it follows that some $\alpha \in \kappa$ is such that $\alpha_n \in \alpha$ for all $n \in \omega$. Clearly for this α, $r \restriction \alpha$ is not a well-ordering and so can not be isomorphic to any $\beta \in \kappa$. From the equivalence of (1) and (2) it follows that well-orderings of type ξ with $\kappa \leq \xi < \kappa^+$ can be singled out from the simple orderings r of κ by sentences of the form (2)

Let the language L contain at least the symbols $\leq$, and an individual constant c_ξ for each $\xi < \kappa$. With only a finite number of additional symbols we can find a set $\mathcal{S}$ of types of $L_{\omega\omega}$ such that any model for L omitting $\mathcal{S}$ will have to satisfy the following:

(i) $c_\xi \leq c_\eta$ if and only if $\xi \leq \eta < \kappa$;

(ii) the field of $\leq$ is precisely the set $\bar{\kappa} = \{c_\xi : \xi < \kappa\}$;

(iii) there is a set R of simple orderings r of $\bar{\kappa}$ such that each $r \in R$ satisfies (2);

(iv) if $r \in R$ and s is an initial segment of r then $s \in R$;

(v) given $r, s \in R$ either r is an initial segment of s or vice-versa;

(vi) if r is an initial segment of s, then $A_r \subset A_s$ (A_r, A_s are sets associated with $r, s \in R$);

(vii) if $r = s + 1$, then $|A_r| \leq 2^{|A_s|}$;

(viii) if $r = \bigcup_{s<r} s$, then $A_r = \bigcup_{s<r} A_s$;

(ix) $\leq$ is the first element of R and $A_{\leq} = 0$;

(x) the universal of the model $A = \bigcup_{r \in R} A_r$.

We see that all assertions other than (i) and (ii) can be said in a single sentence, and $\mathfrak{S}$ is a set of at most κ types. It now follows that

for each ξ, $\kappa \leq \xi \leq \kappa^+$, there is a model $\mathfrak{A}$ of power $\beth_\xi$ which omits $\mathfrak{S}$,

and

no model $\mathfrak{A}$ of power greater than $\beth_{\kappa^+}$ can omit $\mathfrak{S}$.

So $m_\kappa > \beth_{\kappa^+}$.

REMARKS. (a) Once the ordinal κ^+ is fixed by omitting $\mathfrak{S}$ (this is really the point of (i) - (x)) then standard techniques will allow us to fix the ordinals $\kappa^+ + \kappa^+$, $\kappa^+ \cdot \kappa^+$, and $(\kappa^+)^{\kappa^+}$ (ordinal exponentiation) etc, so that m_κ can correspondingly be lifted higher. In fact, one can show that m_κ is greater than any $\beth_\alpha$ where α is an ordinal recursive (in the sense of Kripke) in κ^+, (see [16]).

(b) The abstract of Morley and Morley [16] is stated in terms of κ-logic. For the related results on ω-logic, see Morley [15].

(c) The whole subject of omitting types is an interesting one. Besides the basic paper of Morley [14], we can refer the reader to [3] which contains a general discussion and some open problems.

3. <u>Methods of finding models generated from $\kappa\omega$-indiscernibles</u>.

In this section we assume κ is a regular infinite cardinal and that L has at most κ symbols; we also assume that L has all the required (first-order) Skolem functions. If X is a subset of a model $\mathfrak{A}$, then the Skolem hull of X, denoted by $\mathfrak{H}(X)$, is the closure of X under all the Skolem functions of A. It is well-known that $\mathfrak{H}(X)$ is an elementary submodel of $\mathfrak{A}$. An ordered subset X of a model $\mathfrak{A}$ is a set of $\omega\omega$-indiscernibles, or simply indiscernibles, if for any two finite strictly increasing sequences $x_1 < \cdots < x_n$, $y_1 < \cdots y_n$ of elements of X we have $(\mathfrak{A},x_1,\ldots,x_n) \equiv_{\omega\omega} (\mathfrak{A},y_1,\ldots,y_n)$. This notion was first introduced in Ehrenfeucht-Mostowski [5]. We shall assume throughout that T is a first-order theory of $L_{\omega\omega}$ with infinite models and such that all axioms for Skolem functions are among the axioms of T. Some simple properties of indiscernibles are the following.

(a) (<u>Existence theorem</u>) Given any order type τ, there is a model $\mathfrak{A}$ of T with an ordered subset X of type τ of indiscernibles.

(b) (<u>Automorphism theorem</u>) If X is a set of indiscernibles in $\mathfrak{H}(X)$, then any ordermorphism (order preserving automorphism) of X onto X induces an automorphism of $\mathfrak{H}(X)$ onto $\mathfrak{H}(X)$.

(c) (<u>Subset theorem</u>) If X is a set of indiscernibles in $\mathfrak{H}(X)$ and Y is a subset of X carrying the same ordering as X, then Y is a set of indiscernibles in $\mathfrak{H}(Y)$.

(d) (<u>Extension theorem</u>) If X is a set of indiscernibles in $\mathfrak{H}(X)$ and Y is a superset of X carrying an order which is an extension of the ordering on X, then Y can be extended to a model $\mathfrak{H}(Y)$ such that $\mathfrak{H}(X) \prec \mathfrak{H}(Y)$ and Y is a set of indiscernibles in $\mathfrak{H}(Y)$.

(e) There is a model $\mathfrak{A}$ of T of power κ which is the Skolem hull of a set of indiscernibles.

From (a) - (e) the following result of Ehrenfeucht follows easily:

(f) (<u>Tower theorem</u>) There are a set $\mathfrak{S}$ of at most κ complete types of $L_{\omega\omega}$ and an $\omega\omega$-elementary tower $\mathfrak{H}(X_\lambda)$, λ a cardinal $\geq \kappa$, of models of T such that each X_λ is of type λ, $X_\lambda \subset X_\nu$ if $\lambda \leq \nu$, X_λ is a set of indiscernibles in $\mathfrak{H}(X_\lambda)$, and each $\mathfrak{H}(X_\lambda)$ realizes only the complete types in $\mathfrak{S}$. Clearly the smallest member of the tower has power κ, and each model $\mathfrak{H}(X_\lambda)$ has power exactly λ.

In this section we are interested in finding suitable analogs of (a) through (f) for higher notions of indiscernibility. The reader will see that we have succeeded in recapturing (most of) (a) through (f) in the case of $\kappa\omega$- indiscernibility, but in general the question is almost completely untouched.

An ordered subset X of a model A is a set of $\kappa\omega$-indiscernibles if for any two finite increasing sequences $x_1 < \cdots < x_n$, $y_1 < \cdots <_n$ from X we have

$$(A, x_1, \ldots, x_n) \equiv_{\kappa\omega} (A, y_1, \ldots, y_n).$$

This notion is the natural generalization of $\omega\omega$-indiscernibility. We shall use the results (a) - (f) as a measure of our success or failure at finding models with $\kappa\omega$-indiscernibles. So, it is understood that Skolem function shall remain first-order Skolem functions, but that we wish to replace $L_{\omega\omega}$ by $L_{\kappa\omega}$, $<_{\omega\omega}$ by $<_{\kappa\omega}$, and $\omega\omega$-indiscernibility by $\kappa\omega$-indiscernibility in (a) - (f). We are also allowed to extend L by adding new symbols if necessary.

Our first remark is that if X is a set of $\omega\omega$-indiscernibles in $\mathfrak{A}$ and Y and Z are two subsets of X which are order isomorphic, then

$$(\mathfrak{A}, y)_{y \in Y} \equiv_{\omega\omega} (\mathfrak{A}, z)_{z \in Z},$$

where the elements of Y and Z are displayed in their natural order. This is because syntactic notions in the language $L_{\omega\omega}$ always have the property that something holds if and only if it holds for all finite subsets. On the other hand, because $\kappa\omega$-equivalence does not have this property if $\kappa > \omega$, the above will fail if we replace $\omega\omega$ by $\kappa\omega$ everywhere.

Our second remark is that if X is a set of $\omega\omega$-indiscernibles in $\mathfrak{H}(X)$ and X is finitely transitive, i.e. given any two finite increasing sequences $x_1 < \cdots < x_n$ and $y_1 < \cdots < y_n$ from X there is an ordermorphism of X onto X taking x_i onto y_i, $1 \leq i \leq n$, then the set X is already a set of $\infty\omega$-indiscernibles in $\mathfrak{H}(X)$. This is because the ordermorphism of X onto X induces an automorphism of $\mathfrak{H}(X)$ onto $\mathfrak{H}(X)$, so that

$$(\mathfrak{H}(X), x_1, \ldots, x_n) \cong (\mathfrak{H}(X), y_1, \ldots, y_n).$$

This naturally implies $(\mathfrak{H}(X), x_1, \ldots, x_n) \equiv_{\infty\omega} (\mathfrak{H}(X), y_1, \ldots, y_n)$. This way of finding $\kappa\omega$-indiscernibles (indeed, $\infty\omega$-indiscernibles) is subject to the limitation that the order type τ of X must be finitely transitive. So that (a), (c), and (d) will not hold.

If we analyze the argument above we see that if X is a set of $\omega\omega$-indiscernibles in $\mathfrak{H}(X)$ and Y is a subset of X which is finitely transitive over X, i.e., given any $x_1 < \cdots < x_n$ and $y_1 < \cdots < y_n$ from Y, there is an ordermorphism of X onto X taking x_i onto y_i, then the set Y is a set of $\infty\omega$-indiscernibles in $\mathfrak{H}(X)$. Of course we can not expect Y to be a set of $\infty\omega$-indiscernibles in $\mathfrak{H}(Y)$, so (b) and (e) fail. However, this way of finding $\kappa\omega$-indiscernibles leads to the following remark. We see that Y being finitely transitive over X means that

(1) given $x_1 < \cdots < x_n$, $y_1 < \cdots < y_n$ from Y, we have

$$(X, \leq, x_1,\ldots,x_n) \cong (X, \leq, y_1,\ldots,y_n).$$

From (1) we concluded above that Y is a set of $\kappa\omega$-indiscernibles in $\mathfrak{H}(X)$. It turns out that (1) can be drastically weakened, namely, $(X, \leq, x_1,\ldots,x_n) \equiv'_{\kappa\omega} (X, \leq, y_1,\ldots,y_n)$, and we are still able to show that Y is a set of $\kappa\omega$-indiscernibles in $\mathfrak{H}(X)$. This method is explored in Chang [4]. Even though the method in [4] suffers from the defect that the models are not generated from the $\kappa\omega$-indiscernibles it is at present the only non-trivial way of finding, say, $\omega_1\omega_1$-indiscernibles.

We next discuss a method which satisfies a large number of the requirements (a) - (f). This is really a combination of Morley's technique in (A) and the transformation introduced in (D) in §2. The idea is to express $\kappa\omega$-indiscernibility in terms of ordinary $\omega\omega$-indiscernibility in an expanded language. We outline in what follows the first step of the construction. Specifically, we first proceed as in the proof of (D), §2. For each formula σ of $L_{\kappa\omega}$, we introduce a new predicate symbol P_σ with the correct number of places. Let the language L' contain all the symbols of L plus all such P_σ, σ a formula of $L_{\kappa\omega}$, and a sufficient number of Skolem functions. That is, L' not only contains all the new P_σ's, but for each formula φ of $L'_{\omega\omega}$ there is a Skolem function symbol corresponding to it. Let Δ be the collection of all sentences (i) - (vi), as in (D), §2, as σ ranges over all formulas of $L_{\kappa\omega}$. (Note the absence of (vii).) Clearly the set of sentences $T \cup \Delta$ has arbitrarily large models. This is because T has arbitrarily large models, and every model of T can be expanded in a unique manner to a model for L' which satisfies $T \cup \Delta$. We now transform the set $T \cup \Delta$ into a set $\mathfrak{S}$ of types of $L'_{\omega\omega}$ such that for any model $\mathfrak{A}$ for L',

$$\mathfrak{A} \models T \cup \Delta \text{ if and only if } \mathfrak{A} \text{ omits } \mathfrak{S}.$$

So $\mathfrak{S}$ has arbitrarily large models omitting it. Now regardless of the power of the set $\mathfrak{S}$, we can find by using Morley's technique as outlined in (E), §2, models generated from indiscernibles which omits $\mathfrak{S}$. Furthermore we can choose indiscernibles of any order type τ. Let X be a set of indiscernibles in $\mathfrak{H}(X)$ where $\mathfrak{H}(X)$ omits $\mathfrak{S}$. (Note that $\mathfrak{H}(X)$ is taken with respect to all Skolem functions in L'.) We see that because the expansion of a model of T to a model of $T \cup \Delta$ is uniquely determined, the set X is a set of $\kappa\omega$-indiscernibles in $\mathfrak{H}(X)$ in the sense of L.

Some remarks about this method of construction are in order.

First of all, instead of a theory T of $L_{\omega\omega}$ with infinite models, we can deal equally well with a single sentence φ of $L_{\kappa^+\omega}$ with arbitrarily large models. This only involves a slight enlargement of the sets Δ and $\mathfrak{S}$ and causes no difficulties.

Clearly the existence and automorphism theorems hold for this construction. The rest of the properties (c) - (f) will fail. As we shall discuss in the next two remarks, there are ways in which we can amend the statements of (c) and (d) or of our construction so that (c) and (d) will hold. However this method will generally never yield a model of power κ. The reason is because the number of formulas σ of $L_{\kappa\omega}$ may be greater than κ in the absence of the GCH. So the language L' may have more than κ symbols and this causes (e) to fail.

Returning to an earlier remark, by the method of [4], if we replace the simple inclusion relation $\subset$ between sets of indiscernibles X and Y by the relation $<'_{\kappa\omega}$ (see [4] for notation), then we are able to obtain (c) and (d) with $<'_{\kappa\omega}$ between the Skolem hulls $\mathfrak{H}(X)$ nad $\mathfrak{H}(Y)$.

Finally, notice that the indiscernibles X are $\kappa\omega$-indiscernibles in $\mathfrak{H}(X)$ in the sense of L. They are not $\kappa\omega$-indiscernible in $\mathfrak{H}(X)$ in the sense of the expanded language L'. This however can be fixed up simply by repeating the expansion from L to L' a large number, say κ, of times. Let us say that $\bar{L} = \bigcup_{\xi<\kappa} L^\xi$, where L^ξ is an increasing sequence of languages defined by:

$$L^0 = L;$$

$$L^{\xi+1} = (L^\xi)', \quad \text{for } \xi < \kappa;$$

$$L^\eta = \bigcup_{\xi<\eta} L^\xi, \quad \text{for limit ordinals } 0 < \eta < \kappa.$$

As before let us consider the sets $T \cup \Delta$ and $\mathfrak{S}$ obtained in an analogous manner. Then because κ is regular, any set X of $\omega\omega$-indiscernibles (in the language $\bar{L}$) whose hull $\mathfrak{H}(X)$ omits $\mathfrak{S}$ is also $\kappa\omega$-indiscernible (in the language $\bar{L}$). It is also automatic that for models which omit $\mathcal{S}$ $\prec_{\omega\omega}$ in the sense of $\bar{L}$ will become $\prec_{\kappa\omega}$ in the sense of $\bar{L}$. In this way then we see that except for the restriction on the cardinality of the models, all properties (a) through (d) hold.

The above is a bare outline of the idea of pursuing Morley's technique of constructing indiscernibles. We see from the last remark that except for (e) and (f) all other properties can be satisfied. It turns out that if we combine Morley's technique with the technique of constructing Scott sentences in §1, then we can make (a), (b), (e), (f) hold while sacrificing very little on (c) and (d). This then is the last method of construction we discuss here. Since it does involve at least one new idea, we shall describe the construction in some detail.

We start with a language L with κ symbols and a theory T of L having infinite models. We add to L κ new symbols: Let $\bar{L}$ contain all symbols of L together with

E_n^ξ : a $2(n+1)$-placed predicate symbol for each $n<\omega$, $\xi<\kappa$;

F^ξ : a finitely many placed function symbol for each $\xi<\kappa$.

We have not specified the number of places of the functions F^ξ because we want to arrange matters in such a way that the following is true:

Every formula φ of $\bar{L}_{\omega\omega}$ has one of the F^ξ as its Skolem function. Furthermore, the index ξ of F^ξ is greater than any η such that E_n^η or F^η occurs in φ.

This can be done in a simple manner as the total number of $\bar{L}_{\omega\omega}$ formulas is κ and each such formula contains only a finite number of symbols. We now extend T to a set $\bar{T}$ of sentences of $\bar{L}_{\omega\omega}$ as follows: (If x, y are $(n+1)$-tuples of variables, we write $xE_n^\xi y$ for $E_n^\xi(x\frown y)$.) $\bar{T}$ contains all of the sentences listed below.

(1) All axioms that say: F^ξ is the Skolem function of the corresponding formula φ.

(2) $\forall x,y\,(xE_n^\xi y \to xE_n^\eta y)$, all $n<\omega$, $\eta\le\xi<\kappa$.

(3) $\forall x,y\,(xE_n^{\xi+1} y \to \forall u\,\exists v\,(x\frown u\, E_{n+1}^\xi\, y\frown v) \wedge \forall u\,\exists v\,(x\frown v\, E_{n+1}^\xi\, y\frown u))$, all $n<\omega$, $\xi<\kappa$.

(4) $\forall x,y\,(xE_n^0 y \to (\varphi(x) \leftrightarrow \varphi(y)))$, all $n<\omega$, and all formulas φ of $L_{\omega\omega}$ in the variables $v_0,\ldots,v_n$.

(5) $\forall x,y\,(xE_n^{\xi+1} y \to (\varphi(x) \leftrightarrow \varphi(y)))$, all $n<\omega$, $\xi<\kappa$, and all formulas φ of

$L \cup \{E_n^\eta : \eta \leq \xi,\ n < \omega\} \cup \{F^\eta : \eta \leq \xi\}$ in the variables $v_0, \ldots, v_n$.

Note first of all that all sentences of $\overline{T}$ are first-order sentences of $\overline{L}$.

THEOREM. (i) Every model $\mathfrak{A}$ of T can be expanded to a model $\overline{\mathfrak{A}}$ of $\overline{T}$ in such a way that each E_n^ξ is an equivalence relation on the set of all $(n+1)$-tuples of elements of $\overline{\mathfrak{A}}$; furthermore the number of equivalence classes of E_n^ξ is at most $\beth_{\xi+1}(\kappa)$. ($\beth_\xi(\kappa)$ is the iteration of the beth function starting from κ)

(ii) If T has infinite models, then the set of sentences Δ of $\overline{L} \cup \{c_0, \ldots, c_n, \ldots\}$

$$\Delta = \overline{T} \cup \{c_{p_0} \cdots c_{p_n} E_n^\xi c_{q_0} \cdots c_{q_n} : n < \omega,\ \xi < \kappa,\ p_0 < \cdots < p_n,\ q_0 < \cdots < q_n\}$$

is consistent.

(iii) If $\overline{\mathfrak{A}}$ is a model of $\overline{T}$ and for all $\xi < \kappa$ $\overline{\mathfrak{A}} \models \overline{a} E_n^\xi \overline{b}$, then the $(n+1)$-tuples $\overline{a}$ and $\overline{b}$ are $\kappa\omega$-equivalence; in other words $(\overline{\mathfrak{A}}, \overline{a}) \equiv_{\kappa\omega} (\overline{\mathfrak{A}}, \overline{b})$, where the equivalence is taken in the sense of $\overline{L}_{\kappa\omega}$.

Proof. (i) Let $\mathfrak{A}$ be a model of T. We shall define by transfinite induction an (increasing) sequence of expansions $\mathfrak{A}_\xi$, $\xi < \kappa$, of $\mathfrak{A}$ as follows: each $\mathfrak{A}_\xi$ is supposed to be an expansion of $\mathfrak{A}$ that contains the relations R_n^η and functions G^η for all $\eta < \xi$. Let $\mathfrak{A}_0 = \mathfrak{A}$. Suppose that $\mathfrak{A}_\eta$ for all $\eta < \xi$ have already been defined. We now defined $\mathfrak{A}_\xi$. First assume ξ is a successor ordinal, $\xi = \eta + 1$. For each $n < \omega$, define

$$x\, R_n^\xi\, y \text{ if and only if } \begin{cases} x\, R_n^\eta\, x, \text{ and} \\ (\mathfrak{A}_\eta, x) \equiv (\mathfrak{A}_\eta, y), \text{ and} \\ \forall u \exists v\, x^\frown u\, R_{n+1}^\eta\, y^\frown v, \text{ and} \\ \forall u \exists v\, x^\frown v\, R_{n+1}^\eta\, y^\frown u. \end{cases}$$

Since R_{n+1}^η is an equivalence relation it is easy to verify that so is R_n^ξ. Similarly, if there are at most $\beth_{\eta+1}(\kappa)$ number of equivalence classes of R_{n+1}^η, then there are at most $\beth_{\xi+1}(\kappa)$ number of equivalence classes of R_n^ξ. If F^ξ is supposed to be the Skolem function of some formula φ whose interpretation is already well-defined in $\mathfrak{A}_\eta$, then pick G^ξ to be a Skolem function of the interpretation of φ in $\mathfrak{A}_\eta$; otherwise pick G^ξ arbitrary. Suppose ξ is a limit ordinal. Then define

$$x\, R_n^\xi\, y \text{ if and only if } x\, R_n^\eta\, y \text{ for all } \eta < \xi.$$

Then R_n^ξ is an equivalence relation with at most $\beth_{\xi+1}(\kappa)$ equivalence classes. We pick G^ξ as above. The induction is now complete. We let $\bar{\mathfrak{A}}$ be the union of all the expansions $\mathfrak{A}_\xi$, $\xi < \kappa$. It should be quite evident that $\bar{\mathfrak{A}}$ is a model of T and all sentences of the form (1) - (5) hold in $\bar{\mathfrak{A}}$, so $\bar{\mathfrak{A}}$ is a model of $\bar{T}$.

(ii) Since T has arbitrarily large models, by (i) $\bar{T}$ also has arbitrarily large models. Using the fact that the number of equivalence classes of R_n^ξ is at most $\beth_{\xi+1}(\kappa)$, we can easily show that the set Δ is consistent. (See [12] for details.)

(iii) Let $\bar{\mathfrak{A}}$ be any model of $\bar{T}$ and suppose $\bar{a}, \bar{b}$ are $(n + 1)$-tuples of elements of $\bar{\mathfrak{A}}$ such that $\bar{a}\, R_n^\xi\, \bar{b}$ for all $\xi < \kappa$. Let $\varphi(v_0 \cdots v_n)$ be any formula of $\bar{L}_{\kappa\omega}$. Because κ is regular, there is an ordinal $\alpha < \kappa$ such that φ only contains E_n^η, F^η for $\eta < \alpha$. We now prove by induction on the ordinal ξ that: for all $(m + 1)$-tuples $\bar{c}, \bar{d}$ from $\bar{\mathfrak{A}}$, if $\bar{c}\, R_m^{\alpha+\xi}\, \bar{d}$, then $\bar{c}$ and $\bar{d}$ satisfy exactly the same formulas $\psi(v_0 \cdots v_m)$ of $\bar{L}_{\kappa\omega}$ such that

ψ contains no symbols E_n^η, F^η for $n < \omega$, $\alpha \leq \eta < \kappa$,

and

ψ has quantifier-rank at most ξ.

This induction is very similar to the one in Proposition 7, §1. Since the quantifier rank of φ is less than κ, it follows that

$$\bar{\mathfrak{A}} \models \varphi[\bar{a}] \text{ if and only if } \bar{\mathfrak{A}} \models \varphi[\bar{b}].$$

Since this is true for every φ, we have $(\bar{\mathfrak{A}},\bar{a}) \equiv_{\kappa\omega} (\bar{\mathfrak{A}},\bar{b})$.

Using the theorem, we see that we can easily find indiscernibles X in $\mathfrak{H}(X)$ which are in fact $\kappa\omega$-indiscernibles. It is trivial that (a) and (b) hold. (e) of course holds becuase $\bar{L}$ has at most κ symbols. To see that (f) also holds, we need the following notion. Let Y be a subset of an ordered set X. We say that Y is a nice subset of X (and X a nice extension of Y) if the following holds: Given any finite subset Z of X, there is an order preserving map of Z into Y

which leaves every element of $Z \cap Y$ fixed. (This is equivalent to saying that every existential first-order formula in $\leq$ which holds in X with parameters in Y also holds in Y.)

LEMMA. Suppose X is a set of $\kappa\omega$-indiscernibles in $\mathfrak{H}(X)$ constructed from the set Δ as in the theorem, and suppose Y is a nice subset of X, then Y is a set of $\kappa\omega$-indiscernibles in $\mathfrak{H}(Y)$ and furthermore $\mathfrak{H}(Y) <_{\kappa\omega} \mathfrak{H}(X)$.

Proof. It is obvious that Y is a set of $\kappa\omega$-indiscernibles in $\mathfrak{H}(Y)$. What is not obvious is that finite increasing sequences from Y should satisfy the same $\overline{L}_{\kappa\omega}$ formulas in $\mathfrak{H}(Y)$ as in $\mathfrak{H}(X)$. (This is perhaps the main difference between this construction which uses κ symbols and the iterated construction of Morley, as, there, it is always true that $\mathfrak{H}(Y) <_{\kappa\omega} \mathfrak{H}(X)$ with no restriction on Y.)

We shall prove by induction on the quantifier rank ξ that

$$\mathfrak{H}(Y) <_{\xi\omega} \mathfrak{H}(X), \quad \text{for all } \xi < \kappa.$$

There is only one step of the induction which is worrisome, namely the step from $\varphi(x)$ to $\exists x\, \varphi(x)$. So let $\varphi(v_0 \cdots v_n)$ be a formula of $L_{\kappa\omega}$ of quantifier rank ξ, and let $a_1,\ldots,a_n \in \mathfrak{H}(Y)$. Suppose that $\mathfrak{H}(X) \models \exists v_0\, \varphi[a_1 \cdots a_n]$, we wish to prove that $\mathfrak{H}(Y) \models \exists v_0\, \varphi[a_1 \cdots a_n]$. Let a_0 be an element in $H(X)$ such that $\mathfrak{H}(X) \models \varphi[a_0 a_1 \cdots a_n]$. By switching to the generators, we see that there are Skolem functions $f_0,\ldots,f_n$ and elements $y_1,\ldots,y_p$ in Y and $x_1,\ldots,x_q$ in X such that

$$a_0 = f_0(x_1,\ldots,x_q;\, y_1,\ldots,y_p)$$

$$a_i = f_i(y_1,\ldots,y_p) \qquad 1 \leq i \leq n$$

and

$$\mathfrak{H}(X) \models \varphi[f_0(x_1 \cdots y_1 \cdots),\, f_1(y_1 \cdots),\ldots,f_n(y_1 \cdots)].$$

By the property of being a nice subset, we now slide the elements $x_1 \cdots x_q$ into $y_1',\ldots,y_q'$ in Y so that the ordering is not disturbed. By $\kappa\omega$-indiscernibility we have

$$\mathfrak{H}(X) \models \varphi[f_0(y_1' \cdots y_1 \cdots),\, f_1(y_1 \cdots),\ldots,f_n(y_1 \cdots)].$$

But $f_0(y_1' \cdots y_1 \cdots) \in H(Y)$, so using the induction hypothesis on φ, we have $\mathfrak{H}(Y) \models \exists v_0\, \varphi[a_1 \cdots a_n]$. This finishes the induction.

From the Lemma we see immediately that (c) and (d) hold for nice subsets and nice extensions. As for (f), if λ and μ are cardinals such that $\kappa \leq \lambda < \mu$, then λ is a nice subset of μ. Thus the require elementary tower can be constructed. It is obvious how to define the set $\mathfrak{S}$ of complete types of $\overline{L}_{\kappa\omega}$. A simple sliding argument will insure that each member of the tower realizes only those types in $\mathfrak{S}$.

As a very last remark, I take this opportunity to mention that this last construction is the method I had in mind when I wrote the last sentence of [4]. It is indeed the only method known to me where the models are generated from the $\kappa\omega$-indiscernibles. Just recently, I noticed that if we are willing to give up the notion of models generated from the $\kappa\omega$ -indiscernible within it then one can prove the assertion in the last line of [4] by the method of [4] in a very simple manner. In fact, the problem posed in the next to the last line in [4] can also be solved affirmatively. However, it is still an open problem whether we can create $\omega_1\omega_1$-indiscernibles of any order type.

REFERENCES

[1] Barwise, J., Remarks on universal sentences of $L_{\omega_1\omega}$, to appear.

[2] Chang, C. C., Two refinements of Morley's method on omitting types of elements, Notices Amer. Math. Soc., 11 (1964), 679.

[3] __________, Omitting types of prenex formulas, J. of Symbolic Logic, 32 (1967), 61-74.

[4] __________, Infinitary properties of models generated from indiscernibles, to appear in the Proceedings of the 1967 Int'l Cong. Logic Meth. and Phil. Sci. held in Amsterdam.

[5] Ehrenfeucht, A. and Mostowski, A., Models of axiomatic theories admitting automorphisms, Fund. Math., 43 (1956), 50-68.

[6] Hanf, W., Incompleteness in languages with infinitely long expressions, Fund. Math., 53 (1964), 309-323.

[7] Helling, M., Hanf numbers for some generalizations of first-order language, Notices Amer. Math. Soc., 11 (1964), 679.

[8] Karp, C., Languages with expressions of infinite length, (North-Holland Publ. Co., Amsterdam, 1964).

[9] Karp, C., Finite quantifier equivalence, in Theory of Models, (North-Holland Publ. Co., Amsterdam, 1965), 407-412.

[10] Kueker, D., Definability, automorphisms, and infinitary languages, this volume.

[11] Lopez-Escobar, E. G. K., Universal formulas in the infinitary language $L_{\alpha,\beta}$, Bull. de l'Acad. Polon. des Sci. Ser. math. astr. phys., XIII (1965), 383-388.

[12] ________, On defining well-orderings, Fund. Math., LIX (1966), 13-21.

[13] Malitz, J., The Hanf number for complete $L_{\omega_1\omega}$ sentences, this volume.

[14] Morley, M., Omitting classes of elements, in Theory of Models, (North-Holland Publ. Co., Amsterdam, 1965), 265-273.

[15] ________, The Hanf number for ω-logic, (abstract), J. Symbolic Logic, 32 (1967), 437.

[16] Morley, M. and Morley, V., The Hanf number for κ-logic, Notices Amer. Math. Soc., 14 (1967), 556.

[17] Scott, D., Logic with denumerably long formulas and finite strings of quantifiers, in Theory of Models, (North-Holland Publ. Co., Amsterdam, 1965), 329-341.

[18] Silver, J., Some applications of model theory in set theory, Ph.D. Thesis, Univ. of Calif., Berkeley, 1966.

[19] Vaught, R., The Löwenheim-Skolem theorem, in Proc. of the 1964 Int'l Cong. Logic Meth. and Phil. Sci. held in Jerusalem, 81-89.

UNIVERSITY OF CALIFORNIA, LOS ANGELES

REMARKS ON THE THEORY OF GEOMETRICAL CONSTRUCTIONS

ERWIN ENGELER*

It is astonishing that a field historically as close to the foundations of mathematics as the theory of geometrical constructions has received so little attention by logicians and has been left so largely untouched by the methods of formalization and axiomatization. From the point of view of contemporary mathematics, this has left the field with some gaping ambiguities and inadequacies in the formulation of the most basic notions and results.

Quite apart from these feelings of regret in the state of one of the oldest and historically most important branches of mathematics, there are several reasons for devoting renewed interest to this field. First, in connection with the topic of this conference, the theory allows and motivates interesting applications of infinitary logic without which a formalization could probably not have been obtained. The second reason is psychological: Our intuition in the constructive theory of real numbers is not as weak as we might be led to believe by recursive analysis, for example. We feel, on the contrary, that an adequate (and more nearly standard) constructive analysis could be developed on the basis of computationally closed fields (section 2) rather than on the set of computable real numbers. Finally, there is a didactic reason: In this era of programmed computers, a treatment of geometrical construction programs might serve as a very welcome didactic tool at an early stage of the training of our students.

For people unsympathetic with infinitely long formulas, let it be remarked that the kind of formulas used in our formalization can easily be encoded in expressions of finite length. The detour through general infinitely long formulas is merely a useful technical device.

*Work supported in part by NSF Grant GP-5434.

1. Construction programs in plane Euclidean geometry.

The need for a programming language in which to formulate directions for geometrical constructions has not arisen in the past because these directions were addressed to budding mathematicians rather than to stupid machines. Also, for the rather narrow part of theory of geometrical constructions that has received algebraic treatment as an application of Galois theory, the description of constructions could indeed be left informal. For a more comprehensive approach, however, we do need a clarification of several of the fundamental notions. The groundwork for doing this was laid in [1] to which we refer for the proof of Theorem 1.4 and for the clarification of some of the features below that are not self-explanatory.

Consider the Euclidean plane as a collection of three kinds of entities: points (for which we will use variables $P_0,P_1,P_2,\dots$), lines (variables $\ell_0,\ell_1,\ell_2,\dots$), and circles (variables $\gamma_0,\gamma_1,\gamma_2,\dots$). Between these entities there are defined certain relations, among them equality ($P_i = P_j, \ell_i = \ell_j, \gamma_i = \gamma_j$), incidence ($I(P_i,\ell_j), I(P_i,\gamma_j), I(\ell_i,\gamma_j), I(\gamma_j,\gamma_k)$), betweeness ($B(P_i,P_j,P_k)$), parallelity ($\ell_i \| \ell_j$), and equidistance ($E(P_i,P_j;P_r,P_s)$). In addition, we assume that two lines, x and y, have been singled out such that x and y are perpendicular (a notion that is definable), intersect in a point O, and that we have selected points E_x on x, E_y on y such that O,E_x and O,E_y are equidistant and $E_x \neq O \neq E_y$.

The basic capabilities of the Euclidean constructor can be expressed as follows: ($i,j,k,s = 0,1,2,\dots$):

A. Operating capabilities.

(1) $\ell_i := (P_j,P_k)$, draw a line ℓ_i through two distinct points P_j,P_k;

(2) $P_i := (\ell_j,\ell_k)$, find the intersection P_i of two non-parallel lines;

(3) $\gamma_i := [P_j,P_k]$, draw the circle γ_i of center P_j and passing through $P_k, P_j \neq P_k$;

(4) $P_i := (\gamma_j,\ell_k)$, choose an intersection point P_i of a circle γ_j and an intersecting line ℓ_k;

(5) $P_i := (P_s,\gamma_j,\ell_k)$, given a line through a peripheral point of a circle, find the other intersection point;

(6) $P_i := (\gamma_j,\gamma_k)$, choose an intersection point of two intersecting circles;

(7) $P_i := (P_s,\gamma_j,\gamma_k)$, given an intersection point of two circles, find the other intersection point.

(8) $P_i := 0$, $P_i := E_x$, $P_i := E_y$, $\ell_i := x$, $\ell_i := y$, $P_i := P_j$, $\ell_i := \ell_j$, $\gamma_i := \gamma_j$.

B. Decision-making capabilities.

(1) $P_i = P_j$, $\ell_i = \ell_j$, $\gamma_i = \gamma_j$, decide whether two points, lines, or circles are equal;

(2) $\ell_i \| \ell_j$, decide whether two lines are parallel;

(3) $I(P_i, \ell_j)$, decide whether P_i lies on ℓ_j;

(4) $I(P_i, \gamma_j)$, decide whether P_i lies on γ_j;

(5) $I(\ell_i, \gamma_j)$, decide whether ℓ_i intersects γ_j;

(6) $I(\gamma_i, \gamma_j)$, decide whether γ_i and γ_j intersect;

(7) $B(P_i, P_j, P_k)$, decide whether P_j lies between P_i and P_k;

(8) $E(P_i, P_j; P_k, P_s)$, decide whether P_i, P_j and P_k, P_s are equidistant.

On the basis of these operations and decisions, we can write programs as in [1], composing them of individual labelled instructions of the following kind.

k: do ψ then go to p, (for operations ψ);

k: if φ then go to p else go to q, (for decision-conditions φ).

There is a slight complication here due to the fact that some of the operations are not toal, i.e., not defined for some of the arguments. However, for each such operation there is a condition among B(1) - B(8) which decides whether the operation is geometrically performable. Thus we agree to replace each instruction

s: do $\ell_i := (P_j, P_k)$ then go to p

by a subroutine

s: if $P_j = P_k$ then go to s else go to s';

s': do $\ell_i := (P_j, P_k)$ then go to p

where s' is a label that did not occur before in the program. Similarly for the other partial operations. The end effect is that we may consider all operations as total; the program does not terminate if at some point a geometrically non-performable operation is called upon. We describe this situation by saying that the partial operations in A(1) - A(8) are definite partial operations.

Clearly, the list of operations and decisions above is highly redundant, and it is an easy exercise in plane Euclidean geometry to reduce this list. This means to provide subroutines to replace some operational instructions or to provide subroutines to replace some conditional instructions. For example, a conditional instruction

k: <u>if</u> $I(P_i,\ell_j)$ <u>then go to</u> p <u>else go to</u> q

can be replaced by

k: <u>if</u> $\ell_j \parallel x$ <u>then go to</u> k_1 <u>else go to</u> k_5;

k_1: <u>do</u> $P_0 := (y,\ell_j)$ <u>then go to</u> k_2;

k_2: <u>if</u> $P_0 = P_i$ <u>then go to</u> p <u>else go to</u> k_3;

k_3: <u>do</u> $\ell_0 := (P_0,P_i)$ <u>then go to</u> k_4;

k_4: <u>if</u> $\ell_0 = \ell_j$ <u>then go to</u> p <u>else go to</u> q;

k_5: <u>do</u> $P_0 := (x,\ell_j)$ <u>then go to</u> k_2.

(We assume that the variables P_0,ℓ_0 do not occur in the original program and that the labels $k_1,\ldots,k_5$ don't occur either, otherwise we simply rename the appropriate variables and labels.)

THEOREM 1.1. <u>The set of all points on the line</u> x <u>forms an euclidean ordered field</u> G <u>in which the operations</u> $+, \cdot, -, ^{-1}, \sqrt{\ }$ <u>and the relation</u> $\leq$ <u>are definable by programs in terms of the capabilities</u> A(1) - B(8).

This theorem is due, in essence, to Hilbert. For an operation, say +, to be definable by a program we mean that there is a program π^+ which contains the variables P_i, P_j and P_k and has the property that π^+ terminates for each assignment of points of x to the variables P_i, P_j. The value of P_k at termination is the result of the operation + . A relation, say $\leq$, is definable by a program, if there is a program $\pi^{\leq}$ with two exits and containing the variables P_i, P_j such that $\pi^{\leq}$ terminates for any assignment of points of x to P_i, P_j either in one exit (if the relation holds), or in the other (if it doesn't). The details for writing these programs are well-known.

Conversely, suppose that an euclidean ordered field is given, say $G = \langle A, \leq, +, \cdot, -, ^{-1}, 0, 1\rangle$, with respect to which we have the following capabilities (using $x_0,x_1,\ldots$ for variables over A):

(a) $x_i := x_j$, $x_i := 0$, $x_i := 1$, $x_i := x_j + x_k$, $x_i := x_j \cdot x_k$, $x_i := - x_j$, $x_i := (x_j)^{-1}$ (for $x_j \neq 0$); $x_i := \sqrt{x_j}$ (for $x_j \geq 0$); $i,j,k = 0,1,2,\ldots$.

(b) $x_i \leq x_j$, $x_i = x_j$, $x_i = 0$, $x_i = 1$; $i,j = 0,1,2,\ldots$.

Consider the plane analytic geometry $G(G)$ over G in which points, lines, and circles, incidence, etc. are defined in the usual way.

THEOREM 1.2. $G(G)$ is a geometry in which the operation A(1) - A(8) and the relations B(1) - B(8) are definable by programs in terms of the capabilities (a) and (b).

This theorem may be ascribed to Descartes and is well-known.

COROLLARY 1.3. If G_1 and G_2 are euclidean ordered fields such that each program over (a), (b) terminates in G_1 if and only if it terminates in G_2, then every program over A(1) - B(8) terminates in $G(G_1)$ if and only if it terminates in $G(G_2)$. Conversely if $\mathcal{G}_1$, $\mathcal{G}_2$ are euclidean planes such that each program over A(1) - B(8) terminates in $\mathcal{G}_1$ if and only if it terminates in $\mathcal{G}_2$, then every program over (a), (b) terminates in the field corresponding to $\mathcal{G}_1$ if and only if it terminates in the field corresponding to $\mathcal{G}_2$. (We say that a program terminates if it terminates for all assignments of initial values to the variables.)

The proof is obvious from 1.1 and 1.2.

Loosely speaking, a construction problem is the problem of finding a program over A(1) - B(8) which constructs certain points, lines, and circles from given data in such a way that the configuration that is obtained has certain geometrical properties. The crucial question is: what kind of property? The most natural answer, in our opinion, is that the property be verifiable by the constructive means at our disposal.[1] This means that we have a program whose favorable outcome is a necessary and sufficient condition for the property to hold. It is clear that by a slight change in the program (leading non-favorable exits into non-terminating subroutines), we can normalize the notion of favorable termination to simply termination. Properties that are in this sense verifiable by programs are called algorithmic properties, and we have the following result:

THEOREM 1.4. To every program π with free variables $x_1,\ldots,x_n$ we can effectively find a

[1] See Remark 2 in Section 3 for a discussion of other views of this notion.

quantifier-free formula $\varphi(x_1,\ldots,x_n)$ in $L_{\omega_1,\omega}$ such that the termination of π for an assignment is equivalent with φ holding for this assignment; [1].

We call two euclidean geometries <u>algorithmically equivalent</u> if, for every program π over A(1) - B(8), π terminates in one geometry for all initial assignments if and only if it does in the other. Similarly for euclidean ordered fields.

From Corollary 1.3 follows at once:

COROLLARY 1.5. <u>Two geometries are algorithmically equivalent if and only if their corresponding fields are algorithmically equivalent</u>.

Thus, if our goal is to axiomatize constructive geometry, we need a characterization of the class of all euclidean fields that are algorithmically equivalent to the field of reals. Once an axiomatic characterization of this concept is found the remainder is straightforward.

2. <u>An axiomatization of the algorithmic theory of real numbers</u>.

The algorithmic basis $\mathcal{B}$ for the field $\mathcal{R}$ of real numbers, as realized by the theory of geometrical constructions, consists of the following:

(a) $x_i := x_j$, $x_i := 0$, $x_i := 1$, $x_i := x_j + x_k$, $x_i := x_j \cdot x_k$, $x_i := -x_j$, $x_i = (x_j)^{-1}$ (for $x_j \neq 0$); $i,j,k = 0,1,2,\ldots$.

(b) $x_i \leq x_j$, $x_i = x_j$, $x_i = 0$, $x_i = 1$; $i,j = 0,1,2,\ldots$.

Note that $x_i := (x_j)^{-1}$ is a partial operation, undefined for $x_j = 0$. If we make the operation total by defining arbitrarily $0^{-1} = 0$ we also have to convert each program into a new one by replacing each instruction

k: <u>do</u> $x_i := (x_j)^{-1}$ <u>then go to</u> p

by the subroutine

k: <u>if</u> $x_j = 0$ <u>then go to</u> k <u>else go to</u> k';

k': <u>do</u> $x_i := (x_j)^{-1}$ <u>then go to</u> p

where k' is a label that did not occur in the original program.

With this slight change the work of [1] applies to the field of real numbers. In particular, we have an effective method for formulating for each program π over $\mathcal{B}$ a sentence in (a fragment of)

$L_{\omega_1,\omega}$ that expresses exactly the property of $\mathfrak{R}$ verified by the termination of π. The algorithmic theory of $\mathfrak{R}$ is the set of all such sentences (or their negations) holding in $\mathfrak{R}$. Our goal is to axiomatize this theory in the framework of $L_{\omega_1,\omega}$.

For this purpose we first collect a few outstanding properties, formulated in $L_{\omega_1,\omega}$, which a structure

$$\mathfrak{G} = \langle A, \leq, +, \cdot, -, {}^{-1}, 0, 1\rangle$$

must have in order to be algorithmically equivalent to $\mathfrak{R}$.

LEMMA 2.1. $\mathfrak{G}$ is an archimedian ordered field.

It is easy to write programs over $\mathfrak{R}$ that verify the finitely-many axioms of archimedian ordered fields. For example, the following program verifies the archimedian property:

1: do $x_0 := 0$ then go to 2;

2: do $x_3 := x_2$ then go to 3;

3: if $x_1 \leq x_0$ then go to 7 else go to 4;

4: if $x_2 \leq x_0$ then go to 7 else go to 5;

5: if $x_1 \leq x_2$ then go to 7 else go to 6;

6: do $x_2 := x_2 + x_3$ then go to 5.

The axioms for an ordered field are first-order and hence a fortiori formulable in $L_{\omega_1,\omega}$; the archimedian property is

$$(\forall x)(\forall y)(x > 0 \wedge y > 0 \rightarrow \bigvee_{i=1}^{\infty} (x \leq \underbrace{y + y + \cdots + y}_{i \text{ times}}))$$

which is a formula in $L_{\omega_1,\omega}$.

LEMMA 2.2. $\mathfrak{G}$ is real-closed.

There are well-known numerical procedures that approximate the square root of a nonnegative real number or approximate a root of a polynomial of odd degree (with given coefficients). These procedures can easily be transformed into programs that compute a sequence of nested intervals (by computing the endpoints) that converge towards the real root. Such a program will look as follows:

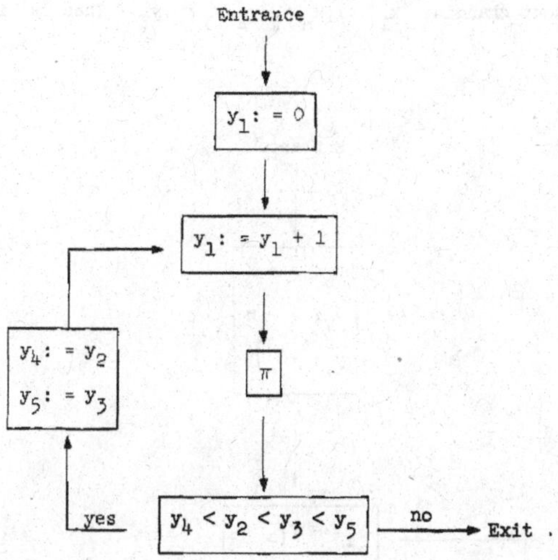

For successive values 1,2,3,... of y_1 the values of y_2, y_3 will form a converging sequence of nested intervals. Convert this program by replacing the box $y_4 < y_2 < y_3 < y_5$ by $y_4 < y_2 < z < y_3 < y_5$ where z did not occur in the original program. The converted program will not terminate if executed in $\mathfrak{R}$ when we assign the real root z as value to z, but will terminate in $\mathfrak{G}$ for all assignments for z in case the root in question does not exist in $\mathfrak{G}$. For $\mathfrak{G}$, being archimedian, is a subfield of $\mathfrak{R}$; if for some z in $\mathfrak{G}$ the program did not terminate in $\mathfrak{G}$ the limit point would equal z would be in $\mathfrak{G}$, and would be a root. Since the notion of a real closed field is formulable in first-order logic, it is, a fortiori, formulable in $L_{\omega_1,\omega}$.

LEMMA 2.3. $\mathfrak{G}$ _is computationally closed._

By this we mean the following: Suppose we are given n elements $a_1,\ldots,a_n$ of $\mathfrak{G}$ and a program π over $\mathfrak{R}$ that computes, in terms of $a_1,\ldots,a_n$ a nested sequence of intervals. Then $\mathfrak{G}$ contains a point in the intersection. We shall need a slightly more general notion of computing a converging sequence of intervals than the one used above. Our more general notion uses the idea of "backtrack".

A program π computes a converging sequence of intervals in the following sense: Let π_0 be a one-exit program which computes y_2 in terms of $x_1,\ldots,x_n,y_1$ without changing $x_1,\ldots,x_n,y_1,y_3,y_4,y_5,y_6,y_7$. Let π_1 be a one-exit program which computes y_4 and y_5 in terms of $x_1,\ldots,x_n,y_1,y_3,y_6,y_7$ without changing $x_1,\ldots,x_n,y_1,y_2,y_3,y_6,y_7$. Then π is of the following form (in terms of flow-diagrams):

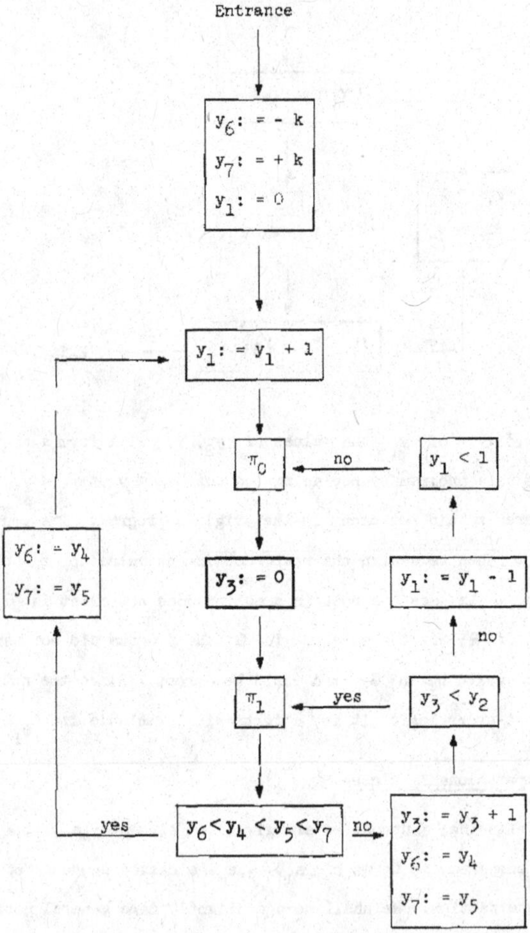

REMARK. To understand the above diagram, consider it first with the right hand side replaced by an exit. The resulting program clearly determines at each turn (i.e., for successive values of y_1) a decreasing sequence of nested intervals. The added right hand side provides for a "backtrack".

The program π computes a converging sequence of intervals if for some k the program does not terminate. If $\varphi(k,x_1,\ldots,x_n)$ expresses that π terminates for $k,x_1,\ldots,x_n$ then $\bigvee_{k=1}^{\infty} \neg \varphi(k,x_1,\ldots,x_n)$ expresses that the program computes a converging sequence of intervals if started on $x_1,\ldots,x_n$.

If in the above program the box $y_6 < y_4 < y_5 < y_7$ is replaced by $y_6 < y_4 < z < y_5 < y_7$ and $\psi(k,z,x_1,\ldots,x_n)$ expresses that the modified program terminates for $k,z,x_1,\ldots,x_n$ then $\bigvee_{k=1}^{\infty} \neg \psi(k,z,x_1,\ldots,x_n)$ expresses that z lies in the intersection of the converging sequence of intervals computed by π. Thus the formula

$$(\forall x_1) \cdots (\forall x_n)\left(\bigvee_{k=1}^{\infty} \neg \varphi(k,x_1,\ldots,x_n) \rightarrow (\exists z) \bigvee_{k=1}^{\infty} \neg \psi(k,z,x_1,\ldots,x_n)\right)$$

states, in the language $L_{\omega_1,\omega}$, that if π computes a converging sequence of intervals for some $x_1,\ldots,x_n$ then there is a point in the intersection. Now, if we associate such a sentence to each program π of the form indicated, we obtain a characterization of the notion of computational closure by a (recursive) set of formulas of $L_{\omega_1,\omega}$.

Results 2.1 and 2.3 together establish:

THEOREM 2.4. <u>If $\mathfrak{A}$ is algorithmically equivalent to $\mathfrak{R}$ then $\mathfrak{A}$ is a computationally closed archimedian ordered field</u>.

The remainder of this section is devoted to a proof of the converse of this theorem.

THEOREM 2.5. <u>If $\mathfrak{A}$ is a computationally closed archimedian ordered field then $\mathfrak{A}$ is algorithmically equivalent to $\mathfrak{R}$</u>.

<u>Proof</u>. Any archimedian ordered field $\mathfrak{F}$ may be considered as a subfield of $\mathfrak{R}$. By the computationally closure of $\mathfrak{F}$ in R we mean the smallest subfield of $\mathfrak{R}$ that contains $\mathfrak{F}$ and is computationally closed. Obviously, this field may be obtained by iterating the adjunction of elements of $\mathfrak{R}$ determined by computable convergent sequences of intervals. In particular, let C be the computational closure of the field of rational numbers. We shall show that all computationally closed archimedian ordered fields $\mathfrak{A}$ are algorithmically equivalent to C.

Note that C is a subfield of $\mathfrak{A}$, hence if φ is an algorithmic property of $\mathfrak{A}$ then it is one of $\mathfrak{C}$ (φ being a universal formula of $L_{\omega_1,\omega}$). It remains to show that, conversely, every algorithmic property of $\mathfrak{C}$ is one of $\mathfrak{A}$. For this it is sufficient to prove that for each algorithmic property $\varphi(x_1,\dots,x_n)$ and for each $a_1,\dots,a_n$ in $\mathfrak{A}$ with $\mathfrak{A} \models \neg\varphi[a_1,\dots,a_n]$ there are $b_1,\dots,b_n$ in $\mathfrak{C}$ such that $\mathfrak{C} \models \neg\varphi[b_1,\dots,b_n]$.

By [1] the formula $\neg\varphi$ is obtained from atomic formulas and negated atomic formulas by the following syntactical operations: conjunction, disjunction with a negated or unnegated atomic formula, primitive substitution $\mathrm{Sub}_k(\psi)$, and conjunctions $\bigwedge_{w\in|\sigma^*|} \mathrm{Sub}_w(\psi)$. First observe that every formula $\neg\varphi$ is logically equivalent to a <u>reduced</u> formula of the form $\bigwedge_{w\in|\sigma|} \mathrm{Sub}_w(\psi)$ where σ is the signature of an appropriate program and ψ is a (finite) Boolean combination of atomic formulas. This is easily established by induction on the structure of $\neg\varphi$, (using the fact that we have infinitely many variables that we can use as dummies). The procedure that determines a point in $\mathfrak{C}^n$ which satisfies a reduced formula $\bigwedge_{w\in|\sigma|} \mathrm{Sub}_w(\psi)$ can be outlined informally as follows: To determine a computable first coordinate consider for each m the formula $(\exists x_2)\cdots(\exists x_n)\psi^{(m)}$, where $\psi^{(m)}$ is the conjunction of the first m conjuncts in $\bigwedge_{w\in|\sigma|} \mathrm{Sub}_w(\psi)$. Since $\mathfrak{A}$ is real closed, Tarski's decision method applies, and the set of all x_1 satisfying this formula is a finite collection of intervals (open, closed, half-open, or degenerated to a point) with algebraic endpoints. The formulas $(\exists x_2)\cdots(\exists x_n)\psi^{(m)}$, $m = 1,2,\dots$, thus determine a convergent sequence of intervals (which can be found by back-tracking of the kind described in the proof of Lemma 2.3). The points of the intersection of such a sequence satisfy $(\exists x_2)\cdots(\exists x_n) \bigwedge_{w\in|\sigma|} \mathrm{Sub}_w(\psi)$. Without going into the tedious details, it is clear that the first coordinate is then obtained by a computable convergent sequence of intervals. Using the first coordinate, we obtain similarly the second, etc.

THEOREM 2.6. <u>The algorithmic theory of $\mathfrak{R}$ is undecidable.</u>

<u>Proof</u>. Since the algorithmic theory of the natural numbers $\mathfrak{n} = \langle N, 0, \text{Successor}\rangle$ is undecidable, [2], it suffices to give a relative interpretation of the algorithmic theory of $\mathfrak{n}$ into that of $\mathfrak{R}$. For this, observe that the program

1: <u>do</u> $x_1 := 0$ <u>then go to</u> 2;

2: <u>do</u> $x_2 := 1$ <u>then go to</u> 3;

3: <u>if</u> $x_1 := x_0$ <u>then go to</u> 5 <u>else go to</u> 4;

4: <u>do</u> $x_1 := x_1 + x_2$ <u>then go to</u> 3,

terminates exactly when started on an assignment of a natural number to x_0. Thus the algorithmic property associated to this program defines the concept of natural number.

Theorems 2.4, 2.5, and 2.6 together establish the main result of this section:

THEOREM 2.8. The algorithmic theory of computationally closed archimedian ordered fields is complete, axiomatizable, and undecidable; its models are exactly those ordered fields that are algorithmically equiavlent to the field of real numbers.

Note that, in contrast, the set of first-order consequences of the set of axioms for computationally closed archimedian ordered fields is decidable (since it is the theory of real closed fields for which we have Tarski's decision procedure). Note also that the requirement of archimedian order may be dropped since it follows easily from computational closure.

3. Miscellaneous remarks.

1. Let us call the algorithmic theory of a geometry its "construction theory." If we now define the notion of a constructively closed Euclidean geometry in the obvious manner analogous to the definition of computationally closed fields, we obtain the following consequence of Theorem 2.7 and Corollary 1.5:

THEOREM. The construction theory of constructively closed Euclidean geometries is complete, axiomatizable, and undecidable; its models are exactly those geometries which are constructively equivalent to the real Euclidean plane.

2. The definition of the notion of construction problem is left implicit in the literature. Most modern writers seem to favor a notion which we would call elementary construction problem and which could be formally defined thus: Suppose that φ is a quantifier-free formula of elementary geometry (i.e., a first-order formula):

$$\varphi\left(P_1,\ldots,P_n,\ell_1,\ldots,\ell_n,\gamma_1,\ldots,\gamma_n,P_{n+1},\ldots,P_{n+m},\ell_{n+1},\ldots,\ell_{n+m},\gamma_{n+1},\ldots,\gamma_{n+m}\right).$$

Then the elementary construction problem associated to φ consists in finding a program π over A(1) - B(8) such that π computes values for $P_{n+1},\ldots,\gamma_{n+m}$ from initial values for $P_1,\ldots,\gamma_n$ such that these values together satisfy the formula φ. Some authors even restrict themselves further by allowing only such φ that, translated into analytical terms, are equivalent to a single algebraic equation. Note that, for example, the problem of the quadrature of the circle is not an elementary construction problem. (It is, however, a construction problem in the sense of the present paper, where we may take for φ any formula of construction theory.) For the deductive theory $\mathcal{E}_2$" which is adequate for the elementary (in the above sense) theory of geometrical constructions see Tarski [5]; it is not adequate for all of traditional construction theory. For example, Kijne [4] shows that the Mohr-Mascheroni construction theorem does not hold in all models of $\mathcal{E}_2$".

3. By the well-known algebraic methods, it is possible to prove that there is a decision procedure for the problem whether a given elementary construction problem can be solved. In contrast, there is no decision procedure for the general construction problem in which φ is an arbitrary formula of construction theory.

4. In a semi-formal exposition of the theory of geometrical constructions, Kijne [4] uses two additional kinds of operations.[2] The selection operation S_1 and the adjunction operation A^g. The need for the first seems to arise there because of the lack of definiteness in the formulation of construction programs. The adjunction operation is more troublesome. Its intuitive counterpart is illustrated by: "... now, let P_i be any point interior to the given triangle, ..." With the aid of the fixed coordinate lines x, y and basis points $0, E_x$ and E_y, we are able to circumvent the use of adjunction operations in all constructions that are independent of the choice of the adjoined element (within the class of possible choices). The proof of this is well-known, [3].

REFERENCES

[1] Engeler, E., Algorithmic properties of structures, Math. Systems Theory, 1 (1967), 183-195.

[2] __________, Formal Languages: Automata and Structures, Markham Publishing Company, Chicago, (1968), viii + 81 pp.

[3] Enriques, F. (edit.), Fragen der Elementargeometrie, vol. 2, B. G. Teuber Verlag, Leipzig (1923), p. 116.

[4] Kijne, D., Plane construction field theory, Ph.D. Thesis, University of Utrecht, (1956), vi + 118 pp.

[5] Tarski, A. What is elementary geometry, The Axiomatic Method, ed. by L. Henkin, P. Suppes, and A. Tarski, North Holland Publishing Company, Amsterdam (1959), 16-29.

UNIVERSITY OF MINNESOTA

[2] We thank Professor H. Guggenheimer for drawing our attention to the work of Kijne.

NOTE ON ADMISSIBLE ORDINALS

HARVEY FRIEDMAN AND RONALD JENSEN

Sacks proved by a forcing argument that every countable admissible ordinal α has the form $\alpha = \omega_1^B$ for some $B \subset \omega$. Later, Jensen proved several generalizations by an alternative forcing argument. In this paper we present a proof of Sacks' result which makes no use of forcing. It appears to be shorter than the earlier proofs. The fact that the theorem may be established by considering appropriate end extensions of L_α was first noticed by Friedman. The present proof contains suggestions due to both authors and is the end result of a series of approximations. The use of induction on α, which is the key to eliminating forcing, was first proposed by Friedman.

We assume that α is the least countable admissible ordinal not of the form ω_1^B. We let $A = L_\alpha$; i.e., the initial segment of L up to α. We associate with A the infinitary language $\mathcal{L}_A$ as defined in Barwise [1].

LEMMA 1. <u>If every α-finite subset of an α-r.e. theory in $\mathcal{L}_A$ is consistent, then so is the theory</u>. The proof is in [1].

We single out a constant symbol, c, and restrict ourselves to $\mathcal{L}_A$ with one 2-ary relation symbol, ϵ, and the constant symbol c. We write this language as $\mathcal{L}$.

We let $\varphi \in \mathcal{L}$ be the conjunction of all formulae in ZF-P (ZF without power set), together with

$$c \text{ is an ordinal and every admissible ordinal} \leq c \text{ is of the form } \omega_1^B$$

$$\text{and } (\forall x)\Big(x \in \omega \rightarrow \bigvee_{i \in \omega} x = i\Big) \text{ and all sets are countable .}$$

We then wish to add a schema in $\mathcal{L}$ to φ which asserts that c is "larger" than all $\beta < \alpha$.

LEMMA 2. There is a natural α-recrusive 1-1 function $f : \alpha \to \mathcal{L}$ so that for all $\beta < \alpha$ and structures M ($= \langle M,R,d \rangle$) we have:

$$M \vdash f(\beta) \text{ if and only if } \{x \mid Rxd\} \text{ is linearly ordered by } R \text{ and has an initial segment of type } \beta .$$

(Here d is the interpretation of c and R is the interpretation of ϵ.) Furthermore, f is 1-1 and the inverse image of any α-finite subset of Range (f) is bounded in α.

Proof. The first half is well-known, and is proven by a standard inductive construction. The second half follows from α being admissible, and f being 1-1.

We define the theory $T \subseteq \mathcal{L}$ by Range $(f) \cup \{\varphi\}$. The second half of Lemma 2 is used to prove that every α-finite subset of T is consistent. By Lemma 2, every α-finite subset of T has only the $f(\beta)$, for some $\beta < \gamma < \alpha$. We define M as HC = hereditarily countable sets, and take $d = \gamma$ (d being the interpretation of c). Then clearly HC has standard integers and satisfies ZF-P and all of the $f(\beta)$, $\beta < \gamma$. In addition, any ordinal $\delta \leq \gamma$ satisfying the formal sentence expressing admissibility must be admissible, and hence (by the choice of α) must be of the form ω_1^B. Now $B \in$ HC (in fact, $B \subseteq \omega$), and so $\omega_1^B = \delta$ must hold in HC.

Now let $N = \langle N,R,d \rangle$ be a model of T. We can, without loss of generality, identify the integers in N with the real integers.

LEMMA 3. No linear ordering of ω of type α occurs in N.

Proof. Suppose x is a linear ordering of type α in N. Then by replacement in N there is some element $\bar{\alpha}$ such that

$$N \models \bar{\alpha} \text{ is an admissible ordinal } \leq d \text{ and } \bar{\alpha} \approx x$$

where we use $\approx$ to indicate the existence of an order isomorphism. Hence

$$N \models \bar{\alpha} \text{ is of the form } \omega_1^D .$$

Choose such a D in N. Then $\omega_1^D \geq \alpha$ since $\{z \mid R z \bar{\alpha}\}$ has order type α when ordered by R. If

$\omega_1^D > \alpha$, then some well-ordering y of type α, recursive in D, occurs in N. But

$$N \models y < \bar{\alpha} \vee y \approx \bar{\alpha} \vee y > \bar{\alpha} .$$

The only possible case is $N \models y \approx \bar{\alpha}$, thus violating $N \models \omega_1^D = \bar{\alpha}$. Thus $\alpha = \omega_1^D$ and we have a contradiction.

To complete the proof, we choose a linear ordering B of ω with $N \models B \approx d$. We can do this, since

$$N \models \text{all sets are countable} .$$

Then $\omega_1^B \geq \alpha$, because B has initial segments of type β, for each $\beta < \alpha$. It now suffices to show $\omega_1^B \leq \alpha$. Supposing $\omega_1^B > \alpha$, we find a D recursive in B, D a well-ordering of type α. But D occurs in N. Thus, we have shown α to be of the form ω_1^B, contrary to assumption.

REFERENCE

[1] Barwise, Jon, Infinitary logic and admissible sets, Thesis, Stanford University, 1967.

STANFORD UNIVERSITY

AN ALGEBRAIC PROOF OF THE BARWISE COMPACTNESS THEOREM[1]

CAROL KARP

The early treatments of infinitary languages had the formulas classified by cardinality (Scott-Tarski [12], 1958, Karp [4], 1959). Formulas of $\underset{\sim}{L}_{\kappa,\lambda}$ were to have conjunctions of length less than κ, quantifier-sequences of length less that λ, where κ and λ were cardinals. Actually there was no good reason for doing things that way as Kreisel rather forcefully pointed out in [8], 1963. If, for instance, one is looking at a logical system for formulas which belong to a set A, what one needs to know about A is that it has closure properties adequate for setting up the logic and that when a formula in A is provable, it has a proof in A. Following Kreisel's suggestions which were, by the way, model-theoretically motivated, Barwise in his thesis [1], 1966, developed infinitary languages and logic along these lines and found, among other things, that the transitive sets A suitable for a natural cut-free deductive system for finite-quantifier formulas are precisely the admissible sets of infinitary recursion theory (Kripke [10]) and their unions. Moreover the admissible sets are suitable for deductions from a $\Sigma_1^{A,+}$-definable set of assumptions which are formulas of A. If we let $\underset{\sim}{L}_A$ be the language with finite-quantifier formulas in A, the old $\underset{\sim}{L}_{\kappa,\omega}$ are equivalent to $\underset{\sim}{L}_{H_\kappa}$, with formulas hereditarily of power less than κ. The constructive infinitary languages introduced by Kino-Takeuti [7], 1963, are equivalent to $\underset{\sim}{L}_{L(\omega_1^c)}$ with formulas in the constructible hierarchy up to the first non-recursive ordinal. But there are many new languages $\underset{\sim}{L}_A$ not studied before, many with better model-theoretic properties than the old ones had. See Kreisel's paper in this volume for a summary.

[1] Prepared with the aid of NSF grant GP-6897.

It is the purpose of this note to show that the Boolean-algebraic methods of Karp [5] still apply to the languages $\underset{\sim}{L}_A$. It is not necessary to use the complicated deductive systems used by Barwise to show

(I) If A is a union of admissible sets, a provable formula in A has a proof in A.

All one needs is a set of rules with a Σ_1^A-definable proof predicate that characterizes Boolean-valued validity. Similarly, it is not necessary to use a complicated deductive system to show

(II) If A is admissible and $S \subseteq A$ is a $\Sigma_1^{A,+}$-definable set of statements, a formula in A is deducible from S if and only if it is deducible from a subset of S which belongs to A.

All one needs to know is that the relation "x is deducible from S" is $\Sigma_1^{A,+}$-definable and that it is equivalent to "every Boolean-valued model of S is a model of x". Of course, it is still true that the deductive systems of [1] give information that Boolean-algebraic methods do not, for example, the important interpolation theorems.

When Boolean logic is adequate for semantic validity as it is for countable formulas, (II) gives rise to a compactness theorem. The Barwise Compactness Theorem says that if A is countable and admissible, and $S \subseteq A$ is a $\Sigma_1^{A,+}$-definable set of statements, then S has a model if and only if every subset of S that belongs to A, has a model. This theorem holds only for countable A, but (II) holds also for additional predicates $R_1,\ldots,R_k$ on A.

(II') If A is $R_1,\ldots,R_k$-admissible and $S \subseteq A$ is a $\Sigma_1^{A,+}$ in $R_1,\ldots,R_k$-definable set of statements, then a formula in A is deducible from S if and only if it is deducible from a subset of S which belongs to A.

Thus when Boolean logic plus a $\Sigma_1^{A,+}$ in $R_1,\ldots,R_k$-definable set of assumptions is adequate for semantic validity on A, then (II') gives rise to a compactness theorem for A. The cases where A has cofinality ω are included here because in these cases the completeness theorems of [5] take a form suitable for the application of (II'). We obtain this theorem which was noticed also by Barwise: If $cf(A) = \omega$ and A is P-admissible, where P is the graph of the power-set operation restricted to A, and closed under the power-set operation, then for a $\Sigma_1^{A,+}$ in P-definable subset $S \subseteq A$ of statements, S has a model if and only if every subset of S that belongs to A has a model.[2] The compactness theorems also hold relative to additional predicates on A.

The work reported here is in an early stage. There was not time to prepare for this publication the weak, but still valuable, form of the Cf(ω) Compactness Theorem that holds for sets not necessarily closed under the power-set operation. Work in progress includes the formulation and study of languages $\underset{\sim}{L}_A$ where A is not transitive. Languages $\underset{\sim}{L}_{H_\kappa}$ do not quite match the old $\underset{\sim}{L}_{\kappa,\omega}$ because symbols for $\underset{\sim}{L}_{H_\kappa}$ must come from H_κ while symbols for a language $\underset{\sim}{L}_{\kappa,\omega}$ could come from anywhere. Thus an arbitrary structure can be symbolized by a language $\underset{\sim}{L}_{\kappa,\omega}$ while only structures that one can map into H_κ can be symbolized by $\underset{\sim}{L}_{H_\kappa}$.

This formulation of infinitary languages depends on the hierarchy of set-theoretically definable predicates first treated systematically by Lévy in his Memoir [11]. A (finite) set-theoretical formula in predicates $\approx$, ε, and possibly extra predicate symbols $\underline{R}_0,\ldots,\underline{R}_k$, is Δ_0 in $\underline{R}_0,\ldots,\underline{R}_k$ if all of its quantifiers appear restricted, i.e., in forms $(w)[w \in v \rightarrow \cdots]$ or $(Ew)[w \in v \wedge \cdots]$. An important property of such formulas is their absoluteness in transitive classes. If $\mathfrak{D}$ is Δ_0 in $\underline{R}_0,\ldots,\underline{R}_k$, $T \subseteq T'$ are transitive classes and $R_0,\ldots,R_k$ are predicates on T, Then $(\forall x_1,\ldots,x_n \in T)(\langle T,\in \upharpoonright T, R_0,\ldots,R_k\rangle \models \mathfrak{D}(x_1,\ldots,x_n) \Leftrightarrow \langle T',\in \upharpoonright T', R_0',\ldots,R_k'\rangle \models \mathfrak{D}(x_1,\ldots,x_n))$ provided the R_i' agree with R_i on T. A formula is generalized Σ_1 in $\underline{R}_0,\ldots,\underline{R}_k$ if it is a finite formula built up from formulas Δ_0 in $\underline{R}_0,\ldots,\underline{R}_k$ by ordinary conjunction, disjunction, existential and bounded universal quantification. Such a formula is Σ_1 in $\underline{R}_0,\ldots,\underline{R}_k$ if it has form $(Ew)\mathfrak{D}$ where $\mathfrak{D}$ is Δ_0 in $\underline{R}_0,\ldots,\underline{R}_k$. Generalized Σ_1 in $\underline{R}_0,\ldots,\underline{R}_k$-formulas reduce to Σ_1 in $\underline{R}_0,\ldots,\underline{R}_k$-formulas by means of these equivalences:

$$(Ew)\mathfrak{A}(w,\ldots) \wedge (Ew')\mathfrak{B}(w',\ldots) \longleftrightarrow (Ew,w')[\mathfrak{A}(w,\ldots) \wedge \mathfrak{B}(w',\ldots)], \quad w \neq w'$$

$$(Ew)\,\mathfrak{A} \vee (Ew')\mathfrak{B} \longleftrightarrow (Ew,w')[\mathfrak{A} \vee \mathfrak{B}]$$

$$(Ew)(Ew')\mathfrak{A} \longleftrightarrow (Ew'')(Ew \in w'')(Ew' \in w'')\mathfrak{A}$$

$$(\forall w \in v)(Ew')\mathfrak{A} \longleftrightarrow (Ew'')(\forall w \in v)(Ew' \in w'')\mathfrak{A}.$$

The definition of admissibility guarantees that whenever a set A is $R_0,\ldots,R_k$-admissible, a

[2] In my lecture for the Symposium, I stated that this theorem held for Σ_1^A-sets of statements provided that A was closed under the power-set operation, had cofinality ω, but was only admissible. This was incorrect. The error was found independently by J. Gregory and J. Barwise.

generalized Σ_1 in $\underline{R}_0,\ldots,\underline{R}_k$-formula reduces to a Σ_1 in $\underline{R}_0,\ldots,\underline{R}_k$-formula on A. Generalized Σ_1 in $\underline{R}_0,\ldots,\underline{R}_k$-formulas are persistent upward from transitive sets. An important property of Σ_1-formulas is their cardinal absoluteness: If G is Σ_1, $\langle V,\in \restriction V\rangle \models G(x) \Leftrightarrow \langle t,\in \restriction t\rangle \models G(x)$, where $t = \{u \mid TC(u) \preceq TC(x)\}$ (TC is the transitive closure).

Let T be a class. An n-place predicate R on T is Σ_1^T in $R_0,\ldots,R_k$-definable if and only if for some Σ_1 in $\underline{R}_0,\ldots,\underline{R}_k$-formula A in n variables,

$$R(x_1,\ldots,x_n) \Leftrightarrow \langle T,\in \restriction T,R_0,\ldots,R_k\rangle \models G(x_1,\ldots,x_n).$$

Moreover R is $\Sigma_1^{T,+}$ in $R_0,\ldots,R_k$-definable if for some $a_1,\ldots,a_p \in T$ and some Σ_1 in $\underline{R}_0,\ldots,\underline{R}_k$-formula in $n + p$ variables, $(\forall x_1,\ldots,x_n \in T)$

$$(R(x_1,\ldots,x_n) \Leftrightarrow \langle T,\in \restriction T,R_0,\ldots,R_k\rangle \models G(x_1,\ldots,x_n,a_1,\ldots,a_p)).$$

The Σ_1^V-definable functions on the universe V (i.e. functions whose graphs are Σ_1^V-definable) have very nice closure properties. Some of the best of these properties are also closure properties for the primitive recursive set functions treated in [3]. Although the terminology of primitive recursive set functions is used in this paper, very little is assumed about them. We list the properties that we do use:

(1) A primitive recursive set function is defined on the whole universe V. The graph of a primitive recursive set function is Σ_1^V-definable. Hence the complement of the graph is also Σ_1^V-definable since $y \neq F(x) \Leftrightarrow (\exists z)(z = F(x)\ \&\ z \neq y)$. If T is transitive and primitive recursively closed, then the graph of F restricted to T is Σ_1^T-definable (in fact, is defined relative to T by its original defining formula).

(2) The following commonly used functions and relations on sets are primitive recursive:

(i) Such relations as $\lambda x,y\ (x \subseteq y)$, λx (x is transitive), λx (x is an ordinal), λx (x is an ordered pair), λx (x is a relation), λx (x is a function), λx (x is a function), $\lambda x\ (x \in \omega)$, $\lambda x\ (x \subseteq \omega)$, λx (x is finite).

(ii) Such functions as $\lambda x(\bigcup x)$, $\lambda x,y\ (x \cup y)$, $\lambda x,y\ (x \cap y)$, $\lambda x,y\ (x - y)$, $\lambda x,y\ (\{x,y\})$, $\lambda x,y\ (\langle x,y\rangle)$, $\lambda x\ (Dm(x))$, $\lambda x\ (Rg(x))$, $\lambda x,y\ (x\text{'}y)$ where $x\text{'}y$ is the value of x at y if x is a function, $\lambda x,y\ (x\text{''}y)$, the image of x on y, $\lambda x,y\ (x \restriction y)$, x restricted to y, $\lambda x,y\,(x \times y)$, $\lambda x,y\,(x^y$ if y is finite, 0 if not), $\lambda x(TC(x))$, $\lambda x(x^{-1})$, $\lambda x(\mathrm{rank}(x))$ where $\mathrm{rank}(x) = \bigcup\{\mathrm{rank}(y) + 1 \mid y \in x\}$.

(3) The primitive recursive set functions are closed under substitution, definition by cases, ordinal recursion and recursion with respect to $\in$, the bounded ordinal min rule. If G and R are primitive recursive, so is F where $F(z,\underset{\sim}{x}) = \{G(u,\underset{\sim}{x}) \mid u \in z \,\&\, R(u,\underset{\sim}{x})\}$.

This supply of initial functions and relations and closure properties is sufficient for the main technical Lemmas 1.1 and 1.2.

§1. Preliminaries.

For this discussion we may suppose that the infinitary finite-quantifier languages are defined set-theoretically along lines of Barwise [1], except that the conjunction is to operate not on a set of formulas but on a function whose range is a set of formulas. This is a technical device designed to insure that the calculations of [6] can be used to guarantee that the syntax of the languages is primitive recursive in the set-theoretical sense [3]. The unnecessary requirement in [6] that the conjunctions be well-ordered has been dropped. In the list that follows, the formulas are the sets appearing on the left, the notation on the right shows how they are to be interpreted.

Individual Symbols:

$(0,x)$, $(2,x)$	individual variables $\ulcorner v_x \urcorner$, $\ulcorner w_x \urcorner$.
$(1,x)$, $(3,x)$, ...	individual constants $\ulcorner c_x \urcorner$, $\ulcorner d_x \urcorner$, ...

Atomic Formulas:

$\langle 0,u,u' \rangle$ where u,u' are individual symbols	$\ulcorner[t \approx t']\urcorner$ where $u = \ulcorner t \urcorner$, $u' = \ulcorner t' \urcorner$.
$\langle 2,u,u' \rangle$ where u,u' are individual symbols	$\ulcorner[t \in t']\urcorner$.
$\langle 2(n+2),u \rangle$ where u is an individual symbol	$\ulcorner[\underline{R}_n t]\urcorner$.

Formulas:

$\langle 1,y \rangle$ where y is a formula	$\ulcorner \neg \mathfrak{a} \urcorner$ where $y = \ulcorner \mathfrak{a} \urcorner$.

$\langle 3,y,z\rangle$ where y, z are formulas.	$\ulcorner \mathfrak{A} \to \mathfrak{B} \urcorner$ where $y = \ulcorner \mathfrak{A} \urcorner$, $z = \ulcorner \mathfrak{B} \urcorner$.
$\langle 5,w\rangle$ where w is a function whose range is a set of formulas.	$\ulcorner \bigwedge_{i\in I} \mathfrak{A}_i \urcorner$ if $I = \mathrm{Dm}(w)$, $w`i = \ulcorner \mathfrak{A}_i \urcorner$.
$\langle 7,x,y\rangle$ where y is a formula.	$\ulcorner (w_x)\mathfrak{A} \urcorner$ where $y = \ulcorner \mathfrak{A} \urcorner$.

Connectives $\wedge$, $\vee$, $\bigvee$, $\leftrightarrow$, and the existential quantifier (Ew_x) are introduced by the customary abbreviations. If S is a set of formulas, $\ulcorner \bigwedge \urcorner S$ is $\langle 5, I \restriction S\rangle$ where I is the identity operation.

An interpretation for the language consists of a system $\mathfrak{A} = \langle A, E, R_0, \ldots, R_n, \ldots, K\rangle$ where the R_n are unary relations on A and K maps the individual constants into A. This broad formulation with convenient names for sets has many advantages. For example, for a set x let $\mathrm{Eq}(x)$ be the sentence $\ulcorner (w_0)\left[w_0 \mathrel{\varepsilon} c_x \leftrightarrow \bigvee_{y\in x} w_0 \approx c_y\right].\urcorner$ Then an easy induction shows that if T is a transitive class, any model of $\{\mathrm{Eq}(x) \mid x \in T\}$ is isomorphic to an end-extension of $\langle T, \in \restriction T\rangle$.

Another advantage is the sharpening of Löwenheim-Skolem arguments by use of a uniform assignment by use of a uniform assignment of witnessing constants for quantifications as in §3.

1.1. LEMMA. <u>The functions and relations listed below are all primitive recursive</u>: Fmla(y), "y <u>is a formula</u>", Neg(y), "y <u>is a negation</u>", Impl(y), "y <u>is an implication</u>", Conj(y), "y <u>is a conjunction</u>", Quan(y), "y <u>is a universal quantification</u>". λy Con(y), <u>the set of individual constants appearing in</u> y <u>if</u> Fmla(y). λy Subfmla(y), <u>the set of subformulas of</u> y <u>if</u> Fmla(y). λy FV(y), <u>the set of variables appearing free in</u> y <u>if</u> Fmla(y). Sent(y), "y <u>is a sentence</u>". $\lambda y,v,u$ SF(y,v,u), <u>the result of substituting</u> u <u>for free occurrences of</u> v <u>in</u> y <u>if</u> Fmla(y), v <u>is an individual variable</u>, u <u>is an individual symbol</u>.

Ordinary Hilbert-style Boolean logic is adequate for our purposes.

<u>Logical Axioms</u>:

(1) Substitutions of the tautologies in $\neg$, $\to$, that one uses to axiomatize ordinary propositional logic.

(2) Formulas of the form $\ulcorner \left[\bigwedge_{i\in I} \mathfrak{A}_i\right] \to \mathfrak{A}_j \urcorner$ where $j \in I$, and $\ulcorner \bigwedge_{i\in I} [\mathfrak{A} \to \mathfrak{B}_i] \to \left[\mathfrak{A} \to \bigwedge_{i\in I} \mathfrak{B}_i\right] \urcorner$.

(3) Formulas of the form $\ulcorner(w_x)\,[\mathfrak{A} \to \mathfrak{B}] \to [\mathfrak{A} \to (w_x)\mathfrak{B}]\urcorner$ and $\ulcorner(w_x)\mathfrak{A} \to \mathfrak{A}(w_x/t)\urcorner$ where t is an individual symbol.

(4) Equality axioms.

Rules of Inference:

Modus ponens

Conjunction: From $\ulcorner\mathfrak{A}_i\urcorner$ for all $i \in I$ infer $\ulcorner\left[\bigwedge_{i\in I} \mathfrak{A}_i\right].\urcorner$

Generalization: From $\ulcorner\mathfrak{A}(v_y)\urcorner$ infer $\ulcorner(w_x)\mathfrak{A}(v_y/w_x)\urcorner$.

Formal Deductions:

A formal deduction from a set S of formulas is a function p whose range consists of formulas having no free occurrence of variables w_x, such that for every $x \in Dm(p)$, either $p'x$ is an axiom or $p'x \in S$ or $p'x$ is the result of applying a rule of inference to formulas in $Rg(p \upharpoonright TC(x))$. A formal deduction from the empty set may also be called a formal proof. If formula y is in the range of a deduction from S we may write "$S \vdash y$". Moreover if y is in the range of a deduction p from S such that $p \in T$, we may write "$S \vdash y$ in T".

The formal development of Boolean logic for these languages is so similar to that in [5] that we do not go into it here. There are however two special properties needed for the theorems below.

1.2. LEMMA. (i) The predicate $Prf(x,y)$, "x is a formal proof and $y \in Rg(x)$" is primitive recursive. The predicate "x is a formal deduction from S" is primitive recursive in S.

(ii) If T is transitive and primitive recursively closed, $S \subseteq T$ and $\ulcorner\left[\bigwedge_{i\in I} \mathfrak{A}_i\right]\urcorner \in T$ and

$(\exists w \in T)(\forall i \in I)(S \vdash \ulcorner\mathfrak{A}_i\urcorner$ in $w)$ then $(\exists w \in T)(S \vdash \ulcorner\left[\bigwedge_{i\in I} \mathfrak{A}_i\right]\urcorner$ in $w)$.

The proof of (i) is similar to that found in [6], Appendix 2. Part (ii) is a routine calculation which involves repositioning the given deductions so that they can be made into a single one in T.

§2. Infinitary Logic and Admissible Sets.

Admissible sets play the same role in infinitary logic that they do in recursion theory. In Kripke's equation calculus, the admissible sets are those that are closed under computations, in infinitary logic the admissible sets are those that are closed under formal proofs and deductions from Σ_1-definable sets of assumptions.

2.1. MAIN LEMMA. <u>Let</u> T <u>be a transitive primitive recursively closed class. Then if</u>

$$(*)_T: \quad \ulcorner\Big[\bigwedge_{i\in I} G_i\Big]\urcorner \in T \;\&\; (\forall i \in I)(\exists p \in T)\mathrm{Prf}(p, \ulcorner G_i\urcorner) \Rightarrow (\exists p' \in T)\mathrm{Prf}(p', \ulcorner \bigwedge_{i\in I} G_i\urcorner)$$

<u>then for any formula</u> $\ulcorner G\urcorner \in T$, $\vdash \ulcorner G\urcorner \Longleftrightarrow \vdash \ulcorner G\urcorner$ <u>in</u> T.

<u>Moreover, if</u> $S \subseteq T$ <u>and if</u>

$$(**)_{S,T}: \quad \ulcorner \bigwedge_{i\in I} G_i\urcorner \in T \;\&\; (\forall i \in I)(\exists s \in T)(s \subseteq S \;\&\; s \vdash \ulcorner G_i\urcorner) \Rightarrow (\exists s' \in T)(s' \subseteq S \;\&\; s' \vdash \ulcorner \bigwedge_{i\in I} G_i\urcorner)$$

<u>then for any formula</u> $\ulcorner G\urcorner \in T$, $S \vdash \ulcorner G\urcorner \Longleftrightarrow (\exists s \in T)(s \subseteq S \;\&\; s \vdash \ulcorner G\urcorner)$.

<u>Proof</u>. Algebraic methods make these proofs straightforward. For instance to prove the second statement, assume that S is a set of statements of T and that $(**)_{S,T}$ holds. We may assume that infinitely many constants, say $(5,n)$, $n \in \omega$, do not appear in statements of S. Suppose further that S is T-consistent, i.e., no contradiction is deducible from any subset of S that belongs to T. Let D be the set of all sentences in T with finitely many constants $(5,n)$ appearing. Form equivalence classes of sentences in D under the relation $\ulcorner G\urcorner \approx \ulcorner \mathcal{B}\urcorner \Longleftrightarrow (\exists s \in T)(s \subseteq S \;\&\; s \vdash \ulcorner G \leftrightarrow \mathcal{B}\urcorner)$. Using the T-consistency of S and the usual properties of Boolean logic it is easy to see that the equivalence classes form a Boolean algebra $\mathfrak{B}$ under the Lindenbaum operations $(\ulcorner G\urcorner/\approx) \cap (\ulcorner \mathcal{B}\urcorner/\approx) = \ulcorner \wedge G,\mathcal{B}\urcorner/\approx$, $-(\ulcorner G\urcorner/\approx) = \ulcorner \neg G\urcorner/\approx$, and so on. By $(**)_{S,T}$, $\inf_{i\in I}(\ulcorner G_i\urcorner/\approx) = \ulcorner \bigwedge_{i\in I} G_i\urcorner/\approx$ whenever the conjunction is in D. Because statements in D have only finitely many $(5,n)$ appearing, $\ulcorner(w_x)G\urcorner/\approx \; = \inf_{n\in\omega}(\mathrm{SF}(\ulcorner G\urcorner,(2,x),(5,n))/\approx) = \inf\{\mathrm{SF}(\ulcorner G\urcorner,(2,x),c)/\approx \mid c$ is a constant in $T\}$.

Let $\mathfrak{B}^*$ be the completion of $\mathfrak{B}$ preserving all existing infs and sups. Define a Boolean-valued interpretation for formulas by taking $\mathfrak{B}^*$ as the algebra of truth values, the set of individual constants in T as domain, and interpreting the equality and membership symbols as $\ulcorner c\urcorner \;\mathrm{EQ}\; \ulcorner c'\urcorner = \ulcorner c \approx c'\urcorner/\approx$ and $\ulcorner c\urcorner \;\mathrm{E}\; \ulcorner c'\urcorner = \ulcorner c \in c'\urcorner/\approx$. Interpret the $\underline{R}_n$ as $R_n(\ulcorner c\urcorner) = \ulcorner \underline{R}_n c\urcorner/\approx$. Assign to individual constants the constant itself it if is in T, $(5,0)$ if not. An easy induction shows that the Boolean value of a statement $\ulcorner G\urcorner \in D$ is $\ulcorner G\urcorner/\approx$. But then for $\ulcorner C\urcorner \in S$, the value of $\ulcorner C\urcorner$ is the unit element of $\mathfrak{B}^*$. Since the axioms are Boolean valid and the rules of inference preserve Boolean validity, it follows that any formula deducible from S is valid in the interpretation. Hence under the assumption that S is T-consistent it follows that no contradiction is deducible from S.

Hence (ii) above.

Part (i) is similar. Suppose that $\ulcorner \mathfrak{A} \urcorner \in T$ has no formal proof in T. We must show not $\vdash \ulcorner \mathfrak{A} \urcorner$. Assume that constants $(5,n)$, $n \in \omega$, do not appear in $\ulcorner \mathfrak{A} \urcorner$ and define an equivalence relation on formulas of D as above, by $\ulcorner \mathfrak{B} \urcorner \approx \ulcorner \mathfrak{B}' \urcorner \leftrightarrow (\exists p \in T)Prf(p, \ulcorner \mathfrak{A} \rightarrow [\mathfrak{B} \leftrightarrow \mathfrak{B}'] \urcorner)$. Then proceed as before.

The notion of admissibility gives precisely $(**)_{S,T}$ for generalized Σ_1-subsets of T, whence the Main Lemma can be reformulated as Theorem 2.3 below.

2.2. DEFINITION. Let $R_1,\ldots,R_k$ be relations on a set T. Then T is $R_1,\ldots,R_k$-admissible if and only if T is transitive, primitive recursively closed, and satisfies the Σ_1^T in $R_1,\ldots,R_k$ reflection principle; i.e., for any $n+2$-place formula $\mathfrak{D}$ which is Δ_0 in $\underline{R}_1,\ldots,\underline{R}_k$, $(\forall a_1,\ldots,a_n,x \in T)\ (\forall u \in x)\ (\exists v \in T) \models \mathfrak{D}(u,v,a_1,\ldots,a_n) \Rightarrow (\exists w \in T)\ (\forall u \in x)\ (\exists v \in w) \models \mathfrak{D}(u,v,a_1,\ldots,a_n)$, where $\models$ denotes satisfaction in $\langle T, \in \restriction T, R_1,\ldots,R_k \rangle$.

An ordinal α is admissible just in case $L(\alpha)$, the set of constructible sets up to level α, is admissible.

In the definition $\mathfrak{D}$ can be taken to be Σ_1 in $\underline{R}_1,\ldots,\underline{R}_k$ since adjacent existential quantifiers can be collapsed in primitive recursively closed sets. Moreover, it then follows that the rules for reducing generalized Σ_1 to Σ_1-formulas hold in an admissible set, so $\mathfrak{D}$ can be taken to be generalized Σ_1 in $\underline{R}_1,\ldots,\underline{R}_k$.

2.3. THEOREM. (Barwise) (i) <u>If</u> T <u>is a union of admissible sets, then whenever</u> $\ulcorner \mathfrak{A} \urcorner$ <u>is a formula in</u> T <u>and</u> $\vdash \ulcorner \mathfrak{A} \urcorner$, $\ulcorner \mathfrak{A} \urcorner$ <u>has a formal proof in</u> T.

(ii) <u>If</u> T <u>is</u> $R_1,\ldots,R_k$<u>-admissible and</u> $S \subseteq T$ <u>is a</u> $\Sigma_1^{T,+}$ <u>in</u> $R_1,\ldots,R_k$<u>-definable set of statements, then any formula</u> $\ulcorner \mathfrak{A} \urcorner \in T$ <u>such that</u> $S \vdash \ulcorner \mathfrak{A} \urcorner$, <u>is deducible from a subset of</u> S <u>in</u> T.

<u>Proof</u>. It suffices to prove (i) for admissible T. Condition $(*)_T$ is a consequence of Lemma 1.2 and the Σ_1^T reflection principle since primitive recursive predicates are generalized Σ_1-definable on any transitive primitive recursively closed set. Thus (i) follows from the Main Lemma.

Under the assumptions of (ii), part (i) of Lemma 1.2 shows that the condition $(s \subseteq S \ \&\ s \vdash \ulcorner \mathfrak{A}_i \urcorner)$ is a $\Sigma_1^{T,+}$ in $R_1,\ldots,R_k$-definable property of $s, i, \ulcorner \bigwedge_{i \in I} \mathfrak{A}_i \urcorner$. Thus the Σ_1^T in $R_1,\ldots,R_k$ reflection principle implies $(**)_{S,T}$. Thus (ii) follows from the Main Lemma.

Barwise shows for his formulation of the deductive systems, that the converse of (i) holds for countable segments of the constructible hierarchy and that the converse of (ii) holds for transitive primitive recursively closed sets generally. In fact, it holds for S fixed as in (iii) below. The function Eq in §1 is primitive recursive, so S in (iii) is primitive recursive in $R_1,\ldots,R_k$.

2.4. THEOREM. For transitive, primitive recursively closed sets T the following are equivalent:

(i) T is $R_1,\ldots,R_k$-admissible.

(ii) If $S \subseteq T$ is a $\Sigma_1^{T,+}$ in $R_1,\ldots,R_k$-definable set of statements, then any formula $\ulcorner \mathfrak{A} \urcorner \in T$ such that $S \vdash \ulcorner \mathfrak{A} \urcorner$, is deducible from some $s \in T$, $s \subseteq S$.

(iii) Let $S = \{Eq(x) \mid x \in T\} \cup \{\ulcorner \underline{R}_i c_x \urcorner \mid 1 \leq i \leq k \;\&\; R_i(x)\} \cup \{\ulcorner \neg \underline{R}_i c_x \urcorner \mid 1 \leq i \leq k \;\&\; \sim R_i(x)\}$ with $Eq(x)$ as defined in §1. Then any formula $\ulcorner \mathfrak{A} \urcorner \in T$ such that $S \vdash \ulcorner \mathfrak{A} \urcorner$, is deducible from some $s \in T$, $s \subseteq S$.

Proof. It suffices to show (iii) $\Rightarrow$ (i). First prove as a lemma that if $\mathfrak{D}(v_1,\ldots,v_n)$ is Δ_0 in $\underline{R}_1,\ldots,\underline{R}_k$, then for $x_1,\ldots,x_n \in T$,

$$\langle T, \in \restriction T, R_1,\ldots,R_k \rangle \models \mathfrak{D}(x_1,\ldots,x_n) \iff S \vdash \ulcorner \mathfrak{D}(c_{x_1},\ldots,c_{x_n}) \urcorner,$$

with S as in (iii). Then suppose that for $a_1,\ldots,a_n$, $x \in T$, the hypothesis of the reflection principle is satisfied, i.e., $(\forall u \in x)(\exists v \in T) \models \mathfrak{D}(u,v,a_1,\ldots,a_n)$. Then according to the lemma, $(\forall u \in x)(\exists v \in T) S \vdash \ulcorner \mathfrak{D}(c_u,c_v,c_{a_1},\ldots,c_{a_n}) \urcorner$. So $S \vdash \ulcorner \bigwedge_{u \in x} (Ew_0)\mathfrak{D}(c_u,w_0,c_{a_1},\ldots,c_{a_n}) \urcorner$. Choose s as in (iii) and let $w = TC\{u \mid u \in x$ or $u \in \{a_1,\ldots,a_n\}$ or $(\exists y \in s) \ulcorner c_u \urcorner \in Con(y)\}$. Then $w \in T$ and formulas of s are true in the interpretation $\langle w, \in \restriction w, R_1,\ldots,R_k, K \rangle$ where $K(\ulcorner c_u \urcorner) = u$. Hence so is the conclusion. Hence $(\forall u \in x)(\exists v \in w) \models \mathfrak{D}(u,v,a_1,\ldots,a_n)$.

§3. Compactness Theorems.

The Barwise Compactness Theorem is an immediate consequence of Theorem 2.3 and the completeness of Boolean logic for countable formulas. For if T is countable, $S \subseteq T$ and $\ulcorner \mathfrak{A} \urcorner \in T$, then $S \vdash \ulcorner \mathfrak{A} \urcorner \iff$ every model of S is a model of $\mathfrak{A}$.

3.1. BARWISE COMPACTNESS THEOREM. Suppose T is countable and $R_1,\ldots,R_k$-admissible. Then if $S \subseteq T$ is a $\Sigma_1^{T,+}$ in $R_1,\ldots,R_k$-definable set of statements, S has a model if and only if every subset of S in T has a model.

Once one sees this, it is natural to try to use the author's confinality-ω completeness theorem in the same way in order to obtain a compactness theorem for certain uncountable sets. It is necessary to strengthen the Boolean logic by adding in distributive laws, and this is where the power-set operation $\mathcal{P}$ enters the picture. One obtains the following result (noticed also by Barwise):

3.2. Cf(ω) COMPACTNESS THEOREM. Let P be the graph of the power-set operation on T, suppose T is closed under $\mathcal{P}$, is $P,R_1,\ldots,R_k$-admissible, and has form $T = \bigcup_{n\in\omega} T_n$, with the $T_n \in T$. Then if $S \subseteq T$ is a $\Sigma_1^{T,+}$ in $P,R_1,\ldots,R_k$-definable set of statements, S has a model if and only if every subset of S in T has a model.

Actually this result uses the cf(ω) completeness theorem in a weak form that is much easier to prove than the result in [5]. We give a proof that is further simplified by use of a uniform assignment of witnessing constants for quantifiers.

The proof of the Cf(ω) Completeness Theorem depends on a combinatorial characterization of those sets S of statements that have models. We may suppose without loss of generality that constants $(3,x) = \ulcorner d_x \urcorner$ do not appear in S. Operations C, D on S are defined by recursion as follows:

$$D(S,0) = \{x \mid x \in \text{Subfmla}^*(S) \ \&\ \text{Sent}(x)\}$$

$$C(S,0) = \text{Con}^*(S).$$

By Subfmla*(S) we mean $\bigcup_{y\in S} \text{Subfmla}(y)$ and by Con*(S), $\bigcup_{y\in S} \text{Con}(y)$.

$$D(S,n+1) = \{x \mid x \in \text{Subfmla}^* D(S,n) \ \&\ \text{Sent}(x)\} \cup$$
$$\{\text{SF}(\ulcorner \mathfrak{a} \urcorner, v, u) \mid \ulcorner \mathfrak{a} \urcorner \in \text{Subfmla}^* D(S,n) \ \&\ \text{FV}(\ulcorner \mathfrak{a} \urcorner) = \{v\} \ \&\ u \in C(S,n)\} \cup$$
$$\{\langle i,u,u' \rangle \mid u,u' \in C(S,n) \ \&\ i \in \{0,2\}\} \cup \{\langle 2(n+2),u \rangle \mid u \in C(S,n) \ \&\ n \in \omega\}.$$

$$C(S,n+1) = C(S,n) \cup \{(3,\ulcorner \mathfrak{a} \urcorner) \mid \ulcorner \mathfrak{a} \urcorner \in D(S,n) \ \&\ \text{Quan}(\ulcorner \mathfrak{a} \urcorner)\}.$$

Let $D(S) = \bigcup_{n\in\omega} D(S,n)$, $C(S) = \bigcup_{n\in\omega} C(S,n)$. Then $D(S)$ is closed under subformulas that are sentences and $\ulcorner (w_x)\mathfrak{a} \urcorner \in D(S)$ and $u \in C(S)$ implies $\text{SF}(\ulcorner \mathfrak{a} \urcorner,(2,x),u) \in D(S)$. Every formula in $D(S)$ is a sentence whose constants either appear in S or are witnessing constants $(3,\ulcorner \mathfrak{a} \urcorner)$. All these constants are in $C(S)$.

3.3. LEMMA. Let S be a set of sentences in which constants $(3,x)$ do not appear. Then S has a model if and only if there is a class S^* such that $S \subseteq S^*$ and the following conditions hold:

(1) (Negation) $(\forall \ulcorner \mathfrak{a} \urcorner \in D(S))(\ulcorner \mathfrak{a} \urcorner \in S^*$ or $\ulcorner \neg \mathfrak{a} \urcorner \in S^*) \ \& \sim (\exists \ulcorner \mathfrak{a} \urcorner \in D(S))(\ulcorner \mathfrak{a} \urcorner \in S^* \ \& \ \ulcorner \neg \mathfrak{a} \urcorner \in S^*)$.

(2) (Implication) $(\forall \ulcorner \mathfrak{a} \rightarrow \mathfrak{B} \urcorner \in D(S))(\ulcorner \mathfrak{a} \rightarrow \mathfrak{B} \urcorner \in S^* \Longleftrightarrow \ulcorner \mathfrak{a} \urcorner \notin S^*$ or $\ulcorner \mathfrak{B} \urcorner \in S^*)$.

(3) (Conjunction) $(\forall \ulcorner \bigwedge_{i \in I} \mathfrak{a}_i \urcorner \in D(S))(\ulcorner \bigwedge_{i \in I} \mathfrak{a}_i \urcorner \in S^* \Longleftrightarrow (\forall i \in I)(\ulcorner \mathfrak{a}_i \urcorner \in S^*))$.

(4) (Quantification) $(\forall \ulcorner (w_x)\mathfrak{a} \urcorner \in D(S)) \, [(\ulcorner (w_x)\mathfrak{a} \urcorner \in S^* \Rightarrow (\forall u \in C(S)) SF(\ulcorner \mathfrak{a} \urcorner,(2,x),u) \in S^*) \ \&$
$(SF(\ulcorner \mathfrak{a} \urcorner,(2,x),(3,\ulcorner (w_x)\mathfrak{a} \urcorner)) \in S^* \Rightarrow \ulcorner (w_x)\mathfrak{a} \urcorner \in S^*)]$.

(5) (Equality) $(\forall u \in C(S))(\langle 0,u,u \rangle \in S^*) \ \& \ (\forall u_1,u_2,u_3,u_4 \in C(S))(\forall i \in \{0,2\})((\langle 0,u_1,u_2 \rangle \in S^* \ \&$
$\langle 0,u_3,u_4 \rangle \in S^* \ \& \ \langle i,u_1,u_3 \rangle \in S^*) \Rightarrow \langle i,u_2,u_4 \rangle \in S^*) \ \& \ (\forall u,u' \in C(S))(\forall n \in \omega)((\langle 0,u,u' \rangle \in$
$S^* \ \& \ \langle 2(n+2),u \rangle \in S^*) \Rightarrow \langle 2(n+2),u' \rangle \in S^*)$.

Proof. Similar to the corresponding proof in [5]. Given a set S^* satisfying conditions (1) - (5) one defines the model on constants $C(S)$ that is always used in Henkin-style completeness proofs.

Conversely, for a model $\mathfrak{A} = \langle A,E,R_1,\ldots,K \rangle$ of S, the set S^* of true statements always satisfies all of the conditions (1) - (5) except (4). The assignment K to constants may be modified to satisfy (4) as follows. The witnessing constants $(3,\ulcorner \mathfrak{a} \urcorner)$ are not in S, so let $K_0 = K \upharpoonright C(S,0)$, the constants in S. Given K_n defined on $C(S,n)$, extend K_n to K_{n+1} on $C(S,n+1)$ in such a way that whenever $(3,\ulcorner (w_x)\mathfrak{a} \urcorner) \in C(S,n+1) - C(S,n)$, and $\langle A,E,R_1,\ldots,K \rangle \models \ulcorner \neg (w_x)\mathfrak{a} \urcorner$, $K_{n+1}((3,\ulcorner (w_x)\mathfrak{a} \urcorner))$ denotes an element $a \in A$ such that $\langle A,E,R_1,\ldots,K \rangle \models \ulcorner \neg \mathfrak{a} \urcorner(a)$. The axiom of choice is needed. Finally let S^* be the set of statements true in $\langle A,E,R_1,\ldots,\bigcup_{n \in \omega} K_n \rangle$. Then S^* satisfies (1) - (5).

3.4. REMARKS. (1) If $S \subseteq T$, T transitive and Prim-closed, then $C(S) \subseteq T$ and $D(S) \subseteq T$. So the proof just given shows that if S has a model, then S has a model whose universe is a subclass of T. This gives a sharpened version of the Löwenheim-Skolem Theorem as described in [3].

(2) Let $W(S)$ be the set of witnessing formulas for $D(S)$, i.e.,

$$W(S) = \{\langle 3, SF(\ulcorner \mathfrak{a} \urcorner,(2,x),(3,\ulcorner (w_x)\mathfrak{a} \urcorner)), \ulcorner (w_x)\mathfrak{a} \urcorner \rangle \mid \ulcorner (w_x)\mathfrak{a} \urcorner \in D(S)\}.$$

Then the proof just given shows that any model of S is the restriction to constants in S of a model of $S \cup W(S)$. Moreover, if S' is any set of statements having only constants in S, then any model of S' is the restriction to constants in S of a model of $S' \cup W(S)$.

3.5. DEFINITION. Let S be a set of statements. Then Array(S) is the union of the following sets of subsets of $D(S) \cup \{\ulcorner \neg \mathfrak{A} \urcorner \mid \ulcorner \mathfrak{A} \urcorner \in D(S)\}$:

(0) $\{\{\ulcorner \mathfrak{A} \urcorner\} \mid \ulcorner \mathfrak{A} \urcorner \in S\}$.

(1) $\{\{\ulcorner \mathfrak{A} \urcorner, \ulcorner \neg \mathfrak{A} \urcorner\} \mid \ulcorner \mathfrak{A} \urcorner \in D(S)\}$.

(2) $\{\{\ulcorner \mathfrak{A} \rightarrow \mathfrak{B} \urcorner, \ulcorner \mathfrak{A} \urcorner\} \mid \ulcorner \mathfrak{A} \rightarrow \mathfrak{B} \urcorner \leftarrow D(S)\} \cup \{\{\ulcorner \mathfrak{A} \rightarrow \mathfrak{B} \urcorner, \ulcorner \neg \mathfrak{B} \urcorner\} \mid \ulcorner \mathfrak{A} \rightarrow \mathfrak{B} \urcorner \leftarrow D(S)\} \cup$

$\{\{\ulcorner \neg [\mathfrak{A} \rightarrow \mathfrak{B}] \urcorner, \ulcorner \neg \mathfrak{A} \urcorner, \ulcorner \mathfrak{B} \urcorner\} \mid \ulcorner \mathfrak{A} \rightarrow \mathfrak{B} \urcorner \in D(S)\}$.

(3) $$\left\{\left\{\ulcorner \neg \bigwedge_{i \in I} \mathfrak{A}_i \urcorner, \ulcorner \mathfrak{A}_j \urcorner\right\} \middle| \ulcorner \bigwedge_{i \in I} \mathfrak{A}_i \urcorner \in D(S) \ \& \ j \in I\right\} \cup$$
$$\left\{\left\{\ulcorner \bigwedge_{i \in I} \mathfrak{A}_i \urcorner\right\} \cup \left\{\ulcorner \neg \mathfrak{A}_i \urcorner \middle| i \in I\right\} \middle| \ulcorner \bigwedge_{i \in I} \mathfrak{A}_i \urcorner \in D(S)\right\}.$$

(4) $\{\{\ulcorner \neg (w_x)\mathfrak{A} \urcorner, SF(\ulcorner \mathfrak{A} \urcorner, (2,x), u)\} \mid \ulcorner (w_x)\mathfrak{A} \urcorner \in D(S) \ \& \ u \in C(S)\} \cup$

$\{\{\ulcorner (w_x)\mathfrak{A} \urcorner, \langle 1, SF(\ulcorner \mathfrak{A} \urcorner, (2,x), (3, \ulcorner (w_x)\mathfrak{A} \urcorner))\rangle\} \mid \ulcorner (w_x)\mathfrak{A} \urcorner \in D(S)\}$.

(5) $\{\{\langle 0,u,u \rangle\} \mid u \in C(S)\} \cup \{\{\langle 1, \langle 0,u_1,u_2 \rangle\rangle, \langle 1, \langle 0,u_3,u_4 \rangle\rangle,$

$\langle 1, \langle i,u_1,u_3 \rangle\rangle, \langle i,u_2,u_4 \rangle\} \mid u_1,u_2,u_3,u_4 \in C(S) \ \& \ i \in \{0,2\}\} \cup$

$\{\{\langle 1, \langle 0,u,u' \rangle\rangle, \langle 1, \langle 2(n+2),u \rangle\rangle, \langle 2(n+2),u' \rangle\} \mid u,u' \in C(S) \ \& \ n \in \omega\}$.

3.6. REMARKS. (1) Let T be transitive and Prim-closed. Then if $S \subseteq T$, then $\text{Array}(S) \subseteq T$.

(2) If $Z \in \text{Array}(S)$, then the disjunction of Z is deducible from $S \cup W(S)$.

3.7. LEMMA. <u>Let</u> S <u>be a set of statements in which constants</u> $(3,x)$ <u>do not appear. Then</u> S <u>has a model if and only if</u> Array(S) <u>has a non-contradictory choice set, i.e., a choice set containing no pair</u> $\ulcorner \mathfrak{A} \urcorner$, $\ulcorner \neg \mathfrak{A} \urcorner$.

<u>Proof</u>. This is just a reformulation of 3.3 designed to bring out the role of the distributive laws.

3.8. DEFINITION. For a set Z of sets of formulas, let Dist(Z) be the

Z, i.e., $\text{Dist}(Z) = \ulcorner \bigwedge_{X \in Z} \bigvee X \rightarrow \bigvee_{f \in C} \bigwedge_{X \in Z} f`x \urcorner$ where $C = \prod_{X \in Z} X$, the Cartesian product of the sets in Z.

3.9. REMARKS. Let T be transitive, Prim-closed, and closed under the power-set operation $\mathcal{P}$. Then (1) T is closed under Dist, and

(2) Dist is a Σ_1^T in P-definable function on T, where P is the graph of $\mathcal{P}$.

Proof. This follows from the form of the definition of Dist.

$$\mathrm{Dist}(Z) = \langle 3,\langle 5,\langle\langle 1,\langle 5,\langle\langle 1,x\rangle \mid x \in X\rangle\rangle\rangle \mid X \in Z\rangle\rangle,\langle 1,\langle 5,\langle\langle 1,\langle 5,\langle f`X \mid X \in Z\rangle\rangle\rangle \mid f \in C\rangle\rangle\rangle\rangle$$

where $C = \{f \in \mathcal{P}(Z \cup \bigcup Z) \mid \mathrm{Fcn}(f) \ \&\ (\forall X \in Z)f`X \in X\}$.

3.10. $\mathrm{Cf}(\omega)$ COMPLETENESS THEOREM (Weak Form). Suppose T is transitive, Prim-closed, and closed under the power-set operation. Suppose also $T = \bigcup_{n\in\omega} T_n$ where each $T_n \in T$. Then if $S \subseteq T$ is a set of statements and $\ulcorner A \urcorner \in T$, then $S \models \ulcorner \mathcal{Q} \urcorner \Leftrightarrow S \cup W(S) \cup \{\mathrm{Dist}(\mathrm{Array}(S) \cap T_n) \mid n \in \omega\} \vdash \ulcorner \mathcal{Q} \urcorner$. (Note: It is not claimed that the indicated deduction is in T.)

Proof. The right-to-left implication follows from 3.4(2) and the fact that with the axiom of choice, the distributive laws are valid. The left-to-right implication is a consequence of the Baire Category Theorem, but it is simple enough to give a direct proof. We may suppose without loss of generality that S has no constants $(3,x)$ so that Lemma 3.7 applies. Let $A_n = \mathrm{Array}(S) \cap T_n$. Then by 3.6(1) each $A_n \in T$ and it suffices to show that if $S' = S \cup W(S) \cup \{\mathrm{Dist}(A_n) \mid n \in \omega\}$ is formally consistent with respect to $\vdash$, then S has a model. So suppose S' is consistent. By 3.6(2), $S \cup W(S) \vdash \bigwedge_{X\in A_n} \bigvee X$ for every n. So for each n,

$$(*)_n\colon\ S' \vdash \ulcorner \bigvee_{f\in C_n} \ \bigwedge_{X\in A_n} f`X \urcorner \quad \text{where} \quad C_n = \prod_{X\in A_n} X.$$

Suppose for $i < n$, we have chosen $f_i \in C_i$ such that $S'_n = S' \cup \bigcup_{i<n} \{f_i`X \mid X \in A_i\}$ is consistent. If then for every $f \in C_n$, $S'_n \cup \{f`X \mid X \in A_n\}$ is inconsistent, we must have $S'_n \vdash \ulcorner \neg \bigwedge_{X\in A_n} f`X \urcorner$ for every $f \in C_n$. So $S'_n \vdash \ulcorner \bigwedge_{f\in C_n} \neg \bigwedge_{X\in A_n} f`X \urcorner$. But then by $(*)_n$, S'_n is inconsistent. Thus if S'_n is consistent, there is $f_n \in C_n$ such that $S'_n \cup \{f_n`X \mid X \in A_n\}$ is consistent.

Using the axiom of choice, there is a sequence $\langle f_n \mid n \in \omega\rangle$ such that each $f_n \in C_n$ and

$\bigcup_{n\in\omega} \{f_n{}^{\prime}X \mid X \in A_n\}$ contains no complementary pair. Since this gives a noncontradictory choice set for Array(S), S has a model by Lemma 3.7.

3.11. Cf(ω) COMPACTNESS THEOREM.

Proof. Assume that T is closed under $\mathcal{P}$, is $P,R_1,\ldots,R_k$-admissible, and has form $T = \bigcup_{n\in\omega} T_n$, with the $T_n \in T$. Let $S \subseteq T$ be a $\Sigma_1^{T,+}$ in $P,R_1,\ldots,R_k$-definable set of statements such that every subset of S which belongs to T, has a model. We must show that S has a model. Again we may suppose without loss of generality that constants (3,x) do not appear in S. By the Cf(ω) Completeness Theorem, it suffices to show that $S' = S \cup W(S) \cup \{\mathrm{Dist}(Z) \mid Z \in T \;\&\; \mathrm{Con}(Z) \subseteq C(S)\}$ is consistent. But since Dist is Σ_1^T in P-definable, S' is $\Sigma_1^{T,+}$ in $P,R_1,\ldots,R_k$-definable. By Theorem 2.3, if S' were inconsistent, S' would have a subset $s' \in T$ such that s' were inconsistent. But $s = s' \cap S$ has a model. This can be extended to a model of $s \cup W(S)$ as was pointed out in 3.4(2). Since the distributive laws are valid, this model shows that s' is consistent, thus completing the proof.

REMARKS. If T is $R_1,\ldots,R_k$-admissible, then a Δ_0 in $P,R_1,\ldots,R_k$-definable predicate on T can be expressed both in Σ_2 in $R_1,\ldots,R_k$ and in Π_2 in $R_1,\ldots,R_k$-forms. So an $R_1,\ldots,R_k$-admissible set that satisfies the Σ_2 in $R_1,\ldots,R_k$-reflection principle is $P,R_1,\ldots,R_k$-admissible. If the extra predicates $R_1,\ldots,R_k$ are missing, one can then use methods of Levy [11] to show in set theory that P-admissible sets exist. A Löwenheim-Skolem argument will then show that P-admissible sets of cf(ω) exist. So these sets are large, but exist far below any inaccessible.

Incidentally, if T is P-admissible, any Σ_2^T-definable predicate is Σ_1^T in P-definable. One uses the cardinal absoluteness of Σ_1-formulas.

REFERENCES

[1] Barwise, J., Infinitary logic and admissible sets, Ph.D. Dissertation, Stanford University, 1966.

[2] __________, Infinitary logic and admissible sets. To appear in The Journal of Symbolic Logic.

[3] Jensen, R. and Karp, C., Primitive recursive set functions. To appear in the Proceedings of the Set Theory Institute, UCLA, Summer, 1967.

[4] Karp, C., Languages with expressions of infinite length, Ph.D. Dissertation, University of Southern California, 1959.

[5] __________, Languages with expressions of infinite length, North-Holland Publishing Company, Amsterdam, 1964, 183 pp.

[6] __________, Nonaxiomatizability results for infinitary systems, The Journal of Symbolic Logic, vol. 32 (1967), pp. 367-384.

[7] Kino, A. and Takeuti, G., On predicates with infinitely long expressions, Journal of the Mathematical Society of Japan, vol. 15 (1963), pp. 176-190.

[8] Kreisel, G., Model-theoretic invariants: Applications to recursive and hyperarithmetic operations, The Theory of Models, North-Holland Publishing Company, Amsterdam, 1965, pp. 190-206.

[9] __________, Choice of infinitary languages by means of definability criteria: Generalized recursion theory. This volume.

[10] Kripke, S., Transfinite Recursions on admissible ordinals (abstract). The Journal of Symbolic Logic, vol. 29 (1964), p. 161.

[11] Levy, A., A hierarchy of formulas in set theory, Memoirs of the American Mathematical Society, No. 57 (1965), 76 pp.

[12] Scott, D. and Tarski, A., The sentential calculus with infinitely long expressions, Colloquium Mathematicum, vol. 6 (1958), pp. 166-170.

UNIVERSITY OF MARYLAND

FORMULAS WITH LINEARLY ORDERED QUANTIFIERS

H. J. KEISLER

In this article we shall continue our study, begun in [5], [10], of logic with infinite alternations of quantifiers. A typical quantifier of this kind is

$$(\exists v_0)(\forall v_1)(\exists v_2)(\forall v_3) \cdots .$$

These quantifiers were introduced by Henkin in [4]. They are closely related to games with infinitely many moves, and the definition of truth is best stated in terms of a player having a winning strategy for a certain game. Our main results in this paper, in §2, are a series of "equivalence theorems" of the following kind: If $\mathfrak{A}$ and $\mathfrak{B}$ are elementary equivalent with respect to a language L_1 then they are elementarily equivalent with respect to a (larger) language L_2. Typically, L_1 will be a language with only finite alternations of quantifiers, and L_2 will be a language allowing infinite alternations of quantifiers. Such theorems show that the power of expression of L_2 is not very much greater than that of L_1. In §3 we shall give a series of "approximations theorems" which are sharper forms of the equivalence theorems. Before stating them we define an appropriate notion of an "approximation" of a formula, and they have the following form: If a sentence φ of L_2 holds in $\mathfrak{A}$, and if every "approximation" of φ in L_1 which holds in $\mathfrak{A}$ holds in $\mathfrak{B}$, then φ holds in $\mathfrak{B}$. In §3 we only state our results and omit the proofs, which are routine modifications of the proofs of the corresponding equivalence theorems. Our approximation theorems here are in the same spirit as those in [5], and answer some questions left open in [5]. In the paper [5] we proved approximation theorems in the case that the models $\mathfrak{A}$ and $\mathfrak{B}$ are saturated, the formula φ

This research was supported in part by National Science Foundation grants GP-5913, and GP-6726. The author is an Alfred P. Sloan fellow.

has well-ordered quantifiers, and each negation symbol in φ has finite scope. In this paper we shall deal with much broader classes of models and formulas.

1. Preliminaries.

We begin with a finitary first order predicate language L with an identity symbol, a set S of predicate and function symbols, and an unlimited supply of individual variables (say one variable v_α for each ordinal number α). By the power of L we mean the maximum of ω and the power of the set S.

Suppose κ, λ are infinite cardinals, $\kappa \geq \lambda$. The (infinitary) logic $L_{\kappa\lambda}$ is formed in the usual way. Namely, we give $L_{\kappa\lambda}$ the negation symbol $\neg$, the infinite conjunction $\bigwedge$, infinite disjunction $\bigvee$ and infinite quantifiers $\forall, \exists$. Then the set of formulas $L_{\kappa\lambda}$ is the least class which contains the atomic formulas, is closed under negation, is closed under conjunctions and disjunctions of sets of fewer than κ formulas, and is closed under existential quantification and universal quantification of sets of fewer than λ variables. Thus:

If $\Phi \subset L_{\kappa\lambda}$ and $|\Phi| < \kappa$, then $\bigwedge\Phi \in L_{\kappa\lambda}$, $\bigvee\Phi \in L_{\kappa\lambda}$.

If $\varphi \in L_{\kappa\lambda}$, and W is a set of variables $|W| < \lambda$, then $(\exists W)\varphi \in L_{\kappa\lambda}$, $(\forall W)\varphi \in L_{\kappa\lambda}$, and $(\neg\varphi) \in L_{\kappa\lambda}$.

Since the class of all variables of L is a proper class, the classes $L_{\kappa\lambda}$ are all proper classes. However, for each set W of variables, we can say the following.

LEMMA 1.1. (1) For each set W of variables, the class $L_{\kappa\lambda}(W)$ of all $\varphi \in L_{\kappa\lambda}$ such that every variable occurring in φ belong to W is a set.

(2) Let W be a set of variables of power greater than κ. Then every formula $\psi \in L_{\kappa\lambda}$ all of whose free variables belong to W is logically equivalent to a formula $\varphi \in L_{\kappa\lambda}(W)$.

The language $L_{\omega\omega}$ is just the ordinary finitary logic. We also write

$$L_{\infty\lambda} = \bigcup_{\kappa} L_{\kappa\lambda}, \qquad L_{\infty\infty} = \bigcup_{\lambda} L_{\infty\lambda}.$$

Thus $L_{\infty\lambda}$ allows conjunctions and disjunctions of arbitrary sets of formulas, and $L_{\infty\infty}$ allows existential and universal quantification of arbitrary sets of variables.

We now extend the language $L_{\infty\infty}$ still further by adding <u>linearly ordered</u> quantifiers $(T,U,<)$. We say that $(T,U,<)$ is a linearly ordered quantifier if T, U are two disjoint sets of variables of L, and the relation $<$ linearly orders the set $T \cup U$. The variables $t \in T$ are said to be <u>universally quantified</u>, and the variables $u \in U$ are said to be <u>existentially quantified</u>, in the quantifier $(T,U,<)$. We shall denote by $L(\infty)$ the least class of formulas which contains the atomic formulas, is closed under all the connectives and quantifiers of $L_{\infty\infty}$, and is also closed under linearly ordered quantifiers:

> If $\varphi \in L(\infty)$ and $(T,U,<)$ is a linearly ordered quantifier, then $(T,U,<)\varphi \in L(\infty)$.

Notice that $L_{\infty\infty} \subset L(\infty)$.

We do not admit empty conjunctions, disjunctions, or quantifiers. Thus whenever we write $\bigwedge\Theta$, $\bigvee\Theta$, $(\exists W)\varphi$, $(\forall W)\varphi$, or $(T,U,<)\varphi$, we shall tacitly assume that the sets Θ, W, and $T \cup U$ are non-empty.

There are several ways in which we can classify the formulas of $L(\infty)$, e.g. by the size of the conjunctions and disjunctions, number of free variables, length of quantifiers, and order types of quantifiers. We shall not commit ourselves at this point on how to classify the formulas. Instead, we shall introduce various subclasses of $L(\infty)$ as we need them for our results.

The notion of truth for formulas in $L(\infty)$ is explained in Henkin [ ]. The models for $L(\infty)$ are the same as the models for L. The only part of the notion of truth which requires any explanation is the step involving a quantifier $(T,U,<)$. Consider a formula $(T,U,<)\varphi$ of $L(\infty)$, a model $\mathfrak{A}$, and an interpretation $a(v)$ of each free variable v occuring in $(T,U,<)\varphi$. Thus the interpretation a is a function on the set of all free variables of $(T,U,<)\varphi$ into A. We imagine a game played by two players, $\forall$ and $\exists$. $\forall$ chooses interpretations $b(t) \in A$ for each variable $t \in T$, and $\exists$ chooses interpretations $c(u) \in A$ for each $u \in U$. When $\forall$ chooses $b(t)$, he knows only the values of $c(u)$ for $u < t$, and when $\exists$ chooses $c(u)$ he knows only the values of $b(t)$ for $t < u$. If the resulting interpretation $a \cup b \cup c$ satisfies φ in $\mathfrak{A}$, then $\exists$ wins the game; otherwise $\forall$ wins. We say that the formula $(T,U,<)\varphi$ is <u>satisfied</u> in $\mathfrak{A}$ by the original interpretation a if and only if the player $\exists$ has a winning strategy for this game. Formally, a <u>winning strategy</u> for $\exists$ is a function

$$f : U \times {}^{T}A \to A$$

(where ^{T}A = the set of all functions on T into A) such that:

(1) For each $u \in U$, $b \in {}^{T}A$, the value of $f(u,b)$ depends only on the values of $b(t)$ where $t < u$, so we may write $f(u,b) = f(u,\langle b(t) : t < u\rangle)$.

(2) For each $b \in {}^{T}A$, if we let $c(u) = f(u,b)$ for all $u \in U$ then the interpretation $a \cup b \cup c$ satisfies φ in $\mathfrak{A}$.

With the above definition of winning strategy, the notion of $(T,U,<)\varphi$ being satisfied in $\mathfrak{A}$ by a is made perfectly precise. We shall say that φ holds in $(\mathfrak{A},a)$ when a satisfies φ in $\mathfrak{A}$. If a, b are interpretations of two disjoint sets of variables in $\mathfrak{A}$, we may write $(\mathfrak{A},a,b)$ for $(\mathfrak{A},a)$.

As far as satisfaction goes, the quantifiers $(\exists W)$, $(\forall W)$ may be regarded as special cases of the linearly ordered quantifiers $(T,U,<)$. In fact, if $<$ is any linear ordering of T, then the formulas

$$(\forall T)\varphi, \quad (T,0,<)\varphi$$

are logically equivalent. If $<$ is any linear ordering of U, then

$$(\exists U)\varphi, \quad (0,U,<)\varphi$$

are logically equivalent.

Now let $C \subset L(\infty)$ be an arbitrary class of formulas in $L(\infty)$. We shall say that two models $\mathfrak{A}$, $\mathfrak{B}$ are equivalent with respect to C,

$$\mathfrak{A} \equiv \mathfrak{B}(C),$$

if for every sentence $\varphi \in C$, φ holds in $\mathfrak{A}$ if and only if φ holds in $\mathfrak{B}$. Similarly, if $C \subset L(\infty)$ is a class of formulas, W is a set of variables, and if a, b are interpretations of W in $\mathfrak{A}$, $\mathfrak{B}$ respectively, we shall write

$$(\mathfrak{A},a) \equiv (\mathfrak{B},b) \ (C)$$

if for every $\varphi \in C$ whose free variables are in W,

φ holds in $(\mathfrak{A},a)$ if and only if φ holds in $(\mathfrak{B},b)$.

All of our main results in this paper involve the notion of equivalence with respect to C. It is trivially true that if $\mathfrak{A}$, $\mathfrak{B}$ are two models of power $< \kappa$ and

$$\mathfrak{A} \equiv \mathfrak{B}(L_{\kappa\kappa}),$$

then $\mathfrak{A}$ and $\mathfrak{B}$ are isomorphic.

We shall prove a series of "equivalence theorems" of the following type:

If $\mathfrak{A} \equiv \mathfrak{B}(C)$ then $\mathfrak{A} \equiv \mathfrak{B}(D)$, where C, D are particular subclasses of $L(\infty)$. Usually, C will be included in $L_{\infty\infty}$ while D will not. In Chang [1], a number of equivalence theorems are stated for classes C, D which are both included in $L_{\infty\infty}$. We shall state two results from [1] which will be particularly relevant to this paper.

THEOREM 1.2. (Scott [14] for countable languages, Chang [1].) _If_ $\mathfrak{A}$, $\mathfrak{B}$ _are countable models and_

$$\mathfrak{A} \equiv \mathfrak{B}(L_{\omega_1\omega}),$$

then $\mathfrak{A}$ _and_ $\mathfrak{B}$ _are isomorphic_.

THEOREM 1.3. (Chang [1].) _For every model_ $\mathfrak{A}$ _and every infinite cardinal_ λ, _there exists a single sentence_ $\varphi \in L_{\infty\lambda}$ _such that for all models_ $\mathfrak{B}$,

φ _holds in_ $\mathfrak{B}$ _if and only if_ $\mathfrak{A} \equiv \mathfrak{B}(L_{\infty\lambda})$.

More generally, for every $\mathfrak{A}$, λ, _and interpretation_ a _of a set_ W _of variables in_ $\mathfrak{A}$, _there exists a single formula_ $\varphi \in L_{\infty\lambda}$ _such that for all_ $(\mathfrak{B},b)$,

φ _holds in_ $(\mathfrak{B},b)$ _if and only if_ $(\mathfrak{A},a) \equiv (\mathfrak{B},b)(L_{\infty\lambda})$.

Chang also gives a cardinal κ such that the formula φ in the above theorem may be taken in $L_{\kappa\lambda}$. κ depends on the power of $\mathfrak{A}$, the power of L, the power of W, and on λ. In particular, if all these things are countable, then $\kappa = \omega_1$ and $\varphi \in L_{\omega_1\omega}$. In this case Theorems 1.2 and 1.3 combine to give a stronger result of Scott [14]:

If $\mathfrak{A}$ and L are countable, then $\mathfrak{A}$ can be characterized up to isomorphism by a single sentence of $L_{\omega_1\omega}$.

In the general case, one still has:

COROLLARY 1.4. (Chang [1].) <u>Every countable model</u> $\mathfrak{A}$ <u>can be characterized up to isomorphism by a single sentence of</u> $L_{\infty\omega}$.

2. <u>Equivalence theorems</u>.

In this section we shall prove theorems of the following type:

If $\mathfrak{A} \equiv \mathfrak{B}(C)$ then $\mathfrak{A} \equiv \mathfrak{B}(D)$, where C, D are particular classes of formulas and D is "much larger" than C. For our first theorem we shall introduce a subclass $L(\omega) \subset L(\infty)$.

We define $L(\omega)$ to be the least class of formulas such that:

(A) Every atomic formula belongs to $L(\omega)$;

(B) If $\varphi \in L(\omega)$, then $(\neg \varphi) \in L(\omega)$;

(C) If Φ is a subset of $L(\omega)$, then $\bigwedge\Phi \in L(\omega)$, $\bigvee\Phi \in L(\omega)$;

(D) If $\varphi \in L(\omega)$, W is a finite or countable set of variables, and $(\exists W)\varphi$ has only finitely many free variables, then

$$(\exists W)\varphi \in L(\omega), \quad (\forall W)\varphi \in L(\omega);$$

(E) If $\varphi \in L(\omega)$, $(T,U,<)$ is a well-ordered quantifier of order type at most ω, and $(T,U,<)\varphi$ has only finitely many free variables, then

$$(T,U,<)\varphi \in L(\omega).$$

For example, every sentence of $L_{\infty\omega}$ belongs to $L(\omega)$, and also the sentence which says "R is well-founded,"

$$\neg(\exists v_0 v_1 v_2 \cdots)(\bigwedge_{n<\omega} R(v_{n+1}v_n)),$$

belongs to $L(\omega)$. Our first theorem indicates that the power of expression of the language $L(\omega)$ is no greater than that of $L_{\infty\omega}$ if we allow disjunctions of proper classes of formulas (cf. Corollary 2.2.).

THEOREM 2.1. <u>If</u> $\mathfrak{A} \equiv \mathfrak{B}(L_{\infty\omega})$, <u>then</u> $\mathfrak{A} \equiv \mathfrak{B}(L(\omega))$.

<u>Proof</u>. By induction on the complexity of formulas we shall show that each $\varphi \in L(\omega)$ has the following property P:

For all interpretations a, a' of the free variables of φ in $\mathfrak{A}$, $\mathfrak{B}$ respectively, if

(1) $$(\mathfrak{A},a) \equiv (\mathfrak{B},a')(L_{\infty\omega})$$

then

$$(\mathfrak{A},a) \equiv (\mathfrak{B},a')\{\varphi\}.$$

From this the theorem will follow by taking φ to be a sentence.

All atomic formulas have the property P since they belong to $L_{\infty\omega}$. The property P is obviously preserved by negation, infinite conjunction, and infinite disjunction. Suppose that a formula $\psi \in L(\omega)$ has the property P, let $(T,U,<)$ be a well-ordered quantifier of order type at most ω, and suppose the formula

$$\varphi = (T,U,<)\psi$$

has only finitely many free variables. We shall show that φ has the property P. Suppose (1) holds and

(2) $$\varphi \text{ holds in } (\mathfrak{A},a).$$

Here a, a' are interpretations of the finite set $\{v_0,\ldots,v_n\}$ of free variables occurring in the formula φ. As an illustration, take the case where $(T,U,<)$ has order type ω and T and U take turns, so

$$(T,U,<) = (\forall t_0 \exists u_0 \forall t_1 \exists u_1 \cdots).$$

Let f be a winning strategy for $\exists$ in (2). To show that

(3) $$\varphi \text{ holds in } (\mathfrak{B},a')$$

we shall describe a winning strategy g for the player $\exists$. Suppose player $\forall$ chooses an element $b_0' \in B$ to interpret t_0. Let $\theta_0(v_0 \cdots v_n t_0) \in L_{\infty\omega}$ be a formula such that θ_0 holds in any $(\mathfrak{C},a'',b_0'')$ if and only if

$$(\mathfrak{B},a',b_0) \equiv (\mathfrak{C},a'',b_0'')(L_{\infty\omega}).$$

Theorem 1.3 shows that there is such a θ_0. Then the formula $(\exists t_0)\theta_0$ belongs to $L_{\infty\omega}$ and holds in $(\mathfrak{B},a')$. Thus by (1), $(\exists t_0)\theta_0$ holds in $(\mathfrak{A},a)$.

Choose $b_0 \in A$ such that θ_0 holds in $(\mathfrak{A},a,b_0)$. It follows that

(4) $$(\mathfrak{A},a,b_0) \equiv (\mathfrak{B},a',b_0')(L_{\infty\omega}).$$

We now go back the other way. By (2), the strategy f picks out an element

$$c_0 = f(u_0,b_0) \in A.$$

Using Theorem 1.3 again, there is a formula

$$\theta_1(v_0 \cdots v_n t_0 u_0) \in L_{\infty\omega}$$

such that θ_1 holds in any $(\mathfrak{C},a'',b_0'',c_0'')$ if and only if

$$(\mathfrak{A},a,b_0,c_0) \equiv (\mathfrak{C},a'',b_0'',c_0'')(L_{\infty\omega}).$$

The formula $(\exists u_0)\theta_1$ holds in $(\mathfrak{A},a,b_0)$, and by (4) it also holds in $(\mathfrak{B},a',b_0')$. We now define the first move of the strategy g by choosing

$$g(u_0,b_0') = c_0' \text{ where } \theta_1 \text{ holds in } (\mathfrak{B},a',b_0',c_0').$$

Then

$$(\mathfrak{A},a,b_0,c_0) \equiv (\mathfrak{B},a',b_0',c_0')(L_{\infty\omega}).$$

We now continue this process back and forth countably many times, and in this way we define a strategy g for $\exists$ such that for all $b \in {}^{T}A$, $b' \in {}^{T}B$, and each $n < \omega$,

$$(\mathfrak{A},a,b_0,\ldots,b_n,c_0,\ldots,c_n) \equiv (\mathfrak{B},a',b_0',\ldots,b_n',c_0',\ldots,c_n')(L_{\infty\omega}),$$

where

$$c_i = f(u_i,b) \quad \text{and} \quad c_i' = g(u_i,b').$$

It follows that

$$(\mathfrak{A},a,b,c) \equiv (\mathfrak{B},a',b',c')(L_{\infty\omega}).$$

Since f is a winning strategy for $\exists$, ψ holds in $(\mathfrak{A},a,b,c)$. But ψ is assumed to have the property P, so it also holds in $(\mathfrak{B},a',b',c')$. The elements c_n' were chosen according to the strategy g, hence g is a winning strategy for $\exists$, and φ holds in $(\mathfrak{B},a)$. This proves that φ has the property P.

Except for notation, exactly the same proof will work for any other well-ordered quantifier $(T,U,<)$ of order type at most ω. The same proof also works for quantifiers $\exists W$, $\forall W$ where W is a finite or countable set of variables, since we may well-order the set W with order type at most ω. It follows that all $\varphi \in L(\omega)$ have the property P.

The following consequence of Theorem 2.1 was already known: If $\mathfrak{A} \equiv \mathfrak{B}(L_{\infty\omega})$ and a relation R of $\mathfrak{A}$ is well-founded, then the corresponding relation of $\mathfrak{B}$ is well-founded. More generally, it follows from 2.1 that if $\mathfrak{A} \equiv \mathfrak{B}(L_{\infty\omega})$ then $\mathfrak{A} \equiv \mathfrak{B}(C)$ where C is the class of all formulas $\varphi \in L_{\infty\omega_1}$ built up using only (A)-(D).

COROLLARY 2.2. (i) <u>Let $\mathfrak{A}$ be a model of power $\leq \kappa$ and let the language L have power $\leq \kappa$. Then there is a sentence $\theta \in L_{\kappa^+\omega}$ such that for every model $\mathfrak{B}$,</u>

$$\theta \text{ holds in } \mathfrak{B} \text{ if and only if } \mathfrak{A} \equiv \mathfrak{B}(L(\omega)).$$

(ii) <u>For every sentence $\varphi \in L(\omega)$ there exists a class of sentences $\Phi \subset L_{\infty\omega}$ such that φ is logically equivalent to $\bigvee\Phi$.</u>

Proof. (i) By Theorem 1.3 there is a sentence $\theta \in L_{\infty\omega}$ such that for every model $\mathfrak{B}$,

(1) $$\theta \text{ holds in } \mathfrak{B} \text{ if and only if } \mathfrak{A} \equiv \mathfrak{B}(L_{\infty\omega}).$$

From the more complete discussion of Theorem 1.3 given in [1], we see that θ may be taken to be in $L_{\kappa^+\omega}$. By Theorem 2.1,

$$\mathfrak{A} \equiv \mathfrak{B}(L_{\infty\omega}) \text{ if and only if } \mathfrak{A} \equiv \mathfrak{B}(L(\omega)).$$

and (i) follows.

(ii) For each model $\mathfrak{A}$ of φ, let $\theta_{\mathfrak{A}}$ be a sentence in $L_{\infty\omega}$ such that for all $\mathfrak{B}$, (1) holds for $\theta = \theta_{\mathfrak{A}}$. Let Φ be the class of all sentences $\theta_{\mathfrak{A}}$, where $\mathfrak{A}$ is a model of φ. Then each model $\mathfrak{A}$ of φ satisfies $\theta_{\mathfrak{A}}$, and hence satisfies $\bigvee\Phi$. Let $\mathfrak{B}$ be any model of $\bigvee\Phi$. Then for some model $\mathfrak{A}$ of φ, $\mathfrak{B}$ satisfies $\theta_{\mathfrak{A}}$. Hence

$$\mathfrak{A} \equiv \mathfrak{B}(L_{\infty\omega}),$$

and by 2.1,

$$\mathfrak{A} \equiv \mathfrak{B}(L(\omega)).$$

Thus $\mathfrak{B}$ is a model of φ. This verifies (ii).

In the above proof, it looks as if the axiom of choice for classes is needed. However, the axiom of choice for classes can easily be avoided by taking the family $\varphi_{\mathfrak{A}}$ in a systematic way, for example

$$\varphi_{\mathfrak{A}} = \bigwedge \{\theta : \theta \in L_{\infty\omega} \cap R(\kappa^+), \theta \text{ holds in } \mathfrak{A}\},$$

where κ is the maximum of the cardinals of $\mathfrak{A}$ and L.

In the above corollary, the class Φ cannot in general be taken to be a set, for this would imply that every sentence $\varphi \in L(\omega)$ is logically equivalent to a sentence in $L_{\infty\omega}$. However, it is known from [11] that the sentence which says "R is a well-ordering" is not logically equivalent to a sentence in $L_{\infty\omega}$, even using predicates other than R. Takenti [15] has shown that there exists a sentence $\varphi \in L(\omega)$ which is not even logically equivalent to a sentence in $L_{\infty\infty}$.

Our next project is to generalize Theorem 2.1 to higher cardinals. The main difficulty is that the proof of the theorem breaks down at limit ordinals when we have a well-ordered quantifier of order type greater than ω. To get around this difficulty we must pay a high price - we shall have to make a very strong assumption about the models $\mathfrak{A}$ and $\mathfrak{B}$. On the other hand, we shall be able to deal with quantifiers somewhat more general than the well-ordered quantifiers. We shall now define a subclass $L(\lambda) \subset L(\infty)$ which is the analogue of $L(\omega)$ for an arbitrary infinite cardinal λ.

We shall say that a linearly ordered quantifier $(T,U,<)$ is <u>at most</u> <u>λ-like</u> if every variable $v \in T \cup U$ has fewer than λ predecessors in the ordering $<$. Thus a quantifier is at most ω-like if and only if it is a well-ordered quantifier of order type at most ω. In general, every well-ordered quantifier of order type at most λ is an at most λ-like quantifier. However, for $\lambda > \omega$, there exist quantifiers which are at most λ-like but are not well-ordered. For example, a quantifier whose order type is ω_1 copies of the rationals is at most ω_1-like, and so is any countable linearly ordered quantifier, since each variable has only countably many predecessors. λ-like orderings are important in model theory, e.g. see Fuhrken [3].

Now define $L(\lambda)$ to be the least class of formulas such that:

(A') Every atomic formula belongs to $L(\lambda)$;

(B') If $\varphi \in L(\varphi)$, then $(\neg \varphi) \in L(\lambda)$;

(C') If Φ is a subset of $L(\lambda)$, then $\bigwedge\Phi \in L(\lambda)$, $\bigvee\Phi \in L(\lambda)$;

(D') If $\varphi \in L(\lambda)$, W is a set of variables of power at most λ, and $(\exists W)\varphi$ has fewer than λ free variables, then

$$(\exists W)\varphi \in L(\lambda), \qquad (\forall W)\varphi \in L(\lambda);$$

(E') If $\varphi \in L(\lambda)$, $(T,U,<)$ is an at most λ-like quantifier, and $(T,U,<)\varphi$ has fewer than λ free variables, then

$$(T,U,<)\varphi \in L(\lambda).$$

The conditions (D') and (E') make sure that if $\varphi \in L(\lambda)$, then every "subformula" of φ which begins with a quantifier has fewer than λ free variables. Here subformulas are obtained in the usual ways, and also by "chopping off" an initial part of a linearly ordered quantifier which is at the beginning of a formula.

THEOREM 2.3. <u>Let</u> $\mathfrak{A}$, $\mathfrak{B}$ <u>be two models for the language</u> L. <u>Suppose that for each set</u> $X \subset \lambda$ <u>of</u>

<u>power</u> $|X| < \lambda$, <u>there is a relation</u> $\sim$ <u>between</u> ${}^{X}A$ <u>and</u> ${}^{X}B$ <u>such that</u>

(i) $0 \sim 0$;

(ii) <u>If</u> $a \sim a'$, <u>then</u> $(\mathfrak{A},a) \equiv (\mathfrak{B},a')(L_{\omega\omega})$;

(iii) <u>If</u> $a \sim a'$, <u>then for all</u> $\alpha \in \lambda - X$ <u>and</u> $b \in A$ <u>there exists</u> $b' \in B$ <u>such that</u>

$$a \cup \langle \alpha,b \rangle \sim a' \cup \langle \alpha,b' \rangle;$$

(iv) <u>If</u> $a \sim a'$, <u>then for all</u> $\alpha \in \lambda - X$ <u>and</u> $b' \in B$ <u>there exists</u> $b \in A$ <u>such that</u>

$$a \cup \langle \alpha,b \rangle \sim a' \cup \langle \alpha,b' \rangle;$$

(v) $a \sim a'$ <u>if and only if for all finite</u> $Y \subset X$, $(a|Y) \sim (a'|Y)$.

<u>Then</u>

$$\mathfrak{A} \equiv \mathfrak{B}(L(\lambda)).$$

<u>Proof</u>. It is easily seen that every sentence $\varphi \in L(\lambda)$ is logically equivalent to a sentence $\varphi' \in L(\lambda)$ in which at most λ different variables occur. For example, we can prove by induction on the complexity of formulas that each formula $\varphi \in L(\lambda)$ is logically equivalent to a formula $\varphi' \in L(\lambda)$ in which at most λ bound variables occur. Consider a particular set $\{v_\alpha : \alpha < \lambda\}$ of λ different variables. Let $L_0(\lambda)$ be the class of all formulas $\varphi \in L(\lambda)$ such that all variables occurring in φ are among v_α, $\alpha < \lambda$. To prove the theorem it suffices to prove that

$$\mathfrak{A} \equiv \mathfrak{B}(L_0(\lambda)).$$

For an arbitrary set $X \subset \lambda$, possibly of power λ, let us write $a \sim a'$ if $a \in {}^{X}A$, $a' \in {}^{X}B$, and $a|Y \sim a'|Y$ for all finite $Y \subset X$. To simplify notation we shall identify each variable v_α with the ordinal α. We shall prove by induction on the complexity of formulas that if $\varphi \in L_0(\lambda)$ and X is the set of all free variables of φ, then φ has the following property P:

For all $a \in {}^{X}A$, $a' \in {}^{X}B$ such that $a \sim a'$,

$$(\mathfrak{A},a) \equiv (\mathfrak{B},a')\{\varphi\}$$

Since $0 \sim 0$, it will follow at once that

$$\mathfrak{A} \equiv \mathfrak{B}(L_0(\lambda)).$$

By (ii), every atomic formula in the variables v_α, $\alpha < \lambda$, has the property P. Again, it is obvious that P is preserved by negation, infinite conjunction, and infinite disjunction. Let $(T,U,<)\varphi$ be a member of $L_0(\lambda)$ such that φ has the property P. Thus $(T,U,<)$ is an at most λ-like quantifier. In the case that $(T,U,<)$ is a well-ordered quantifier, we can prove that $(T,U,<)\varphi$ has the property P by a straightforward "back and forth" argument which is like the proof of Theorem 2.1. The conditions (iii) and (iv) are used at successor stages, and (v) gets us past the limit stages in the well-ordered quantifier $(T,U,<)$. However, for the general case the "back and forth" argument doesn't work and we must do something else. Our argument will be similar to the proof of Theorem C in [10].

Before going any further, we shall use (v) to get improved versions of (iii) and (iv). Suppose X, Y are two disjoint subsets of λ of power less than λ, $a \in {}^X A$, $a' \in {}^X B$, and $a \sim a'$. Then:

(iii') For all $b \in {}^Y A$ there exists $b' \in {}^Y B$ such that $a \cup b \sim a' \cup b'$.

(iv') For all $b' \in {}^Y B$ there exists $b \in {}^Y A$ such that $a \cup b \sim a' \cup b'$.

To prove (iii'), we define b' by transfinite recursion. Suppose $\gamma \in Y$ and we have defined $b'(\beta)$, $\beta \in Y \cap \gamma$, so that

$$a \cup (b|Y \cap \gamma) \sim a' \cup (b'|Y \cap \gamma).$$

By (iii) we can then choose $b'(\gamma)$ so that

$$a \cup (b|Y \cap \gamma + 1) \sim a' \cup (b'|Y \cap \gamma + 1).$$

If $\gamma < \lambda$ is a limit ordinal, it follows from (v) that if

$$a \cup (b|Y \cap \beta) \sim a' \cup (b'|Y \cap \beta)$$

for all $\beta < \gamma$, then

$$a \cup (b|Y \cap \gamma) \sim a' \cup (b'|Y \cap \gamma).$$

This shows that (iii') holds. The proof of (iv') is similar.

Now consider a formula $(T,U,<)\varphi$ in $L_0(\lambda)$ such that φ has the property P. The set Z of all free variables of $(T,U,<)\varphi$ has power $|Z| < \lambda$, and we have $Z \subset \lambda$. Let $a \in {}^Z A$, $a' \in {}^Z B$, and $a \sim a'$. Assume that

(1) $$(T,U,<)\varphi \text{ holds in } (\mathfrak{A},a).$$

We wish to prove

(2) $$(T,U,<)\varphi \text{ holds in } (\mathfrak{B},a').$$

By the definition of truth of a formula of $L(\infty)$, there exists a winning strategy f for $\exists$ in (1), and we must construct a winning strategy g for $\exists$ in (2).

First we well-order the set of all pairs (W,b'), where $|W| < \lambda$, W is an initial segment of $T \cup U$ in the ordering $<$, and $b' \in {}^{(T\cap W)}B$. We arrange them in a list, (W_η, b'_η), $\eta < \delta$. It doesn't matter how large δ is. For each $b' \in {}^T B$ and $v \in T \cup U$, let us write,

$$b'|v = \langle b'(t) : t \in T \text{ and } t \leq v\rangle$$

for the restriction of b' to the set of all predecessors of v. We shall define two sequences of functions

$$b_\eta \in {}^{(T\cap W_\eta)}A, \quad c'_\eta \in {}^{(U\cap W_\eta)}B, \quad \eta < \delta,$$

such that:

(3) If $t \in T \cap W_\eta \cap W_\xi$ and $b'_\eta|t = b'_\xi|t$, then $b_\eta(t) = b_\xi(t)$;

(4) If $u \leq u_\eta$, $u \leq u_\xi$, $u \in U$, and $b'_\eta|u = b'_\xi|u$, then $c'_\eta(u) = c'_\xi(u)$;

(5) For each $\eta < \delta$, if we let $c_\eta = \langle f(u,b_\eta) : u \in U \cap W_\eta\rangle$,

then

$$a \cup b_\eta \cup c_\eta \sim a' \cup b'_\eta \cup c'_\eta .$$

Suppose we have established (3) - (5). Let g be the function on $U \times {}^T B$ into B given by $g(\mu,b') = c'_\eta(u)$, η the first ordinal with $u \in W_\eta$, and $b'_\eta \subset b'$. It follows from (4) that g is a strategy for $\exists$, becuase if

$$u \in U \cap W_\eta \cap W_\xi, \quad b'_\eta|u = b'_\xi|u,$$

then

$$g(u,b'_\eta) = c'_\eta(u) = c'_\xi(u) = g(u,b'_\xi).$$

We show now that g is a winning strategy for $\exists$ in (2). Consider any $b' \in {}^T B$. It follows from (3) that there is a unique function $b \in {}^T A$ such that for all $\eta < \delta$, $t \in T$,

$$\text{if } b'|t = b'_\eta|t \text{ and } t \in W_\eta, \text{ then } b(t) = b_\eta(t).$$

Let $c \in {}^U A$, $c' \in {}^U B$ be given by

$$c(u) = f(u,b), \quad c'(u) = g(u,b').$$

Since f is a winning strategy for $\exists$ in (1), φ holds in $(\mathfrak{A}, a \cup b \cup c)$. If the quantifier $(T,U,<)$ has fewer than λ variables, $|T \cup U| < \lambda$, then there is an $\eta < \delta$ with $W_\eta = T \cup U$, $b'_\eta = b'$. Then $c = c_\eta$, $c' = c'_\eta$, $b = b_\eta$. It follows from (5) that

(6) $$a \cup b \cup c \sim a' \cup b' \cup c'.$$

If $|T \cup U| = \lambda$, we can still prove (6) but we must use condition (v). Let $Y \subset T \cup U$ be finite. Choose $\eta < \delta$ so that $U \cap Y \subset W_\eta$ and $b'_\eta \subset b'$. Then $b_\eta \subset b$, and $c'_\eta \subset c'$. Such a segment W_η exists because the quantifier $(T,U,<)$ is at most λ-like. Therefore

$$(a \cup b \cup c)|Y = (a \cup b_\eta \cup c_\eta)|Y,$$

and

$$(a' \cup b' \cup c')|Y = (a' \cup b'_\eta \cup c'_\eta)|Y.$$

By (5),

$$a \cup b_\eta \cup c_\eta \sim a' \cup b'_\eta \cup c'_\eta \,.$$

Therefore

$$(a \cup b \cup c)|Y \sim (a' \cup b' \cup c')|Y.$$

Since this holds for all finite $Y \subset \lambda$, it follows from (v) that (6) holds, that is,

$$a \cup b \cup c \sim a' \cup b' \cup c'.$$

The formula φ is assumed to have the property P and therefore φ holds in $(\mathfrak{B}, a' \cup b' \cup c')$. It follows that g is a winning strategy for (2), and therefore (2) holds.

It remains to construct functions satisfying (3) - (5). Suppose $\alpha < \delta$ and we have already constructed b_η, c'_η for all $\eta < \alpha$, with (3) - (5) satisfied. We shall construct b_α, c_α. Let W be the largest initial segment of W_α such that for all $v \in W_\alpha$ there exists $\eta < \alpha$ with

(7) $$v \in W_\eta \quad \text{and} \quad b'_\eta|v = b'_\alpha|v.$$

For $t \in T \cap W$ and $u \in U \cap W$, choose $\eta < \alpha$ such that (7) holds with v the maximum of t and u, and define

$$b_\alpha(t) = b_\eta(t), \qquad c'_\alpha(u) = c'_\eta(u).$$

By (3) and (4) it doesn't matter which such $\eta < \alpha$ we choose. No matter how we define $b_\alpha(t)$ and $c'_\alpha(u)$ for $t, u \in W_\alpha - W$, (3) and (4) will hold for all $\eta, \xi \leq \alpha$.

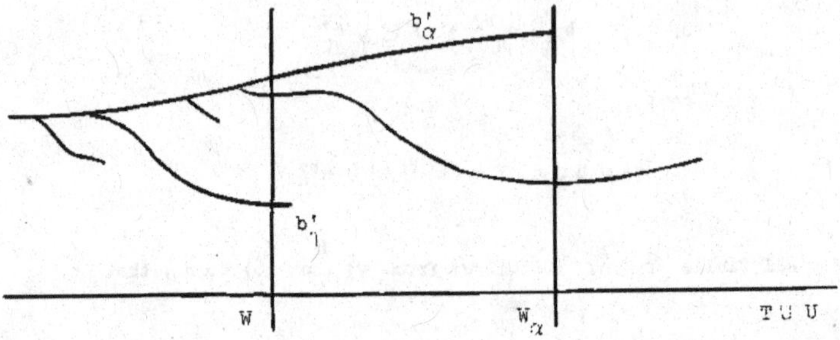

Since $b_\alpha|W$ is defined, $c_\alpha|W$ may be defined as in (5). We may now apply (v) and (5) for $\eta < \alpha$ to get

$$a \cup (b_\alpha|W) \cup (c_\alpha|W) \sim a' \cup (b_\alpha|W) \cup (c'_\alpha|W).$$

Let $W' = W_\alpha - W$. The sets Z, V and W' all have power less than λ, and we are already given $b'_\alpha(u)$ for $u \in U \cap W'$. Therefore by (iv') we may choose $b_\alpha(u) \in A$, $u \in U \cap W'$, so that

$$a \cup b_\alpha \cup (c_\alpha|W) \sim a' \cup b'_\alpha \cup (c'_\alpha|W).$$

Once b_α is defined, c_α is defined as in (5). By (iii') we may choose $c'_\alpha(u)$, $u \in U \cap W'$, so that

$$a \cup b_\alpha \cup c_\alpha \sim a' \cup b'_\alpha \cup c'_\alpha.$$

This shows that (5) holds for α. By induction we see that if φ has the property P, then so does $(T,U,<)\varphi$.

A similar argument shows that if φ has the property P, $W \subset \lambda$, and $(\exists W)\varphi$ has fewer than λ free variables, then the formulas $(\exists W)\varphi$ and $(\forall W)\varphi$ have property P. Our proof is complete.

There are several special cases in which we have a relation $\sim$ satisfying the conditions (i)-(v) of Theorem 2.3, and in each of these cases, Theorem 2.3 tells us that $\mathfrak{A} \equiv \mathfrak{B}(L(\lambda))$. As a first example, let $\lambda = \omega$ and suppose $\mathfrak{A} \equiv \mathfrak{B}(L_{\infty\omega})$. Let $a \sim a'$ mean that $(\mathfrak{A},a) \equiv (\mathfrak{B},a')(L_{\infty\omega})$. Then it is easily seen (using 1.3) that $\sim$ satisfies (i)-(v), so by Theorem 2.3, $\mathfrak{A} \equiv \mathfrak{B}(L(\omega))$. This shows that Theorem 2.1 is a consequence of Theorem 2.3. We give two other examples in the next two corollaries.

A model $\mathfrak{A}$ is said to be λ-saturated if for every set $\Phi \subset L_{\omega\omega}$ having fewer than λ free variables, and each variable v, $\mathfrak{A}$ is a model of the universal closure of the formula

$$\bigwedge\{(\exists v) \wedge \Psi : \Psi \subset \Phi \text{ and } \Psi \text{ finite}\} \rightarrow (\exists v) \wedge \Phi.$$

(See [9], [13].)

COROLLARY 2.4. Suppose $\mathfrak{A}, \mathfrak{B}$ are λ-saturated and $\mathfrak{A} \equiv \mathfrak{B}(L_{\omega\omega})$. Then $\mathfrak{A} \equiv \mathfrak{B}(L(\lambda))$.

Proof. The relation $\sim$ such that $a \sim a'$ if and only if $(\mathfrak{A},a) \equiv (\mathfrak{B},a')(L_{\omega\omega})$ satisfies the conditions (i)-(v).

A model $\mathfrak{A}$ is said to be λ-homogeneous (see [13]) if for all $\alpha < \lambda$, all $a,b \in {}^{\alpha}A$, and all $c \in A$, if

$$(\mathfrak{A},a) \equiv (\mathfrak{A},b)(L_{\omega\omega})$$

then there exists $d \in A$ such that

$$(\mathfrak{A},a,c) \equiv (\mathfrak{A},b,d)(L_{\omega\omega}).$$

It is shown in [13] that every λ-saturated model is λ-homogeneous.

COROLLARY 2.5. Suppose $\mathfrak{A}, \mathfrak{B}$ are λ-homogeneous and $\mathfrak{A} \equiv \mathfrak{B}(L_{\kappa^+\omega})$, where κ is the power of L. Then $\mathfrak{A} \equiv \mathfrak{B}(L(\lambda))$.

Proof. It is shown in [16] that if $\mathfrak{A}, \mathfrak{B}$ are λ-homogeneous and $\mathfrak{A} \equiv \mathfrak{B}(L_{\kappa^+\omega})$, then for all $\alpha < \lambda$:

(1) For all $a \in {}^{\alpha}A$ there exists $a' \in {}^{\alpha}B$ such that $(\mathfrak{A},a) \equiv (\mathfrak{B},a')(L_{\omega\omega})$;

(2) For all $a' \in {}^{\alpha}B$ there exists $a \in {}^{\alpha}A$ such that $(\mathfrak{A},a) \equiv (\mathfrak{B},a')(L_{\omega\omega})$.

Using (1) and (2), we easily see that the relation $\sim$ given by

$$a \sim a' \text{ if and only if } (\mathfrak{A},a) \equiv (\mathfrak{B},a')(L_{\omega\omega})$$

satisfies (i)-(v).

Theorem 2.3 has a slightly more general form which we shall now state to prepare for another application.

THEOREM 2.3'. Let $\mathfrak{A}'$, $\mathfrak{B}'$ be two models for the language L. Let A, B be two sets and let F, G be functions mapping A onto A' and B onto B', respectively. Suppose that for each set $X \subset \lambda$ of power $|X| < \lambda$ there is a relation $\sim$ between ${}^{X}A$ and ${}^{X}B$ such that conditions (i), and (iii)-(v) of Theorem 2.3 hold, and moreover

(ii') If $a \sim a'$, then $(\mathfrak{A}',F \circ a) \equiv (\mathfrak{B}',G \circ a')(L_{\omega\omega})$.

Then

$$\mathfrak{A}' \equiv \mathfrak{B}'(L(\lambda)).$$

The proof is essentially the same as the proof of Theorem 2.3. We show that for all $\varphi \in L(\lambda)$, $a \sim a'$ implies $(\mathfrak{A}',F \circ a) \equiv (\mathfrak{B}'G \circ a')\{\varphi\}$. Choose once and for all an element $F^*(a)$ for each $a \in A'$ such that $F(F^*(a)) = a$, and similarly for B'. The functions c_η in the proof are defined by

$$c_\eta = \langle F^*(f(u,Fb_\eta)) : u \in U \cap W_\eta \rangle.$$

Then from the function g constructed in the proof we can define a winning strategy g' for $\exists$ by

$$g'(u,b') = G(g(u,G^* \circ b')).$$

Theorem 2.3' can be used, for example, to show that 2.3 holds in logic without identity. However, we have another reason for including it.

Using almost the same proof as in Theorem 2.3', we can prove another very similar result, which nevertheless does not appear to be a direct consequence of Theorem 2.3'.

Let $L(<\lambda)$ be the class of all formulas $\varphi \in L(\infty)$ such that each quantifier in φ contains fewer than λ variables. Notice that for sentences,

$$L(<\lambda) \subset L(\lambda) \subset L(<\lambda^+),$$

hence

$$\mathfrak{A} \equiv \mathfrak{B}(L(<\lambda^+)) \Rightarrow \mathfrak{A} \equiv \mathfrak{B}(L(\lambda)) \Rightarrow \mathfrak{A} \equiv \mathfrak{B}(L(<\lambda)).$$

Formulas in $L(<\lambda^+)$ have linearly ordered quantifiers of power at most λ, but such quantifiers need not be at most λ-like. However, they must be at most λ^+-like. The next theorem has both a stronger hypothesis and conclusion than Theorem 2.3.

THEOREM 2.6. Let $\mathfrak{A}'$, $\mathfrak{B}'$ be two models for the language L. Let F, G map A onto A' and B onto B', respectively. Suppose that for each set $X \subset \lambda$, there is a relation $\sim$ between ${}^X A$ and ${}^X B$ such that conditions (i)-(v) of 2.3' hold. Then

$$\mathfrak{A}' \equiv \mathfrak{B}'(L(<\lambda^+)).$$

Proof. Again we need only consider formulas whose variables are among v_α, $\alpha < \lambda$. We argue as in Theorem 2.3. The conditions (iii') and (iv') of that proof can now be shown for all disjoint sets $X, Y \subset \lambda$. Instead of well-ordering the set of all pairs (W, b') where W is an initial segment of $(T \cup U, <)$ of power less than λ, and $b' \in {}^{(T \cap W)}B$, we well-order the set of all functions $b' \in {}^T B$. The argument is as before but simpler, for the references to the initial segments W_η are supressed.

An example of the above theorem is the following result about reduced powers. A proper filter D over a set I is said to be λ-regular if there exists a set $E \subset D$ of power $|E| = \lambda$ such that each $i \in I$ belongs to only finitely many $e \in E$. Regular filters are discussed in [7], [9]. The notion of λ-regular gets stronger as λ increases. It is easily seen that an ultrafilter is ω-regular if and only if it is countably incomplete. Moreover, if I is an infinite set of power λ, then there exist λ-regular ultrafilters over I (but there are no λ^+-regular filters over I). (See [9]).

THEOREM 2.7. Let $\mathfrak{A}_0$, $\mathfrak{B}_0$ be two models for L and let D be a λ-regular filter. Let L have at most λ predicate and function symbols. If $\mathfrak{A}_0 \equiv \mathfrak{B}_0(L_{\omega\omega})$, then the reduced powers have the

<u>relationship</u>

$$\text{D-prod } \mathfrak{A}_0 \equiv \text{D-prod } \mathfrak{B}_0(L(<\lambda^+)).$$

<u>Proof.</u> Let $E \subset D$, $|E| = \lambda$, be such that each $i \in I$ belongs to only finitely many $e \in E$. We may assume without loss of generality that L has no function symbols, for each n-ary function symbol F may be replaced by an $(n+1)$-ary relation symbol R, and the atomic formula $F(X_1 \cdots X_n) = X_{n+1}$ may be replaced by $R(X_1 \cdots X_{n+1})$. Let us call a formula $\varphi \in L_{\omega\omega}$ <u>normal</u> if no variable occurs both free and bound in φ, and furthermore for any two quantifiers Q_1, Q_2 occuring in φ with Q_2 within the scope of Q_1, Q_1 and Q_2 have no variables in common. Let S_0 be any set of predicate symbols and variables of L. We shall say that φ is a S_0-<u>formula</u> if φ is normal and every predicate symbol and variable occurring in φ belongs to S_0. We shall use the following fact, which can be proved by induction on the "quantifier depth" of formulas.

(1) If S_0 is finite, then up to logical equivalence there are only finitely many S_0-formulas.

We also need the following:

(2) Every formula $\varphi \in L_{\omega\omega}$ is logically equivalent to some normal formula.

Now consider the set S consisting of all predicate wymbols of L and all the variables v_α, $\alpha < \lambda + \omega$. Since both S and E have power λ, we may map E onto S, and write S in the form

$$S = \{s_e : e \in E\}.$$

For each $i \in I$, let

$$S_i = \{s_e : i \in e \text{ and } e \in E\}.$$

Then each set S_i is finite, and hence by (1) there are only finitely many S_i-formulas up to logical equivalence. For brevity, let us write

$$\mathfrak{A}' = \text{D-prod } \mathfrak{A}_0, \quad \mathfrak{B}' = \text{D-prod } \mathfrak{B}_0.$$

Let $A = {}^I A_0$, $B = {}^I B_0$, and for $x \in A$, $y \in B$ let $F(x) = x/D$, $G(y) = y/D$. We are now ready to define a relation $\sim$ so that (i)-(v) hold. For $a \in {}^X A$, $i \in I$, let $a_i = \langle a(x)(i) : x \in X \rangle$.

Let $X \subset \lambda$ and $a \in {}^X A$, $a' \in {}^X B$. We shall say that $a \sim a'$ if and only if for all $i \in I$,

$$(\mathfrak{A}_0, a_i) \equiv (\mathfrak{B}_0, a_i')(C(i,X))$$

where $C(i,X)$ is the set of all S_i-formulas all of whose free variables are in X.

By hypothesis, $\mathfrak{A}_0 \equiv \mathfrak{B}_0(L_{\omega\omega})$, and therefore $0 \sim 0$. Thus (i) holds. Let $a \sim a'$. For each $i \in I$, the set of all $j \in I$ such that $S_i \subset S_j$ belongs to D, because this is just the finite intersection $\bigcap \{e : i \in e \in E\}$, and $E \subset D$. Thus the set $\{j : (\mathfrak{A}_0, a_j) \equiv (\mathfrak{B}_0, a_j')(C(i,X))\}$ is in D. It follows from (2) that for every finite set of formulas of $L_{\omega\omega}$ with free variables in X, there is an i such that each of these formulas is equivalent to a formula in the set $C(i,X)$. It follows from a theorem of Feferman and Vaught [17] that

$$(\mathfrak{A}', F \circ a) \equiv (\mathfrak{B}', G \circ a')(L_{\omega\omega}),$$

and thus (ii') holds.

The conditions (iii) and (iv) follow readily from (1) and the definition of the relation $\sim$. Finally, condition (v) follows from the fact that each S_i is finite, and:

$$Y \subset X \text{ implies } C(i,Y) \subset C(i,X),$$

$$S_i \cap X \subset Y \text{ implies } C(i,X) \subset C(i,Y).$$

We may therefore apply Theorem 2.6 and we conclude that

$$\mathfrak{A}' \equiv \mathfrak{B}'(L(< \lambda^+)).$$

Let $\mathfrak{A}$, $\mathfrak{B}$, D be as in Theorem 2.7. Since the notion of a λ^+-saturated model can be expressed by a sentence of $L(< \lambda^+)$ it follows that D-prod $\mathfrak{A}$ is λ^+-saturated if and only if D-prod $\mathfrak{B}$ is. This is a result proved in [9]. Other notions from model theory, such as λ^+-homogeneous, can also be expressed in $L(< \lambda^+)$. Thus D-prod $\mathfrak{A}$ is λ^+-homogeneous if and only if D-prod $\mathfrak{B}$ is. Also, if D is an ultrafilter, D-prod $\mathfrak{A}$ is λ-universal, that is, if $\mathfrak{B} \equiv \mathfrak{A}(L_{\omega\omega})$ and $|B| \leq \lambda$ then $\mathfrak{B}$ is elementary embeddable in D-prod $\mathfrak{A}$ (this is a known result from [2], [9]). Another easy consequence of equivalence with respect to $L(<\lambda^+)$ is the following.

If $\mathfrak{A}_1$, $\mathfrak{B}_1$ are elementary submodels of D-prod $\mathfrak{A}$, D-prod $\mathfrak{B}$ respectively, and $|A_1| \leq \lambda$, $|B_1| \leq \lambda$, then there exist elementary submodels $\mathfrak{A}_2$ of D-prod $\mathfrak{A}$ and $\mathfrak{B}_2$ of D-prod $\mathfrak{B}$ such that $A_1 \subset A_2$, $B_1 \subset B_2$, and $\mathfrak{A}_2$ is isomorphic to $\mathfrak{B}_2$.

Chang has raised the following open question: Under the hypotheses of Theorem 2.7, can we prove that

$$\text{D-prod } \mathfrak{A} \equiv \text{D-prod } \mathfrak{B}(L(\lambda^+)) \;?$$

An affirmative answer would imply affirmative answers to the following two questions which involve only finitary logic. Let L here have power at most λ.

If D is λ-regular, must D-prod $\mathfrak{A}$ be λ^+-universal?

If D is a λ-regular ultrafilter over a set of power λ, $\mathfrak{A} \equiv \mathfrak{B}(L_{\omega\omega})$, $\mathfrak{A}$ and $\mathfrak{B}$ have power at most λ^+, and $\lambda^+ = 2^\lambda$, must D-prod $\mathfrak{A}$ and D-prod $\mathfrak{B}$ be isomorphic?

We know that the answer to the above question is affirmative if D is a good ultrafilter (see [8]), but the question is open for arbitrary λ-regular ultrafilters.

We conclude this section with one more equivalence theorem. We have seen that the main difficulty in generalizing Theorem 2.1 to uncountable quantifiers came at the limit of an increasing sequence of initial segments of a linearly ordered quantifier. However, for <u>inversely well-ordered</u> quantifiers, this difficulty never arises, and we get an equivalence theorem which is very simple to state. Let $L^*(\lambda)$ be the class of all formulas $\varphi \in L(\infty)$ such that every quantifier in φ has fewer then λ variables, and every linearly ordered quantifier is inversely well-ordered (that is, contains no subset of order type ω). Then for sentences,

$$L_{\infty\lambda} \subset L^*(\lambda) \subset L(<\lambda) \subset L(\lambda).$$

Also,

$$L^*(\omega) = L_{\infty\omega},$$

if we identify finite linearly ordered quantifiers with finite strings of universal and existential quantifiers.

THEOREM 2.8. *If*

$$\mathfrak{A} \equiv \mathfrak{B}(L_{\infty\lambda}),$$

then

$$\mathfrak{A} \equiv \mathfrak{B}(L^*(\lambda)).$$

Proof. We need only consider formulas of $L^*(\lambda)$ involving the variables v_α, $\alpha < \lambda$, and as before we identify variables with ordinals. For each set $X \subset \lambda$ of power $|X| < \lambda$, the relation $a \sim a'$ between ${}^XA$ and ${}^XB$ given by

$$a \sim a' \text{ if and only if } (\mathfrak{A},a) \equiv (\mathfrak{B},a')(L_{\infty\lambda})$$

satisfies the conditions (i)-(iv) of Theorem 2.3. The only missing condition is (v). Moreover, every formula φ of $L^*(\lambda)$ which has fewer than λ free variables belongs to $L(\lambda)$, since every "subformula" of φ will also have fewer than λ free variables.

Let us check through the proof of Theorem 2.3 and see where condition (v) was used. It was first used to establish the stronger conditions (iii') and (iv') of that proof. However, in this case we can prove that (iii') and (iv') hold directly. We prove (iii') only, because the proof of (iv') is similar. Let X, Y be two disjoint subsets of λ of power less than λ, $a \in {}^XA$, $a' \in {}^XB$, and $a \sim a'$. Let $b \in {}^YA$. By Theorem 1.3, there is a formula $\theta \in L_{\infty\lambda}$ with free variables $X \cup Y$ such that θ holds in any model $(\mathfrak{C},a'',b'')$ if and only if $(\mathfrak{A},a,b) \equiv (\mathfrak{C},a'',b'')(L_{\infty\lambda})$. Therefore $(\exists Y)\theta$ belongs to $L_{\infty\lambda}$ and holds in $(\mathfrak{A},a)$. It follows that $(\exists Y)\theta$ holds in $(\mathfrak{B},a')$, and there exists $b' \in {}^YB$ such that θ holds in $(\mathfrak{B},a',b')$. Now, we have

$$(\mathfrak{A},a,b) \equiv (\mathfrak{B},a',b')(L_{\infty\lambda}).$$

This means that $a \cup b \sim a' \cup b'$, and (iii') is proved.

We next used (v) to show that the strategy g is a winning strategy $\exists$, in the case that the linearly ordered quantifier $(T,U,<)$ has λ variables. But here the quantifiers have fewer than λ

variables, whence (v) is not needed at that point.

In constructing functions to satisfy (3)-(5), the condition (v) was needed to establish the relation

$$(8) \qquad a \cup (b_\alpha|W) \cup (c_\alpha|W) \sim a' \cup (b'_\alpha|W) \cup (c'_\alpha|W),$$

assuming (3)-(5) holds for all $\eta < \alpha$. Here W denoted the largest initial segment of W_α such that for all $v \in W_\alpha$ there exists $\eta < \alpha$ with

$$v \in W_\eta \quad \text{and} \quad b'_\eta|v = b'_\alpha|v.$$

However, this time the quantifier $(T,U,<)$ is an inversely well-ordered quantifier. It follows that there is a greatest $v \in W_\alpha$ such that for some $\eta < \alpha$, $b'_\eta|v = b'_\alpha|v$. Then W is just the set of all $t \in T$ wuch that $t \leq v$. Let η be as above. Then we have

$$b'_\eta|W = b'_\alpha|W,$$

whence by (3) and (4),

$$b_\alpha|W = b_\eta|W, \quad c_\eta|W = c_\alpha|W, \quad c'_\eta|W = c'_\alpha|W.$$

Since (5) holds for η,

$$a \cup b_\eta \cup c_\eta \sim a' \cup b'_\eta \cup c'_\eta,$$

and restricting to W we obtain (8).

Finally, the condition (v) was needed in passing over the quantifiers $(\exists W)$, $(\forall W)$ when W has power λ. However, (v) was not needed in case W has power less than λ. Thus the proof goes through in the present case.

The above theorem has a corollary analogous to Corollary 2.2.

COROLLARY 2.9. <u>For every sentence</u> $\varphi \in L^*(\lambda)$ <u>there exists a class of sentences</u> $\Phi \subset L_{\infty\lambda}$ <u>such that</u>

$$\varphi \leftrightarrow \bigvee \Phi$$

<u>is valid.</u>

The proof is just like the proof of 2.2, using Theorem 1.3. In case $\lambda = \omega$, the above corollary is trivial, and the class Φ may be replaced by a single sentence in $L_{\infty\omega}$. For $\lambda > \omega$, we do not know whether Φ can be replaced by a single sentence $\psi \in L_{\infty\lambda}$. In other words, we do not know of any sentence of $L^*(\lambda)$ which is not logically equivalent to some sentence of $L_{\infty\lambda}$.

It is obvious that if

$$\mathfrak{A} \equiv \mathfrak{B}(L(\lambda))$$

and $|A| \leq \lambda$, $|B| \leq \lambda$ then $\mathfrak{A}$ and $\mathfrak{B}$ are isomorphic. For consider the sentence

$$(\exists u_\alpha \forall t_\alpha)_{\alpha<\lambda}\ D(\mathfrak{A}) \wedge \bigwedge_{\alpha<\lambda} \bigvee_{\beta<\lambda} t_\alpha = u_\beta,$$

where $D(\mathfrak{A})$ is a diagram of the model $\mathfrak{A}$ using the variables u_α. This sentence φ belongs to $L(\lambda)$ and holds in $\mathfrak{A}$, therefore it holds in $\mathfrak{B}$. In $\mathfrak{B}$, let player $\forall$ choose the t_α's so that they enumerate the set B. Then $\exists$ may choose the u_α's so that the inner part of the formula is satisfied. This means that the u_α's will enumerate B and provide an isomorphism from $\mathfrak{A}$ to $\mathfrak{B}$.

Morley has given an example, for each successor cardinal $\lambda = \kappa^+$, of a pair of models $\mathfrak{A}$, $\mathfrak{B}$ of power λ such that $\mathfrak{A} \equiv \mathfrak{B}(L_{\infty\lambda})$ but $\mathfrak{A}$ and $\mathfrak{B}$ are not isomorphic. Thus Scott's Theorem 1.1 fails for successor cardinals. It follows from Morley's example that Theorem 2.8 becomes false if we replace $L^*(\lambda)$ by $L(\lambda)$. We do not know whether Theorem 2.8 can be improved be replacing $L^*(\lambda)$ by $L(<\lambda)$. We also do not know what happens for most limit cardinals λ. However, for λ cofinal with ω, Chang [1] has shown that an analogue of Scott's Theorem 1.1 holds.

3. <u>Approximation theorems.</u>

In proving an equivalence theorem of the form: "If $\mathfrak{A} \equiv \mathfrak{B}(C)$ then $\mathfrak{A} \equiv \mathfrak{B}(D)$," one can often get additional information by keeping closer track of just which formulas in the class C are needed to take care of a formula in D. Instead of equivalence theorems, one can obtain approximation theorems of the form:

If φ holds in $\mathfrak{A}$, and every approximation of φ which holds in $\mathfrak{A}$ holds in $\mathfrak{B}$, then φ holds in $\mathfrak{B}$.

Various approximation theorems are proved in [5], [10], [12]. In this section we shall state a series of approximation theorems which arise from the proofs of some of the equivalence theorems in §2. We shall omit the proofs because they can be supplied by examining the proofs in §2.

We shall begin with the notion of a finite approximation similar to that in [5]. Using this notion and our present methods we shall improve the results of [5] and answer some questions left open there. However, our answers will be for a different notion of approximation than the one in [5].

Let us consider formulas $\varphi \in L(\lambda)$ and recall the equivalence Theorem 2.4. We wish to obtain a sharper form of this equivalence theorem by weakening the hypothesis $\mathfrak{A} \equiv \mathfrak{B}(L_{\omega\omega})$. What we want to do is to introduce, for each $\varphi \in L(\lambda)$, a class $A(\varphi)$ of "finite approximations" of φ which has the following properties:

I. $0 \neq A(\varphi) \subset L_{\omega\omega}$.

II. The formulas $\theta \in A(\varphi)$ all resemble φ as much as possible in their syntactical appearance.

III. For all λ-saturated models $\mathfrak{A}$ and $\mathfrak{B}$ and all sentences $\varphi \in L(\lambda)$, if φ holds $\mathfrak{A}$ and every $\theta \in A(\varphi)$ which holds in $\mathfrak{A}$ holds in $\mathfrak{B}$, then φ holds in $\mathfrak{B}$.

In [5] we were able to carry out this program for a certain class of formulas $\varphi \in L(\lambda)$, the "admissable" formulas. We shall now prove similar theorems for all formulas of $L(\lambda)$. To do this we must define $A(\varphi)$ in a slightly different way than we did in [5]. The smaller we can make the class $A(\varphi)$ of finite approximations of φ, the stronger our results will be.

Our exposition will be simpler if we don't have too many kinds of quantifiers. So in this section we shall change our format a little and do away with the quantifiers $(\exists W)$, $(\forall W)$. The language $L_{\infty\lambda}$ is now understood as the class of formulas built up using only linearly ordered quantifiers $(T,0,<)$ and $(0,U,<)$ where T and U have power less than λ. These quantifiers have exactly the same interpretation as the quantifiers $(\forall T)$, $(\exists U)$, so we have made no essential change. We shall also sometimes identify a finite linearly ordered quantifier with a string of single quantifiers. Thus $(\exists u)$ means $(0,\{u\},<)$ and $(\forall t)$ means $(\{t\},0,<)$.

We shall say that a formula $\varphi \in L(\infty)$ is <u>conjunctive</u> (cf. [5]) if it contains no infinite disjunctions, and the scope of each occurrence of a negation symbol in φ is finite. In other words, the class of conjunctive formulas is the least class of formulas which contains all the finite formulas and is closed under linearly ordered quantifiers, arbitrary conjunction, and finite disjunction. The reason for introducing conjunctive formulas is that by treating them separately we can make the class $A(\varphi)$ smaller and thus get stronger theorems. For any formula φ, let $V(\varphi)$ be the set of all free variables of φ.

Now suppose $\varphi \in L(\infty)$ is a conjunctive formula. We define the class $A(\varphi)$ of <u>finite approximations</u> of φ recursively as follows:

a1. If $\varphi \in L_{\omega\omega}$, then $A(\varphi) = \{\varphi\}$.

a2. If $\varphi = \theta_1 \wedge \cdots \wedge \theta_n$, then $A(\varphi) = \{\sigma_1 \wedge \cdots \wedge \sigma_n : \sigma_1 \in A(\theta_1),\ldots,\sigma_n \in A(\theta_n)\}$.

Similarly for finite disjunctions.

a3. If $\varphi = \bigwedge \Theta$ where Θ is infinite, then $A(\varphi) = \{\sigma_1 \wedge \cdots \wedge \sigma_n :$ for some distinct $\theta_1,\ldots,\theta_n \in \Theta$, $\sigma_1 \in A(\theta_1),\ldots,$ and $\sigma_n \in A(\theta_n)\}$.

a4. If $\varphi = (T,U,<)\psi$, then $A(\varphi) = \{\theta : V(\theta) \subset V(\varphi)$ and $\theta = (T_0,U_0,<) \wedge \Theta$ for some finite $T_0 \subset T$, $U_0 \subset U$, and $\Theta \subset A(\psi)\}$.

For conjunctive formulas with well-ordered quantifiers this is exactly the class $A(\varphi)$ defined in [5].

LEMMA 3.1. *Suppose* $\varphi \in L(\infty)$ *is conjunctive. Then* $\varphi \rightarrow \bigwedge A(\varphi)$ *is valid. Moreover, for any* $\theta_1,\ldots,\theta_n \in A(\varphi)$ *there exists* $\theta \in A(\varphi)$ *such that* $\theta \rightarrow \theta_1 \wedge \cdots \wedge \theta_n$ *is valid.*

The following result is a special case of a theorem in [10].

THEOREM 3.2. *Suppose* $\varphi \in L(\lambda)$ *and* φ *is conjunctive. Then the formula*

$$\varphi \leftrightarrow \bigwedge A(\varphi)$$

is valid in every λ*-saturated model.*

We now define the class $A(\varphi)$ for arbitrary $\varphi \in L(\infty)$.

A1. If φ is conjunctive, then $A(\varphi)$ is as defined by a1 - a4.

For the following assume that φ is not conjunctive.

A2. If either $\varphi = \bigwedge\Theta$ or $\varphi = \bigvee\Theta$, then $A(\varphi) = \bigcup\{A(\theta) : \theta \in \Theta\}$.

A3. Suppose $\varphi = \neg\psi$. Then $A(\varphi) = \{\neg\theta : \theta \in A(\psi)\}$.

A4. Suppose $\varphi = (T,U,<)\psi$ and the quantifier $(T,U,<)$ is well-ordered. Let Φ be the least subset of $L_{\omega\omega}$ such that:

(i) $A(\psi) \subset \Phi$.

(ii) If $t \in T$, $\theta_1,\ldots,\theta_n \in \Phi$, and every variable $v \in T \cup U$ which occurs free in $\theta_1 \vee \cdots \vee \theta_n$ is $\leq t$, then

$$(\forall t)(\theta_1 \vee \cdots \vee \theta_n) \in \Phi.$$

(iii) If $u \in U$, $\theta_1,\ldots,\theta_n \in \Phi$, and every variable $v \in T \cup U$ which occurs free in $\theta_1 \wedge \cdots \wedge \theta_n$ is $\leq u$, then

$$(\exists u)(\theta_1 \wedge \cdots \wedge \theta_n) \in \Phi.$$

Then we define

$$A(\varphi) = \{\theta : \theta \in \Phi \text{ and } V(\theta) \subset V(\varphi)\}.$$

The last part of the definition takes care of quantifiers which are not well-ordered in formulas which are not conjunctive. The proof of Theorem 2.3 suggests that in this case we want approximations in which the existential quantifiers may have been shifted to the right. We now complete our definition.

A5. Suppose $\varphi = (T,U,<)\psi$, where the quantifier is not well-ordered. We then define $A(\varphi)$ exactly as in A4 except that in the condition (iii) we replace "$v \in T \cup U$" by "$v \in U$."

The main difference between our present definition of $A(\varphi)$ and the definition given in [5] is that in [5] finite disjunctions and conjunctions were treated differently than infinite ones while here they are treated the same. This change allows us to handle a much wider class of formulas.

The difference between A4 and A5 is that in A5 we allow a wider class of approximations obtained by "moving the variables in U to the right" in the quantifier. We list some elementary facts about $A(\varphi)$ in the lemma below.

LEMMA 3.3. Let $\varphi \in L(\infty)$. Then:

(i) $0 \neq A(\varphi) \subset L_{\omega\omega}$.

(ii) If $\theta \in A(\varphi)$, then $V(\theta) \subset V(\varphi)$.

(iii) If $T = T_1 \cup T_2$, $U = U_1 \cup U_2$, and $x < y$ for all $x \in T_1 \cup U_1$, $y \in T_2 \cup U_2$, then

$$A((T_1,U_1,<)(T_2,U_2,<)\varphi) \subset A((T,U,<)\varphi).$$

If the quantifier $(T,U,<)$ is well-ordered then equality holds.

(iv) If φ is positive, or universal, so is every finite approximation of φ. If a variable v is existentially (or universally) quantified in some $\theta \in A(\varphi)$, then v is also existentially (or universally) quantified in φ.

In the above lemma, part (i) is the first of our desired properties for $A(\varphi)$, I. The remaining parts of the lemma help justify our aim II. Part (iv) also holds for many other syntactical properties of φ. If all quantifiers in φ are well-ordered, then the finite approximations of φ also preserve the relative position of each variable in a quantifier in φ.

We now state our main theorem about finite approximations.

THEOREM 3.4. Suppose $\mathfrak{A}$ and $\mathfrak{B}$ are λ-saturated models and $\varphi \in L(\lambda)$. If φ holds in $\mathfrak{A}$ and every finite approximation of φ which holds in $\mathfrak{A}$ holds in $\mathfrak{B}$, then φ holds in $\mathfrak{B}$.

The following corollaries answer questions raised in [5]. Let $A^c(\varphi)$ be the closure of $A(\varphi)$ under finite conjunction and disjunction.

COROLLARY 3.5. (Separation theorem). Suppose that for each $i \in I$, we are given sentences $\sigma_i \in L_{\omega\omega}$, $\varphi_i \in L(\infty)$, such that the following formulas are valid:

$$\neg \bigwedge \{\varphi_i : i \in I\},$$

$$\sigma_i \rightarrow \varphi_i, \quad i \in I.$$

Then there exist $\theta_i \in A^c(\varphi_i)$, $i \in I$, *such that*

$$\neg \bigwedge \{\theta_i : i \in I\},$$

$$\sigma_i \rightarrow \theta_i, \quad i \in I$$

are valid.

COROLLARY 3.6. (i) (Interpolation theorem). *Suppose* σ, η *are sentences in* $L_{\omega\omega}$, φ *is a sentence in* $L(\infty)$, *and the sentences*

$$\sigma \rightarrow \varphi, \quad \varphi \rightarrow \eta$$

are valid. Then there is a sentence $\theta \in A^c(\varphi)$ *such that*

$$\sigma \rightarrow \theta, \quad \theta \rightarrow \eta$$

are valid.

(ii) *Let* ψ *be a sentence of* $L(\infty)$. *If* ψ *is logically equivalent to some sentence* $\sigma \in L_{\omega\omega}$, *then* ψ *is logically equivalent to some sentence* $\theta \in A(\psi)$.

COROLLARY 3.7. *Suppose* $\varphi \in L(\lambda)$. *Then either* φ *is valid in all* λ-*saturated models, or* $\varphi \rightarrow \bigvee A(\varphi)$ *is valid in all* λ-*saturated models*.

We now give a sharper form of Theorem 2.1. For this purpose we need a notion of infinite approximation. For each formula $\varphi \in L(\omega)$ we shall define a class $B(\varphi) \subset L_{\infty\omega}$ of infinite approximations of φ.

Let $\varphi \in L(\omega)$. The class $B(\varphi)$ is defined in the following way:

B1. If φ is finite, then $B(\varphi) = \{\varphi\}$.

For the following assume that φ is infinite, i.e. $\varphi \in L(\omega) - L_{\omega\omega}$.

B2. If $\varphi = \bigwedge \Theta$ or $\varphi = \bigvee \Theta$, then

$$B(\varphi) = \bigcup \{B(\theta) : \theta \in \Theta\}.$$

B3. If $\varphi = \neg \psi$, then

$$B(\varphi) = \{\neg\sigma : \sigma \in B(\psi)\}.$$

B4. Let $\varphi = (T,U,<)\psi$. Let Φ be the least class of formulas in $L_{\infty\omega}$ such that:

(i) $B(\psi) \subset \Phi$.

(ii) If $u \in U$, Θ is a subset of Φ, and every $v \in T \cup U$ which occurs free in $\bigwedge\Theta$ is $\leq u$, then $(\exists u)\bigwedge\Theta \in \Phi$.

(iii) If $t \in T$, $\theta \in \Phi$, and if every $v \in T \cup U$ which occurs free in θ is $\leq t$, then $(\forall t)\theta \in \Phi$.

(iv) If $t \in T$, Θ is a subset of Φ, every $v \in T \cup U$ which occurs free in $\bigvee\Theta$ is $\leq t$, and φ is not conjunctive, then $(\forall t)\bigvee\Theta \in \Phi$.

Then we define

$$B(\varphi) = \{\theta : \theta \in \Phi \text{ and } V(\theta) \subset V(\varphi)\}.$$

Notice that the infinite conjunctions and disjunctions appear in the approximations of φ only at the quantifier steps.

LEMMA 3.8. _Suppose_ $\varphi \in L(\omega)$.

(i) $0 \neq B(\varphi) \subset L_{\infty\omega}$.

(ii) _If_ $\theta \in B(\varphi)$, _then_ $V(\theta) \subset V(\varphi)$ _and_ $V(\theta)$ _is finite_.

(iii) _For each cardinal_ κ, _the class_

$$B(\varphi) \cap L_{\kappa\omega}$$

is a set.

(iv) _If_ φ _is conjunctive, then so is every formula_ $\theta \in B(\varphi)$.

(v) _If_ φ _is conjunctive then_ $\varphi \to \bigwedge B(\varphi)$ _is valid_.

(vi) _If_ $T = T_1 \cup T_2$, $U = U_1 \cup U_2$, _and if_ $x < y$ _for all_ $x \in T_1 \cup U_1$, $y \in T_2 \cup U_2$, _and if_ $\varphi = (T,U,<)\psi$, _then_

$$B((T_1,U_1,<)(T_2,U_2,<)\psi) \subset B((T,U,<)\psi).$$

THEOREM 3.9. Let κ be an infinite cardinal and let $\varphi \in L(\omega)$ be a conjunctive formula. Then the formula

$$\bigwedge(B(\varphi) \cap L_{\kappa^+\omega}) \longleftrightarrow \varphi$$

is valid in all models of power at most κ.

The following corresponds to Theorem 2.1.

THEOREM 3.10. Let κ be an infinite cardinal, let $\mathfrak{A}$, $\mathfrak{B}$ be two models of power at most κ, and let φ be a sentence in $L(\omega)$. If every sentence $\theta \in B(\varphi) \cap L_{\kappa^+\omega}$ which holds in $\mathfrak{A}$ holds in $\mathfrak{B}$, and if φ holds in $\mathfrak{A}$, then φ holds in $\mathfrak{B}$.

We do not have any reasonable approximation theorems corresponding to Theorems 2.3, 2.5, 2.6, or 2.7. We shall conclude our paper by stating an approximation theorem corresponding to Theorem 2.8. For this we need approximations of formulas $\varphi \in L^*(\lambda)$ which belong to $L_{\infty\lambda}$.

Let λ be an infinite cardinal and let $\varphi \in L^*(\lambda)$. We define the class $C(\varphi)$ of all approximations of φ as follows.

C1 - C3 are the same as B1 - B4.

C4. Suppose $\varphi = (T,U,<)\psi$. Let Ψ be the least class of formulas in $L_{\infty\infty}$ such that

(i) $C(\psi) \subset \Phi$.

(ii) If U_0 is an interval in $\langle U,<\rangle$, Θ is a subset of Φ, and for every $v \in U$ which occurs free in $\bigwedge\Phi$, either $v \in U_0$ or $v < U_0$, then $(\exists U_0)\bigwedge\Theta \in \Phi$.

(iii) If T_0 is an interval in $\langle T,<\rangle$, $\theta \in \Phi$, and if for every variable $v \in T \cup U$ free in θ, either $v < T_0$ or $v \in T_0$, then $(\forall T_0)\theta \in \Phi$.

(iv) If T_0 is an interval in $\langle T,<\rangle$, if Θ is a subset of Φ, if for every $v \in T \cup U$ which occurs free in $\bigvee\Theta$ either $v \in T_0$ or $v < T_0$, and if φ is not conjunctive, then $(\forall T_0)\bigvee\Theta \in \Phi$.

We then define

$$C(\varphi) = \{\theta : \theta \in \Phi \text{ and } V(\theta) \subset V(\varphi)\}.$$

LEMMA 3.11. Let $\varphi \in L^*(\lambda)$. Then:

(i) $B(\varphi) \subset C(\varphi) \subset L_{\infty\lambda}$.

(ii) If $\theta \in C(\varphi)$ then $V(\theta) \subset V(\varphi)$.

(iii) For each κ, the class $C(\varphi) \cap L_{\kappa\lambda}$ is a set.

(iv) Parts (iv), (v), (vi) of Lemma 3.7 hold for $C(\varphi)$.

THEOREM 3.12. Let $\varphi \in L^*(\lambda)$ be conjunctive. Then the formula

$$\bigwedge C(\varphi) \cap L_{(\kappa^\lambda)^+\lambda} \leftrightarrow \varphi$$

is valid in all models of power at most κ.

THEOREM 3.13. Let φ be a sentence of $L^*(\lambda)$ and let $\mathfrak{A}$, $\mathfrak{B}$ be any two models. If φ holds in $\mathfrak{A}$ and every sentence $\sigma \in C(\varphi)$ which holds in $\mathfrak{A}$ holds in $\mathfrak{B}$, then φ holds in $\mathfrak{B}$.

The proof follows the method of Theorem 2.8.

The definition of $C(\varphi)$ makes sense for arbitrary formulas $\varphi \in L(\infty)$, but to date this case has not yielded any approximation theorems.

REFERENCES

In the list below, TM is an abbreviation for "The Theory of Models," ed. by J. W. Addison, L. Henkin, and A. Tarski, Amsterdam 1965.

[1] Chang, C. C., this volume.

[2] Frayne, T., Morel, A. and Scott, D., Reduced direct products, Fund. Math., 51 (1962), 195-248.

[3] Fuhrken, G., Languages with the added quantifier "there exist at least $\aleph_\alpha$," TM, 121-131.

[4] Henkin, L., Some remarks on infinitely long formulas, Infinitistic methods, Warsaw 1961, 167-183.

[5] Keisler, H. J., Finite approximations of infinitely long formulas, TM, 158-170.

[6] __________, Some applications of infinitely long formulas, J. Symb. Logic, 30 (1965), 339-349.

[7] __________, On cardinalities of ultraproducts, Bull. Amer. Math. Soc., 70 (1964), 644-647.

[8] __________, Ultraproducts and saturated models, Indag. Math., 26 (1964), 178-186.

[9] __________, Ultraproducts which are not saturated, J. Symb. Logic, 32 (1967), 23-46.

[10] ___________, Infinite quantifiers and continuous games, Symposium on applications of model theory held at Cal. Inst. of Tech., 1967. To appear.

[11] Lopez-Escobar, E., On defining well-orderings, Fund. Math., 59 (1966), 13-21.

[12] Malitz, J., Problems in the model theory of infinite languages. Doctoral dissertation, Univ. of Cal., Berkeley, 1966.

[13] Morley, M. and Vaught, R., Homogeneous universal models, Math. Scand., 11 (1962), 37-57.

[14] Scott, D., Logic with denumerably long formulas and finite strings of quantifiers, TM, 329-341.

[15] Takeuti, G., this volume.

[16] Keisler, H. J. and Morley, M., On the number of homogeneous models of a given power, Israel. J. Math., 5 (1967), 73-78.

[17] Feferman, S. and Vaught, R., First order properties of products of algebraic systems, Fund. Math., 47 (1959), 57-103.

UNIVERSITY OF WISCONSIN

AND

UNIVERSITY OF CALIFORNIA, LOS ANGELES

SOME PROBLEMS IN GROUP THEORY

R. D. KOPPERMAN AND A. R. D. MATHIAS

Except as noted, notation is as defined in [3] group-theoretic notions as in [1].

DEFINITION. A class $\mathcal{A}$ of groups is called <u>bountiful</u> if and only if for every group G with $\omega \leq c(G)$ (ω the first infinite cardinal, $c(G)$ the cardinality of G), if there is an $H \in \mathcal{A}$ such that $G \subset H$, then there is an $H' \in \mathcal{A}$ such that $G \subset H' \subset H$ and $c(G) = c(H')$.

The following are examples of bountiful classes of groups:

(1) $\mathcal{G}$, the class of abelian groups

(2) $\mathcal{S}$, the class of simple groups

(3) $\mathcal{C}$, the class of characteristically simple groups, where G is said to be characteristically simple if and only if $\{e\}$ and G are the only subgroups closed under every automorphism of G

(4) $\mathcal{J}$, the class of groups G for which if $K \lhd H \lhd G$ then $K \lhd G$ (where $A \lhd B$ means that A is a normal subgroup of B, $A < B$ means that A is a subgroup of B)

(5) $\mathfrak{J}$, the class of groups G such that if $H, K < G$ are finitely generated and $H \cong K$, then for some $g \in G$, $g^{-1}Hg = K$.

We also have the following open questions:

(a) Is $\mathcal{I}$, the class of groups G with center $\{e\}$ and such that every automorphism is inner, bountiful?

(b) Can we find a useful general criterion for a class $\mathcal{A}$ of groups to be bountiful?

We note that our first bountiful class $\mathfrak{G} = M(\sigma)$ (the class of models of σ) for some $\sigma \in L$ (the usual [finite] first-order language of groups), and that the Löwenheim-Skolem theorem immediately yields the fact that this class is bountiful. The infinitary downward Löwenheim-Skolem theorem provides a partial answer to open question (b), by telling us that if $\mathfrak{D} = M(\sigma)$ for some $\sigma \in L_{\omega_1\omega}$ (the language of group theory extended by allowing countable conjunctions and disjunctions), then $\mathfrak{D}$ is a bountiful class. We use this partial answer to (b) to give model-theoretic proofs that $\mathfrak{S}$, C, $\mathfrak{J}$, and $\mathfrak{F}$ are bountiful.

LEMMA 1. G <u>is simple if and only if for every</u> $a \in G - \{e\}$, $[\{a\}] = G$, <u>where</u>:

DEFINITION. Let $A \subset G$, G a group. $[A]_G = \bigcap \{H \mid A \subset H \lhd G\}$ (and is called the normal subgroup generated by A in G), and $\langle A \rangle_G = \bigcap \{H \mid A \subset H < G\}$ (and is called the subgroup generated by A in G). If it is apparent which group is meant, the subscript $_G$ may be removed from either of these notations.

<u>Proof of the lemma</u>. Clearly if G is simple and $a \neq e$, $[\{a\}] = G$. Also if G has the property that $[\{a\}] = G$ and $H \neq \{e\}$ is a normal subgroup of G, let $a \in H - \{e\}$. Then $G = [\{a\}] \subset H \subset G$, so $H = G$, thus G is simple.

It will be convenient from now on to denote by $\mathfrak{G}$ the class of all groups, $\mathfrak{G} = M(\gamma)$, where γ is the conjunction of the group axioms, $\gamma \in L$.

COROLLARY 1. $\mathfrak{S} = M(\sigma)$ <u>for some</u> $\sigma \in L_{\omega_1\omega}$. <u>Thus</u> $\mathfrak{S}$ <u>is bountiful</u>.

<u>Proof</u>. It is well-known and simple to check that $[\{a\}] = \{g_1 a^{m_1} g_1^{-1} \cdots g_n a^{m_n} g_n^{-1} \mid n \in \omega,\ m_i \in Z$ (the set of integers) for $1 \leq i \leq n$, $g_i \in G$ for $1 \leq i \leq n\}$. Let

$\tau = (\forall x)(x \neq e \to (\forall y)\bigvee_{i\in\omega-\{0\}} (\exists v_1 \cdots v_i)[\bigvee_{m_1\ldots m_i \in Z} v_1 x^{m_1} v_1^{-1} \cdots v_i x^{m_i} v_i^{-1} = y])$. Clearly $\tau \in L_{\omega_1\omega}$, and if $\sigma = \tau \,\&\, \gamma$, then so is σ. By the lemma, $\mathfrak{S} = M(\sigma)$.

LEMMA 2. <u>Let</u> G <u>be a group</u>, $a \in G$. <u>Set</u> $G_a = \{f_1(a^{m_1}) \cdots f_n(a^{m_n}) \mid f_1,\ldots,f_n$ <u>automorphisms of</u> G, $n \in \omega - \{0\}$, $m_1,\ldots,m_n \in Z\}$. <u>Then</u> G <u>is characteristically simple if and only if for every</u> $a \in G - \{e\}$, $G_a = G$.

<u>Proof</u>. We note that for every $a \in G$, G_a is a subgroup closed under every automorphism of G. G_a is clearly closed under the group operation, and $(f_1(a^{m_1})\cdots f_n(a^{m_n}))^{-1} = f_n(a^{-m_n})\cdots f_1(a^{-m_1}) \in G_a$.

Finally, if f is an automorphism of G, $f(f_1(a^{m_1}) \cdots f_n(a^{m_n})) = (ff_1)(a^{m_1}) \cdots (ff_n)(a^{m_n}) \in G_a$. Thus if G is characteristically simple, $a \in G - \{e\}$, $G_a = G$. Conversely, let H be a subgroup of G closed under every automorphism, $H \neq \{e\}$, let $a \in H - \{e\}$. Then $G = G_a \subset H \subset G$, so $H = G$, and G must be characteristically simple.

The use for the lemma above is not quite so straightforward as that of Lemma 1, since our language $L_{\omega_1\omega}$ lacks the predicates necessary to discuss automorphisms. We therefore consider structures of the form $\langle S,G,M,A,\cdot,e\rangle$, with $\langle G,\cdot,e\rangle$ a group, $G \cup M = S$, $G \cap M = \emptyset$, and A a ternary relation (corresponding to $f(x) = y$) such that,

(1) $(\forall f)(\forall x)(\forall y)(A(f,x,y) \rightarrow M(f) \;\&\; G(x) \;\&\; G(y))$ (A relates a "member" of M and two "members" of G)

(2) $(\forall f)(\forall x)(\forall y)(\forall x')(\forall y')(A(f,x,y) \;\&\; A(f,x',y') \rightarrow (x = x' \leftrightarrow y = y'))$ (each "element" of M is a one-to-one function)

(3) $(\forall f)(\forall y)(G(y) \rightarrow (\exists x)A(f,x,y))$ (each "map" has range $= G$)

(4) $(\forall f)(\forall x)(\forall y)(\forall x')(\forall y')(\forall x'')(\forall y'')(A(f,x,y) \;\&\; A(f,x',y') \;\&\; xx' = x'' \;\&\; yy' = y'' \rightarrow A(f,x'',y''))$ (our "maps" are homomorphisms)

(5) $(\exists g)(\forall x)(G(x) \rightarrow A(g,x,x))$ (the "identity" is in M)

(6) $(\forall f)(\exists g)(\forall x)(\forall y)(A(f,x,y) \rightarrow A(g,y,x))$ ("elements" of M have "inverses" in M)

(7) $(\forall f)(\forall g)(\exists h)(\forall x)(\forall y)(\forall z)(A(f,x,y) \;\&\; A(g,y,z) \rightarrow A(h,x,z))$ (M is closed under "composition").

Every model of (1)-(7) must be a group together with a subgroup of its group of automorphisms, and that all these axioms are in L^t, the (finitary) language of structures of type $t = \langle 1,1,3,3,0\rangle$. In the extension $L^t_{\omega_1\omega}$ of this language by the use of countable conjunctions and disjuntions, we can write: $(\forall y)(\forall x)(x \neq e \rightarrow \bigvee_{i\in\omega-\{0\}} (\exists f_1)(\exists f_2) \cdots (\exists f_i)(\exists z_1) \cdots (\exists z_i)(\exists v_2) \cdots (\exists v_{i-1})[\bigvee_{m_1,\ldots,m_i \in Z} A(f_1,x^{m_1},z_1) \;\&\; \cdots \;\&\; A(f_i,x^{m_i},z_i) \;\&\; z_1 z_2 = v_2 \;\&\; \cdots \;\&\; v_{i-2} z_{i-1} = v_{i-1} \;\&\; v_{i-1} z_i = y] \vee \neg G(y)) = \tau$. Now let $\sigma = \tau \;\&\; \gamma \;\&\; (1) \;\&\; \cdots \;\&\; (7) \in L^t_{\omega_1\omega}$, $A(G)$ stand for the group of automorphisms of G.

COROLLARY 2. C <u>is bountiful</u>.

Proof. Let $G \in \mathcal{G}$ $G \subset H \in \mathcal{C}$. Then $\langle A(H) \cup H, H, A(H), R, \cdot, e\rangle \in M(\sigma)$, where $R(f,x,y) \leftrightarrow f(x) = y$. $G \subset A(H) \cup H$, thus by downward Löwenheim-Skolem $\langle S, S \cap H, S \cap A(H), S^3 \cap R, S^3 \cap \cdot, e\rangle \in M(\sigma)$, with $c(S) = c(G)$, and since $G \subset H$, $G \subset S$, we have $G \subset H \cap S$, and $H \cap S$ must be a characteristically simple group since the automorphisms in $S \cap A(H)$ "already" make it such.

LEMMA 3. $G \in \mathcal{J}$ if and only if (*) for $K \lhd H \lhd G$, with $H = [S]_G$, $K = [S']_H$, S, S' finite, we have $K \lhd G$.

Proof. Clearly for $G \in \mathcal{J}$ (*) holds. Conversely, assume (*). We first show (**) if $K \lhd H \lhd G$, $K = [S]_H$, S finite, then $K \lhd G$. But if $H \lhd G$, $[S]_G \subset H$ for any $S \subset H$. Thus $K \lhd [S]_G \lhd G$, with S finite, so by (*), $K \lhd G$.

Using (**) we now establish the theorem. Let $K = [A]_H$, and assume that the theorem has been established for all $[B]_H$ with $c(B) < c(A)$. We now write A as an ascending union of B_i's with $c(B_i) < c(A)$ (this may be done by putting A in one-to-one correspondence with its cardinal $c(A)$ and setting $B_i = \{a_j \mid j < i\}$). Now note that $K = \bigcup_{i<c(A)}[B_i]_H$, an ascending chain of normal subgroups of G. Thus $K \lhd G$.

COROLLARY 3. $\mathcal{J} = M(\sigma)$ for some $\sigma \in L_{\omega_1\omega}$, thus $\mathcal{J}$ is a bountiful class.

Proof. Let $\tau = \bigwedge_{m,n\in\omega-\{0\}}(\forall a_1)\cdots(\forall a_n)(\forall b_1)\cdots(\forall b_m)(\forall x)(H(a_1,\ldots,a_n,x) \leftrightarrow J(a_1,\ldots,a_n,b_1,\ldots,b_m,x))$, where we define $H(a_1,\ldots,a_n,x) \leftrightarrow \bigvee_{p\in\omega\{0\}}(\exists z_1)\cdots(\exists z_p)\bigvee_{i_{11},\ldots,i_{pn}\in Z}(x = z_1 a_1^{i_{11}} \cdots a_n^{i_{1n}} z_1^{-1} \cdots z_p a_1^{i_{p1}} \cdots a_n^{i_{pn}} z_p^{-1})$ (and this corresponds to $x \in [\{a_1,\ldots,a_n\}]_G$) and we define $J(a_1,\ldots,a_n,\ldots,b_m,x) \leftrightarrow \bigvee_{p\in\omega-\{0\}}(\exists z_1)\cdots(\exists z_p)(H(b_1,\ldots,b_m,z_1) \,\&\, \cdots \,\&\, H(b_1,\ldots,b_m,z_p) \,\&\, [\bigvee_{i_{11},\ldots,i_{pn}\in Z}(x = z_1 a_1^{i_{11}} \cdots a_n^{i_{1n}} z_1^{-1} \cdots z_p a_1^{i_{p1}} \cdots a_n^{i_{pn}} z_p^{-1})])$ (corresponding to $x \in [\{a_1,\ldots,a_n\}]_{[\{b_1,\ldots,b_m\}]}$).

Then by Lemma 3, $\sigma = \tau \,\&\, \gamma$ is our statement since σ states that the normal subgroup generated by $a_1,\ldots,a_n$ in the normal subgroup generated by $b_1,\ldots,b_m$ is the same as the normal subgroup generated by $a_1,\ldots,a_n$ in the entire group. Thus a finitely-generated normal subgroup of a finitely-generated normal subgroup is normal.

PROPOSITION 1. $\mathfrak{F} = M(\sigma)$ for some $\sigma \in L_{\omega_1\omega}$, thus $\mathfrak{F}$ is a bountiful class.

Proof. We define $G(a_1,\ldots,a_n,x) \leftrightarrow \bigvee_{p\in\omega-\{0\},i_{11},\ldots,i_{pn}\in Z}(a_1^{i_{11}} \cdots a_n^{i_{1n}} a_1^{i_{21}} \cdots a_n^{i_{2n}} \cdots a_n^{i_{pn}} = x)$ (corresponding to $x \in \langle\{a_1,\ldots,a_n\}\rangle$), $I(a_1,\ldots,a_n,b_1,\ldots,b_n) \leftrightarrow (\exists z_1) \cdots (\exists z_n)(\forall x)([G(b_1,\ldots,b_m,x) \leftrightarrow G(z_1,\ldots,z_n,x)] \ \& \ \bigwedge_{p\in\omega-\{0\},i_{11},\ldots,i_{pn}\in Z} (z_1^{i_{11}} \cdots z_n^{i_{1n}} z_2^{i_{21}} \cdots z_2^{i_{2n}} \cdots z_n^{i_{pn}} = e \leftrightarrow a_1^{i_{11}} \cdots a_n^{i_{1n}} \cdots a_n^{i_{pn}} = e))$. It is simple to check that $I(a_1,\ldots,a_n,b_1,\ldots,b_m)$ holds if and only if $\langle\{a_1,\ldots,a_n\}\rangle \cong \langle\{b_1,\ldots,b_m\}\rangle$, (where the isomorphism $f_1\langle\{a_1,\ldots,a_n\}\rangle \to \langle\{b_1,\ldots,b_m\}\rangle$ is defined by $f(a_1^{i_{11}} \cdots a_n^{i_{pn}}) = z_1^{i_{11}} \cdots z_n^{i_{pn}}$).

Thus if $\tau = \bigwedge_{m,n\in\omega-\{0\}}(\forall a_1) \cdots (\forall a_n)(\forall b_1) \cdots (\forall b_m)(I(a_1,\ldots,a_n,b_1,\ldots,b_m) \leftrightarrow (\exists g)(\forall x)(G(a_1,\ldots,a_n,x) \leftrightarrow G(b_1,\ldots,b_m,gxg^{-1}))$, then τ says that if any two finitely-generated subgroups are isomorphic, they are conjugate. If $\sigma = \tau \ \& \ \gamma$, then $\mathfrak{F} = M(\sigma)$.

The question of whether $\mathfrak{I}$ is bountiful remains open. A major difficulty seems to be the fact that structures of the form $\langle G \cup A(G),G,A(G),A,\cdot,e\rangle$ cannot be characterized in any language $L^s_{\alpha\beta}$ (for any type s, any regular cardinals α, β). This can be shown by use of infinitary downward Löwenheim-Skolem.

The following considerations suggest a problem on the border between group theory and logic:

Recall that for $a \in G$, $\text{ord}(a) = \begin{cases} \text{the least } n > 0 \text{ such that } a^n = e \\ \infty \text{ if no such } n > 0 \text{ exists} \end{cases}$.

If G is an abelian group and if $\text{ord}(a) \neq 1, 2$, then $a \neq a^{-1}$, thus the map $f: G \to G$ by $f(x) = x^{-1}$, an automorphism in this case, does not leave a fixed. Therefore, for any formula $F \in L$, if $G \models F[a]$, then $G \models F[a^{-1}]$, so it is impossible to distinguish a from all other elements of G by a set of formulas $S \subset L$ with one free variable.

We may now ask if for any group G whether every first-order-definable element of G is of order 1 or 2.

For G non-abelian, $a \notin C(G)$, we have an element $b \in G$ such that $ab \neq ba$, thus $a \neq bab^{-1}$. Therefore the (inner) automorphism $f_b, G \to G$ defined by $f_b(x) = bxb^{-1}$ does not fix a, and since a cannot be distinguished from its image under an automorphism, no $a \notin C(g)$ is first-order-definable. We have shown:

PROPOSITION 2. Let $a \in G$. If

(1) $\mathrm{ord}(a) \neq 1,2$ and G is abelian, or

(2) if $a \notin C(G)$,

then a is not first-order-definable.

The answer to the proposed question, however, is negative. In fact for every $n < \infty$ there are an infinite number of groups with first-order-definable elements of order n. To see this, suppose G is finite. Then two elements $a,b \in G$ cannot be distinguished by a set of first-order-formulas in one free variable if and only if $\langle G,\cdot,e,a\rangle \equiv \langle G,\cdot,e,b\rangle$ in the language L^c of groups with a constant (i.e. if and only if the two structures satisfy the same sentences in that language). But since G is finite this implies $\langle G,\cdot,e,a\rangle \cong \langle G,\cdot,e,b\rangle$, i.e., there is an automorphism f of G such that $f(a) = b$. It will therefore do to show that for every $n > 2$ there is an infinite number of finite groups, each with an element of order n fixed under every automorphism. The following is due to Stephen Meskin:

LEMMA. _If_ $1 < n < \infty$ _then there is an infinite number of finite groups with elements of order_ n _fixed under every automorphism_.

Proof. By Dirichlet's theorem (a statement of which may be found on p. 20 of [4], a proof on p. 49 of [2]) there are infinitely many primes p such that n divides $p - 1$. For each such p, consider the group generated by a, b with $a^p = b^{n^2} = e$ $bab^{-1} = a^t$, where $0 < t < p$ and the order of t in the multiplicative group of z mod p is n (the existence of such a t under these circumstances is shown on p. 54 of [4], the fact that this group exists is shown as on p. 148 of [1]).

The following facts may be established by induction on m:

(1) $b^m a^s b^{-m} = a^{t^m s}$ (thus $b^m a^s = a^{t^m s} b^m$)

(2) $(b^x a^y)^m = a^{(t^x + \cdots + t^{mx})y} b^{mx}$ (the proof of this uses (1))

and (1) can also be used to show that every element of our group can be written uniquely in the form $b^x a^y$ (or $a^r b^s$). Call our group H and let f be an automorphism of H. $f(a)$ must be of order p, but any element of order p must be in the subgroup $\langle\{a\}\rangle$ (since by (1), $\langle\{a\}\rangle$ is normal, and if $g : H \to H/\langle\{a\}\rangle$ is the natural homomorphism, $c \in H$ of order p, then $\mathrm{ord}(g(c))$ divides p, but also must divide the order of the group $H/\langle\{a\}\rangle = n^2$, and since p, n^2 are relatively prime, $\mathrm{ord}(g(c)) = 1$. Thus $g(c) = e$, so $c \in \langle\{a\}\rangle)$. Thus for some k, $f(a) = a^k$ and $(k,p) = 1$ (i.e., k, p are relatively prime), and we may also write $f(b) = b^x a^y$.

Thus $f(bab^{-1}) = f(b)f(a)f(b)^{-1} = b^x a^y a^k a^{-y} b^{-x} = b^x a^k b^{-x} = a^{t^x k}$. But $f(bab^{-1}) = f(a^t) = f(a)^t = a^{kt}$, so $a^{t^x k} = a^{tk}$, thus $t^x k \equiv tk \pmod p$, thus since $(k,p) = 1$, $t^{x-1} \equiv 1 \pmod p$, so $x \equiv 1 \pmod n$. Now note that $\mathrm{ord}(b^n) = n$, and $f(b^n) = f(b)^n = (b^x a^y)^n = a^{(t^x+\cdots+(t^x)^n)y} b^{nx} = a^{(t+\cdots+t^n)y}(b^n)^x = a^{[(1-t^n)/(1-t)]ty}\, b^n = a^{0ty} b^n = b^n$ (the third equation from the end uses the fact that $x \equiv 1 \pmod n$, so $t^x \equiv t \pmod p$). Thus b^n is also fixed under every automorphism.

COROLLARY 4. <u>For any positive</u> n <u>there is a group</u> G <u>with an element</u> c <u>of order</u> n <u>which is first-order-definable</u>.

The question remains open for order ∞.*

The problems in the first part arose in the course of lectures given by Philip Hall in Cambridge in 1966. Hall showed that the classes $\mathfrak{S}$, $\mathfrak{C}$, $\mathfrak{J}$, and $\mathfrak{Z}$ are bountiful, and asked whether the same is true of $\mathfrak{I}$. The term "bountiful" is due to Mathias, who noticed the similarity of Hall's theorems to that of Löwenheim and Skolem, and raised the second question. The answer given here, that the four classes are (with some sleight of hand) axiomatisable in $L_{\omega_1\omega}$, was found by Kopperman in the Westwood Village Delicatessen, Los Angeles, at 7:32 p.m. on December 30, 1967.

The problem of first-order definable elements was raised by Mathias.

---

*This question has been solved in part. Mathias and Meskin have independently discovered a group with an element of infinite order which is first-order-distinguishable from all other elements of the group. It is distinguishable, however, only by an infinite set of sentences, rather than a single sentence. (The example is much like those given in the text, with b "made" infinite in order.) Let D be of order 5, d a generator. Let C be cyclic of infinite order, c a generator. Form the split extension G of D by C defined by setting $d^c = d^2$, $d^{c-1} = d^3$. Then c^4 centralises D. We assert that c^4 is definable in G by the following infinite set of formulae:
$(\exists x)(\exists w)(\forall y)(wy = yw \;\&\; x \neq e \;\&\; x^5 = e \;\&\; x^w = x^2 \;\&\; z = w^4)$ for each $m \in Z$, $m \neq 0$:
$\neg(\exists w)(\forall y)(wy = yw \;\&\; w^{4m+1} = z)$. The argument is similar to that of Meskin, and is omitted.
$(g^h = h^{-1}gh.)$

REFERENCES

[1] Fraleigh, J., A First Course in Abstract Algebra, Addison-Wesley, 1967.

[2] Gelfond, A. and Linnik, Y., Elementary Methods in the Analytic Theory of Numbers, (Translation by D. Brown). M. I. T. Press, 1966.

[3] Kopperman, R., The $L_{\omega_1\omega_1}$-theory of Hilbert spaces, J. Symbolic Logic, 32 (1967), 295-304.

[4] Niven, I. and Zuckerman, H., An Introduction to the Theory of Numbers, (Second Edition), Wiley, 1966.

CHOICE OF INFINITARY LANGUAGES BY MEANS OF DEFINABILITY CRITERIA; GENERALIZED RECURSION THEORY

G. KREISEL*

The use of languages with infinitely long expressions suggests itself at the very beginning of logic. Suppose we start with a finitary ('ordinary') language $\mathcal{L}$ and enumerate its closed formulae $\varphi_1, \varphi_2, \ldots$. Then a truth definition for $\mathcal{L}$ is nothing but the infinite disjunction, with the free number variable x,

$$(x = 1 \wedge \varphi_1) \vee (x = 2 \wedge \varphi_2) \vee \cdots$$

As emphasized by Tarski, in general, there is no formula of $\mathcal{L}$ which is equivalent to this disjunction (for the interpretation of $\mathcal{L}$ considered). Put differently there is no finite explicit truth definition in $\mathcal{L}$ for $\mathcal{L}$. (The existence of finite implicit definitions for the truth predicate or, more generally, for the satisfaction relation, will be important below.)

In the old philosophical literature the meaning of quantifiers over some fixed domain, such as the natural numbers, was explained in terms of infinite conjunctions or disjunctions. More recently, because of a preoccupation with communication by means of concrete symbols, one tended to reverse the procedure. Specifically, the meaning of (technically: the satisfaction relation for) infinite formulae is defined by means of finite formulae; in general the latter contain variables of higher type than those appearing in the infinite formulae considered. Though, taken literally, this device does not apply to the truth definition for set theory, Gödel [5] pointed out a certain modification: he found (finite) formulae in the language $\mathcal{L}_E$ of ordinary set theory, so called axioms of infinity, which

*This note is an expanded version of Section 13 in my paper 'A survey of proof theory,' to appear in JSL 33(1968). The preparation of this note was partially supported by NSF grant GP-6726.

imply all consequences formulated in $\mathcal{L}_E$ of the existence of a truth definition.

Thus, within the framework of full set theory any use of infinite expressions that we know at present is _reducible_ (to formulations in finitary languages with a sufficiently abstract interpretation). However, _technically_ the conception of infinitary languages seems most useful, generally speaking for the following reason. As is well known, one of the major successes of mathematical logic was the discovery that (i) many interesting mathematical structures can be defined in a certain _restricted_ language, namely first order predicate calculus PC, (ii) the structures definable in PC satisfy some remarkably _simple_ laws; evidently just _because_ restrictions have been imposed. Similarly just because infinitary languages and their satisfaction relation can be defined in finite (higher order) languages, there is reason to expect that the theory of such infinitary languages may be simpler than that of the finite higher order languages, i.e., the structures definable by use of the infinitary languages may satisfy simpler laws.

There is a particular defect of finite languages, also stressed in [5], whenever the notion of _natural number_ is definable (for the interpretations of the language considered). By means of diagonalization one shows that the class of definable relations is _not closed_ under the following principle. If the expression fn defines, for each _n_, a definition, say, δ_n of a number d_n (or set D_n) then there is a definition of the relation $F = \{(n,m) : m = d_n\}$ $(\{(n,X) : X = D_n\})$. Natural notions of definability, e.g. recursive definitions, _do_ satisfy this principle; in fact, Kleene's latest characterization of the class of recursive functions in [10] is, literally, based on this principle. Thus, one tries to balance syntax and semantics: one admits those objects in the syntax which can be defined for the (intended) interpretation or interpretations, i.e., the intended semantics, of the language. Note that PC satisfies this condition when _general_ models are considered: only finite structures are definable (up to isomorphism).

From this point of view it is natural to use _definability_ criteria in the choice of language; and if _several_ interpretations are considered, some kind of _invariance_ (w.r.t. all these interpretations) will be required of these definitions. The purpose of the present note is to go over the steps by which such _invariant definability criteria_ were derived. The principal ideas were introduced in [12], developed and applied by Kunen [17] and particularly Barwise [1]. The principal novelty of the present note is the discussion of open problems and of new fields of application.

REMARK. If I overemphasize definability in this _general_ exposition, the reason is simple. I happen to have become interested in infinitary languages in connection with a precise characterization of predicativity where emphasis on definitions is certainly quite in order!

We begin with first order predicate calculus PC as a typical example of a formal language. We use L_α to denote the ramified hierarchy of sets of levels $< \alpha$ and, for initial ordinals (cardinals) α, we use H_α for the collection of sets of hereditary cardinal $< \alpha$. Thus $H_\omega = L_\omega$ = collection of hereditarily finite sets, $H_{\aleph_1}$ = collection of hereditarily countable sets.

(a) Analysis of the problem: syntax and semantics.

(i) General syntax concerns the combination of symbols; thus the syntactic objects of PC are elements of H_ω or finite sequences or simply finite ordinals; the latter when, e.g., one identifies formulae and their Gödel numbers.

The difference between sets and ordinals (or constructible sets) is minor in the finite case because $H_\omega = L_\omega$ and the natural structures on H_ω and ω are interdefinable. But it will matter in the infinite case, e.g. for $H_{\aleph_1}$ and $L_{\aleph_1}$ respectively.

(ii) Semantics assigns a realization to the basic symbols used and thereby determines the two familiar properties of syntactic objects: being a meaningful[1] expression, and: being valid. Further there is the consequence relation (between a formula φ and a class Φ of formulae).

In PC the usual choice of propositional operations $(\neg, \wedge)$ is distinguished by their functional completeness; Mostowski [21] has given an interesting, but as yet inconclusive, criterion of characterizing the quantifiers $(\forall, \exists)$ in terms of the form of their validity predicate.[2] The logical operations in infinitary languages will have to be chosen by more delicate considerations than functional completeness, for instance a language of cardinality $\aleph_1$ cannot contain symbols for all truth functions with $\aleph_0$ arguments.

In PC the set of consequences of an arbitrary Φ is defined from the validity predicate by use of the finiteness theorem and the fact that every finite set of formulae is equivalent to a single one. Note that in the early days of logic one did not consider arbitrary Φ but only formal systems: their sets of consequences are r.e.; it was a (surprising) discovery that the main theorems in model theory

---

[1] Two senses of meaningful are to be distinguished: (i) for all realizations of the language or only (ii) for models of given axioms. For example, different treatments of ι-terms are suitable in the two cases: in (i) one would like every ι-term to have a meaning, in (ii) only those $\iota_x Ax$ for which $\exists ! x Ax$ is a consequence of the axioms. In the case of ι-terms, requirement (i) leads to a recursive set of terms, (ii) only to an r.e. set: I do not know if this case is typical of the general situation.

[2] Specifically, a recursively enumerable validity predicate. The characterization is incomplete as it stands since it only applies to validity in countable structures [25]. It would be interesting to have a general characterization for the role of countable structures, e.g., because of the countability of the languages considered by Mostowski.

of PC apply to arbitrary and not only r.e. sets Φ. Naturally, one exploited this discovery, and alternative proofs not using it were often found surprisingly late, cf. [3] or Chapter 6 of [15]. Finally, note that, for those earlier proofs, PC with non-denumerably many symbols was needed, but not for the later ones; this difference will also be important for the generalization.

(b) Principal proposals. The older work on infinitary languages as presented e.g. in [8] fixed the syntax by cardinality conditions, and chose the logical operations by 'straight' analogy: $\neg$, $\wedge$, $\bigwedge\!\!\bigwedge$, arbitrary strings of quantifiers, or by pragmatic considerations: $\neg$, $\wedge$, $\bigwedge\!\!\bigwedge$, finite strings of quantifiers [23].

As stated in the introduction we choose the syntax and the logical operations by definability, more precisely invariant definability, considerations (which happen to coincide with cardinality conditions in the case of PC). We proceed as follows. In the first place we consider the definition of infinite expressions by means of finite formulae: afterwards we verify that no new infinite formulae can be defined by use of our infinite formulae (closure conditions). Once definability criteria are accepted in the choice of language, it is cogent to use them also e.g. in the generalization of the finiteness theorem, instead of the cardinality conditions in the familiar generalization of the so-called compactness theorem.

Basic principles. We separate our problem into first generalizing the notions used in syntax and semantics to a wide class of structures ($\mathfrak{A}$ in place of H_ω used in PC), and then investigating what additional conditions on $\mathfrak{A}$ are needed in order that the basic results on PC hold for our generalization.

(i) Recursion theoretic notions:

$\mathfrak{A}$-finite = absolutely invariantly definable

$\mathfrak{A}$-rec = invariantly definable on (the universe A of the structure) $\mathfrak{A}$

$\mathfrak{A}$-r.e. = semi-invariantly definable on $\mathfrak{A}$.[3]

An elaborate discussion is in [12]. Regarding finiteness as a recursion theoretic rather than cardinality property in this context is the problematic and fruitful step.

[3] The terminology of [17] differs slightly from [12], e.g. 'strongly' for 'absolutely' or 'A is self definable' for 'the universe A of the structure $\mathfrak{A}$ is $\mathfrak{A}$-finite.' The notion $\mathfrak{A}$-r.e. was introduced in [17], under the name: semi-invariantly implicitly definable, following Mostowski [22]; the other notions are in [12]. Note that, in [12], $\mathfrak{A}$-r.e. was defined to mean the range of an $\mathfrak{A}$-rec function: prima facie this definition is not reasonable for general $\mathfrak{A}$, e.g., not for $A = N^N$, since an r.e. set of functions is Σ_1^0 and not Σ_0^1 (which it would be as the range of a function defined on N^N). What made the definition plausible is that for a recursive relation $R(\alpha,\beta)$ $(\alpha \in N^N, \beta \in N^N)$, there is a recursive $R_1(\alpha,n)$ such that $\forall\alpha[\exists\beta R(\alpha,\beta) \leftrightarrow \exists n R_1(\alpha,n)]$.

(ii) Logical notions. The syntactic objects themselves such as terms or formulae are to be elements of the universe A of our structure $\mathfrak{A}$. Formal derivations, if used at all, are to be $\mathfrak{A}$-finite. The set of valid formulae is to be $\mathfrak{A}$-r.e., and the set of meaningful ones, possibly, $\mathfrak{A}$-rec. The finiteness theorem takes the form:

A consequence of an $\mathfrak{A}$-r.e. set Φ of formulae is also a consequence of an $\mathfrak{A}$-finite subset of Φ.

For any given $\mathfrak{A}$, and $X \subset A$, all the definability notions above can be relativized to: invariant definability from X. Note that the ordinary finiteness theorem for countable A is included in our version since, for all X, H_ω-finite in $X \leftrightarrow$ finite.

(c) Choice of definitions. To apply the principles of (b) one has to specify two things: what (finite) language is to be used for the invariant definitions or, in other words, what structure $\mathfrak{A}$ is to be considered; put differently, if we start with some given set A, what structure is put on A? Also for what class of realizations of this language are the definitions to be invariant? Our aim is to make the notions insensitive to the choice of language: in general, restricting the class of realizations helps because more relations become invariantly definable and, by transitivity, adding them to the structure on A does not alter our notions. Second, one uses implicit rather than explicit definitions. This is plausible because we are using here definitions in a finite language to discover an infinitary language and we know, from the introduction, (i) defects of finite explicit definitions, (ii) implicit definitions of the satisfaction relation which is the central connection between syntax and semantics.[4] In any case, we are proposing to use recursion theoretic notions, and recursive definitions are the most familiar example of implicit (invariant) definitions; cf. also [4].

The considerations above led to the following general scheme [12].

Given a structure $\mathfrak{A}$ and a class $\mathfrak{C}$ of realizations, where $\mathfrak{A} \in \mathfrak{C}$, we consider finite expansions (of the language) of $\mathfrak{A}$, structures which realize some additional relation or function symbols, say $T_0, T_1, \ldots, T_n$. Let φ be a formula of the expanded language.

φ is called an implicit invariant definition of the set $X \subset A$ provided

[4] An instructive case to consider is the (finitary) language of first order arithmetic with variables ranging over the natural numbers. The class of (explicitly) arithmetically definable sets suffers patently from the defect mentioned in the introduction. All hyperarithmetic sets are implicitly definable (by use of auxiliary predicates). In contrast, the class of hyperarithmetic predicates has the property that if a unique set T satisfies a hyperarithmetic condition then T is again hyperarithmetic. The means of definition employed in Kleene's original description of this class in terms of his $H_e (e \in O)$ uses the existential quantifier (or jump), finite iterations of it, and infinite recursive disjunctions for limit numbers $e = 3.5^n$.

$$(\forall \mathfrak{A}' \in \mathfrak{C})[(\mathfrak{A}' \vdash \exists T_0 \ldots \exists T_n \varphi) \rightarrow (\forall X_0 : X_0, \mathfrak{A}' \vdash \exists T_1 \ldots \exists T_n \varphi)(X_0 = X)]$$

and φ is an implicit invariant definition of X on A provided

$$(\forall \mathfrak{A}' \in \mathfrak{C})[(\mathfrak{A}' \vdash \exists T_0 \ldots \exists T_n \varphi) \rightarrow (\forall X_0 : X_0, \mathfrak{A}' \vdash \exists T_1 \ldots \exists T_n \varphi)(X_0 \cap A = X)] .$$

The notion is immediately relativized to definitions from a set $Y \subset A$ (or, more generally, $\subset Y^m$ for some m).

There is an important distinction: should $\mathfrak{A}$ itself satisfy $\exists T_0 \ldots \exists T_n \varphi$ or only some $\mathfrak{A}' \in \mathfrak{C}$? Fortunately, we have a strong stability result: though different formulae φ satisfy the two kinds of conditions, the same class of sets X is obtained.[5]

We now come to the question: given A, what structure $\mathfrak{A}$ should be put on it? For collections A of sets or ordinals the natural structures are $\in$ and $<$ respectively. Beyond this the choice seems, at first, sheer guesswork. The latter is reduced by use of the notion of transitive or end extension, for (binary) $R = \in$ or $<$ restricted to $A \times A$:

(A', R') is an end extension of (A, R) means:

$A \subset A'$ and $\forall x \forall y[(y \in A \wedge xR'y) \rightarrow (xRy \wedge x \in A)]$.

(If (A', R') is not an end extension, even if $A \subset A'$ the object $y \in A \cap A'$ is not the same abstract object in A as in A'; same in the sense made precise by the notion of hard core in [12].)

Note in passing that the familiar Rosser condition in arithmetic

$$\forall x[x < 0^{(n+1)} \leftrightarrow (x = 0 \vee x = 0' \vee \cdots \vee x = 0^{(n)})]$$

ensures that all models are end extensions of $(\omega, <)$.

Principle. In the case of collections A of sets we use the structure $(A, \in)$. So our general

[5] The distinction is explicit in Lacombe [18]; the equivalence is implicit there too, and explicit in Kunen [17]. This distinction expresses neatly and generally the difference which I formulated clumsily in the particular case where $\mathfrak{A}$ is $N \cup M$ and $M \subset \mathfrak{P}(N)$ (with its usual structure), in terms of definability from a property and from a set ([12], p. 197). The connection is this: even if the given $\mathfrak{A}$ satisfies the comprehension axiom $\exists X \forall n(n \in X \leftrightarrow \psi n)$ when ψ does not contain the additional symbols $T_0, \ldots, T_n$, it may fail to do so when these symbols (and their diagrams) are included.

scheme requires that definitions be invariant for classes of end extensions which are finitely axiomatizable, namely those satisfying the 'axiom' $\exists T_0 \ldots \exists T_n \varphi$.

It turns out that the usual set theoretic relations are indeed $\mathfrak{A}$-invariantly definable. Furthermore, for the infinitary languages associated with $\mathfrak{A}$ below, no new relations are so definable even if one uses $\mathfrak{A}$-r.e. sets of infinite formulae.

As far as I know, the situation is less clear cut for collections A of ordinals and the structure $(A,<)$.[6]

Since the universe A determines the structure (A,ϵ) put on it, we shall occasionally replace '$\mathfrak{A}$-finite' or '$\mathfrak{A}$-r.e.' by 'A-finite' or 'A-r.e.'.

(d) Simultaneous choice of $\mathfrak{A}$ and an associated language $\subset A$. All the languages considered will include

$\mathcal{L}_A$: its operations are $\neg$, $\wedge$, $\bigwedge\!\!\!\bigwedge$; finite strings of quantifiers $\forall$, $\exists$.

All the A considered will satisfy elementary closure conditions (needed to ensure that $\mathcal{L}_A \subset A$); for an explicit list of axioms for such a 'rudimentary' set theory, see [1] p. 21.

Problem: What further conditions must A satisfy in order that the basic properties of PC generalize to $\mathcal{L}_A$ for the translation given in (b)? For what extensions of $\mathcal{L}_A$ do these properties persist?

To avoid a circle in the choice of A, one uses properties $\underset{\sim}{P}$ (of a language) which are stable for restrictions of the language: if $\mathcal{L}_A$ doesn't have $\underset{\sim}{P}$, neither does any extension of $\mathcal{L}_A$, e.g. if $\underset{\sim}{P}$ is the property: the validity predicate is $\mathfrak{A}$-r.e. In contrast, e.g. the interpolation property is not stable for restrictions. Trivially, consider ordinary propositional formulae φ and ψ without common variables, such that $\varphi \rightarrow \psi$ holds: $(\neg, \wedge, \top)$ contains an interpolation formula, but $(\neg, \wedge)$ does not.

It is easy to guess reasonably good additional conditions on A: e.g. in [12] p. 194(b) for a recursion theory on L_{ω_1}, where ω_1 is the first non-recursive ordinal, and for ω-logic; independently, in [16], equivalent conditions for a recursion theory on ordinals. Platek considered a modification of

[6] When $\mathfrak{A} = \langle\omega, 0, s\rangle$, where s denotes the successor, exactly the recursive sets are invariantly implicitly definable on ω (with respect to general models), by [4]. One can ensure by means of a single formula of PC that a realization is an end extension. This is not possible for ordinals $> \omega$ in the case of general models.

our axioms, more suited to the case when A does not have an invariantly definable well ordering, and called sets satisfying his axioms: admissible.

The work below (for details see [1] and [17]) bears on the question: admissible for what?

(i) Validity and consequence. In one direction the main results are:

For all admissible A, an $\mathfrak{A}$-r.e. set of formulae in $\mathcal{L}_A$ has an A-r.e. set of consequences [17].

For countable admissible A, $\mathfrak{A}$-r.e. = Σ_1 (over A, in the sense of [19]), $\mathfrak{A}$-rec = Δ_1, $\mathfrak{A}$-finite = being an element of A ([17] and [1]).

But there are uncountable A for which these last results fail, e.g. $A = H_{\aleph_1}$ or $A = H_{\kappa_1}$ where κ_1 is the first inaccessible; so to speak, if one accepts uncountable cardinals at all, $\aleph_1$ and κ_1 play the role of specific natural numbers for the 'finite mind.' This property of κ_1 is not surprising; see footnote 6 of [13], p. 104.

In the opposite direction [1] shows, for countable rudimentary A:

If, uniformly, for all $\Phi \subset \mathcal{L}_A$, the set of consequences of Φ is Σ_1 in Φ, then A is admissible.

There are countable non-admissible unions A of admissible sets, for which the validity predicate is $\mathfrak{A}$-r.e., with refinements when A is of the form L_α and $\alpha < \aleph_1$.

So one properly distinguishes between validity-admissible and consequence-admissible A.

Incidentally, it would be more satisfactory to prove the

Conjecture. For all rudimentary A, if all $\mathfrak{A}$-r.e. sets of formulae $\in \mathcal{L}_A$ have an $\mathfrak{A}$-r.e. set of consequences, then A is admissible, i.e. the result above with $\mathfrak{A}$-r.e. in place of Σ_1.

(ii) Choice of language. First of all, for many (uncountable) admissible A, e.g. $A = H_\kappa$ where κ is inaccessible, the validity predicate is $\mathfrak{A}$-r.e. even if infinite strings of quantifiers are allowed ([17]). So the restriction to finite strings does not seem to be justified generally.[7]

For countable A of the form L_α, by [1]:

For recursively accessible α and $A = L_\alpha$, the validity predicate for $\mathcal{L}_A$ is a complete $\mathfrak{A}$-r.e. predicate; for rec. inaccessible (admissible) α it is $\mathfrak{A}$-rec.

---

[7] More precisely, a good generalization is possible (and therefore desirable) without this restriction. We do not expect all results about PC to generalize! For instance, if a formula of PC has a model it has one of cardinal $\leq \aleph_0$, its Loewenheim number and if it has one of cardinal $\aleph_0$ it has a model of every cardinal $\geq \aleph_0$ and so of arbitrarily large cardinals, its Hanf number.

It seems plausible that this result can be improved to establish the

Conjecture. For rec. accessible α, suppose the extension of $\mathcal{L}_A$ by a new propositional operation π has an $\mathfrak{A}$-r.e. validity predicate; then π is definable in $\mathcal{L}_A$. (For $A = L_{\omega_1}$, this is true.)

Problem. For rec. inaccessible α: what 'natural' propositional operations and/or quantifiers have to be added to $\mathcal{L}_A$ to get a complete $\mathfrak{A}$-r.e. validity predicate?

Other problems of this type will occur to the reader automatically if he goes into the subject.

(iii) $\mathfrak{A}$-finiteness: a controversial issue.[8] Barwise [1] makes a strong case for imposing the following closure condition on A:

$$\forall X (X \text{ is } \mathfrak{A}\text{-finite} \leftrightarrow X \in A) ,$$

satisfied by all countable admissible A. In general terms his idea is this: Let us look for a simple condition $\mathfrak{Z}$ on sets A such that, for countable A, $\mathfrak{Z}$ is equivalent to admissibility and, generally, $\mathfrak{Z}$ implies the closure condition above, the equation $\mathfrak{A}$-r.e. $= \Sigma_1$, and the generalized finiteness theorem. Note that these consequences of $\mathfrak{Z}$ are known in the special case where A is of the form H_α. (In [1] Barwise considers a certain 'indescribability' condition $\mathfrak{Z}$.)

The closure condition has evident consequences for a proof theoretic (inductive) analysis of the validity predicate for $\mathcal{L}_A$. But I should like to know if the condition is necessary! e.g. by finding a counterexample to the generalized finiteness theorem for some, necessarily uncountable, admissible A (which does not satisfy the closure condition - and hence not $\mathfrak{Z}$).

Or, if there is none, to show that every admissible A can be regarded as a fragment of a set A' satisfying $\mathfrak{Z}$; 'fragment' in the sense that:

A' is a transitive extension of A,

$\forall X[(X \subset A \wedge X \text{ is } \mathfrak{A}'\text{-finite}) \leftrightarrow X \text{ is } \mathfrak{A}\text{-finite}]$,

$\forall X[(X \subset \mathfrak{A} \wedge X \text{ is } \mathfrak{A}'\text{-r.e.}) \leftrightarrow X \text{ is } \mathfrak{A}\text{-r.e.}]$.

[8] Several, in general inequivalent, generalizations of finite were considered in [12] pp. 202, 203. Another definition, not considered there and related to Barwise's proposal, is this. For $X \subset A$, X is $\mathfrak{A}$-finite if X is $\mathfrak{A}$-recursive and there is no $\mathfrak{A}$-recursive 1-1 mapping of A onto X. This is a kind of 'relativized' cardinality property.

This situation would be analogous to the use of a finite language for ω-logic where, roughly, $A = L_\omega \cup \{L_\omega\}$, and $A' = L_{\omega_1}$ ('roughly' because this A is, of course, not admissible).

(e) Testing the framework (b). While the work reported in (d) accepted the basic principles of (b), we now wish to consider alternatives to (b).

(i) Old cardinality conditions: here the languages $\mathcal{L}_A$ for countable A, though eminently suited for defining interesting algebraic structures, are excluded from the start. By [7], strong compactness cuts out many A of the form H_κ which satisfy even Barwise's Π_1^1-indescribability criterion. On the one hand, as pointed out at the end of (a), there is at present little evidence that strong compactness is actually needed for generalizing basic properties of PC. On the other, the new version in terms of $\mathfrak{A}$-finite subsets of $\mathfrak{A}$-r.e. sets of axioms, seems to be genuinely useful. For examples, see [12] p. 202, last paragraph of 3.1.[9]

(ii) Alternative recursion theories on sets (Platek). For all admissible $\alpha > \aleph_1$, the set of valid propositional formulae of $\mathcal{L}_{L_\alpha}$ is not constructible, hence not Σ_1 on L_α, and so not generalized r.e. in the sense of Platek, though even L_α-rec by [17]. This result, due to Kunen (see [1], p. 52), uses Ramsey cardinals; it is of course not surprising that large cardinals should have consequences for logical questions: see the lecture (Jan. 1964) [13], bottom of p. 116.

(iii) Restriction to constructible structures. Once again we assume large cardinals and consider $A = L_\alpha$ for α strictly between $\aleph_1^{(L)}$ (the first constructibly uncountable ordinal) and $\aleph_1$. By [1], the set of formulae of $\mathcal{L}_A$ valid in all models is L_α-r.e. and hence Σ_1 on L_α; but, relativizing [9], there are many such α for which the set of formulae valid in all constructible models is not L_α-r.e.

---

[9] I do not know 'applications' in the strict sense of the word, namely consequences of the generalized finiteness theorem which are formulated without any definability notion. But we certainly have applications in the sense that the generalized finiteness theorem was used to answer precisely stated open problems, particularly in ω-logic. Thus my first application [11] was used to refute a conjecture of [6] by constructing a Π_1^1-set of axioms which is not equivalent, in ω-models, to any hyperarithmetic (sub)set. The principle was to consider $A = L_{\omega_1}$ (or, more precisely, in [11] the universe $N \cup HYP$ where HYP is the collection of hyperarithmetic sets of natural numbers), when Π_1^1 sets are $\mathfrak{A}$-r.e., and Δ_1^1-sets are $\mathfrak{A}$-finite; then I simply transferred the familiar construction of an infinite r.e. sets which is not logically equivalent to any finite subset. Other applications (mentioned in [12]) depend on transferring the well known principle for constructing non-standard models of arithmetic by adding the r.e. set $a \neq 0$, $a \neq 1$, $a \neq 2$, etc. and observing that every finite subset is satisfied trivially in the standard model. Since the set of ordinal notations is Π_1^1, i.e. $\mathfrak{A}$-r.e. for the $\mathfrak{A}$ above, the corresponding method applies. See also the postscript following the bibliography.

Here we have a striking parallel to Vaught's result [24] for PC, if we compare constructive (in the sense of [24]) for PC with constructible (in the sense of Gödel) for L_A, when $A = L_\alpha$ and $\aleph_1^{(L)} < \alpha < \aleph_1$.

(iv) Interpolation lemma. As pointed out in (d), the interpolation property is not stable for restrictions. Therefore 'counterexamples' have to be judged by whether a suitable choice of language was made. The same principle applies (even more) to a failure of the definability theorem, for, if a relation is invariantly implicitly, but not explicitly definable, then the language in question is simply inadequate. Incidentally I have the impression that Malitz's ingenious counterexamples [20] would be avoided by adding to the $\mathcal{L}_A$ considered certain quantifiers with zero arguments (expressing something about the cardinality of the universe) in such a way that the extended language still has an $\mathfrak{A}$-r.e. validity predicate.

Another 'counterexample' to the interpolation lemma is given in [2] for the case of finitary higher order languages: defects are discussed in the review of [2]. In my opinion, higher order languages present the best potential test for the ideas of (b):

Clearly the notions of invariant definability make perfect sense also for higher order languages. Now, as is well known, finite higher (even: second order formulae) permit the absolutely invariant definition of large structures. So suppose one accepts the generalization of finiteness in (b) and suppose that Barwise's condition on the class A, from which formulae are taken, is satisfied. Then a second order language and, a fortiori, higher order languages should contain 'huge' formulae. Put differently, the class of finite second order formulae cannot be expected to satisfy anything like the closure conditions needed for a smooth theory.[10]

[10] This point is, perhaps, relevant to an analysis of the common malaise about second order languages, despite the fact that the meaning of a second order formula is formulated in ordinary set theoretic terms (also used in defining the meaning of first order formulae). A familiar reason for the malaise is that the theory of second order validity depends sensitively on the existence of large cardinals ([14], p. 157, para. 2), while first order validity is reduced, by Skolem-Loewenheim, to validity in countable domains. Now, if my view is right, we have an additional reason in that the class of finite second order formulae is not even a passable approximation to the 'full' second order 'language.'

BIBLIOGRAPHY

[1] Barwise, J., Thesis, Stanford University (1967).

[2] Craig, W., Satisfaction for nth order language defined in nth order, JSL 30 (1965), 13-21, reviewed MR 33 (1967), 659-660, #3883.

[3] Feferman, S., Persistent and invariant formulas for outer extensions, Compositio 20 (1968).

[4] Fraissé, R., Une notion de récursivité relative, pp. 323-328 in: Infinitistic Methods, Warsaw, 1961.

[5] Gödel, K., Remarks before the Princeton Bicentennial Conference on problems in mathematics, pp. 84-88 in: The Undecidable, ed. M. Davis, N.Y., 1955.

[6] Grzegorczyk, A., A. Mostowski, and S. Ryll-Nardzewski, The classical and ω-complete arithmetic, JSL 23 (1958), 188-206.

[7] Hanf, W., Incompactness in languages with infinitely long expressions, F.M. 53 (1964), 309-324.

[8] Karp, C. R., Languages with expressions of infinite length, Amsterdam (1964).

[9] ________, Primitive recursive set functions: a formulation with applications to infinitary formal systems, JSL 31 (1966), 294.

[10] Kleene, S. C., Recursive functionals and quantifiers of finite types, Trans. Amer. Math. Soc. 91 (1959), 1-52.

[11] Kreisel, G., Set theoretic problems suggested by the notion of potential totality, Proceedings of the Symposium on Infinitistic Methods in the Foundations of Mathematics, Warsaw, Sept. 2-8, 1959 (1961), 103-140.

[12] ________, Model-theoretic invariants; applications to recursive and hyperarithmetic operations, The theory of Models (1965), 190-205.

[13] ________, Mathematical Logic, in Lectures on modern mathematics, vol. III, ed. Saaty (1965), 95-195.

[14] ________, Informal rigour and completeness proofs, pp. 138-171 of: Problems in the Philosophy of Mathematics, Amsterdam, (1967).

[15] Kreisel, G., and J. L. Krivine, Elements of mathematical logic; theory of models, North Holland Publishing Co., (1967).

[16] Kripke, S., Transfinite recursions on admissible ordinals I, II, JSL, 29 (1964), 161-162.

[17] Kunen, K., Implicit definability and infinitary languages, JSL 33 (1968).

[18] Lacombe, D., Deux généralisations de la notion de récursivité, C. R. Acad. Sc. Paris 258 (1964), 3141-3143 and 3410-3413.

[19] Levy, A., A hierarchy of formulas in set theory, Memoirs of the Amer. Math. Soc. 57 (1965).

[20] Malitz, J. I., Thesis, University of California, Berkeley 1966.

[21] Mostowski, A., On a generalization of quantifiers, F.M. 44 (1957), 12-36.

[22] ________, Representability of sets in formal systems, Recursive function theory, Proc. Symposia in Pure Mathematics, 5 (1962), 29-48.

[23] Scott, D. S., Logic with denumerably long formulas and finite strings of quantifiers, pp. 329-341 of Theory of Models, Amsterdam, 1965.

[24] Vaught, R. L., Sentences true in all constructive models, JSL 24 (1959), 1-15.

[25] ________, The completeness theorem of logic with the added quantifier 'there are uncountably many' F.M. 54 (1964), 303-304.

Postscript (to footnote 9). Quite recently a more impressive application was made by Friedman and Jensen, in their contribution to this volume, to give a new proof of a theorem of Sacks characterizing the ordinals of the form ω_1^x for $x \subseteq \omega$.

DEFINABILITY, AUTOMORPHISMS, AND INFINITARY LANGUAGES

DAVID W. KUEKER

1. Introduction.

Let τ be a (finitary) similarity type. We consider the languages $L_{\kappa\lambda}(\tau)$ (defined elsewhere in this volume) whose only non-logical symbols are those occurring in τ. Models of type τ (and thus models for the languages $L_{\kappa\lambda}(\tau)$) will be denoted by the symbols $\mathfrak{A}$ and $\mathfrak{B}$. We will follow the convention that the universe of $\mathfrak{A}$ is A and that of $\mathfrak{B}$ is B.

If $\underline{P}$ is a k-place predicate symbol not occurring in τ $(k \in \omega)$, then we shall also consider models of similarity type $\tau \cup \{\underline{P}\}$. Such models will be written as $(\mathfrak{A},P)$, where $\mathfrak{A}$ is a model of type τ and P is a k-place relation on A. Also, if $\mathfrak{A}$ is a model of type τ and $a_0,\ldots,a_{n-1} \in A$ then $(\mathfrak{A},a_0,\ldots,a_{n-1})$, or $(\mathfrak{A},a_i)_{i<n}$, is a model of τ together with n additional individual constant symbols.

We shall supress all reference to the similarity type τ. Thus, we assume throughout that we have a fixed type τ, and write simply $L_{\kappa\lambda}$ instead of $L_{\kappa\lambda}(\tau)$. All models $\mathfrak{A}$ and $\mathfrak{B}$ are assumed to be of this type τ. We further assume throughout the following

Convention. $\underline{P}$ is a unary predicate symbol which does not occur in τ.
Thus, no formula of any language $L_{\kappa\lambda}$ contains the symbol $\underline{P}$.

We assume familiarity with the basic concepts of the infinitary languages $L_{\kappa\lambda}$ and of model theory. Unless otherwise noted, we employ standard terminology and notation. In particular, $\cong$

Most of the results in this paper appeared in the author's doctoral dissertation written under the direction of Professor C. C. Chang.

denotes the relation of isomorphism between models; the cardinality of a set X is denoted by $|X|$; by the cardinality of a model $\mathfrak{A}$ we mean $|A|$; and $\equiv_{\kappa\lambda}$ denotes the relation of elementary equivalence with respect to sentences of $L_{\kappa\lambda}(\tau')$ (for the appropriate type τ').

In this paper we present several results concerning infinitary definability, particularly in the language $L_{\omega_1\omega}$. Our results all have the form of an equivalence between certain model-theortic condition on a model and some infinitary (syntactical) definability condition on the model. The proofs are all model-theoretic rather than syntactical. The best known result of this sort is the following Countable Definability Theorem of Scott [12].

THEOREM 1.1. Let $\mathfrak{A}$ be a countable model and let $P \subset A$. Then the following are equivalent:

(i) For any $Q \subset A$, if $(\mathfrak{A},P) \cong (\mathfrak{A},Q)$ then $P = Q$.

(ii) There is some formula $\varphi(x)$ of $L_{\omega_1\omega}$ such that

$$(\mathfrak{A},P) \models \forall x\, [\underline{P}(x) \leftrightarrow \varphi(x)].$$

This theorem can be considered as an $L_{\omega_1\omega}$ analogue for countable models of the well-known definability results of Beth [1] and Svenonuis [13]. As Lopez-Escobar showed in [5], Beth's Theorem holds for theories given by a single sentence of $L_{\omega_1\omega}$. Lopez-Escobar's result, however, appears to be neither weaker nor stronger than Theorem 1.1. On the one hand, there is no way known of obtaining Lopez-Escobar's theorem from Scott's result, or indeed, in any model-theoretic fashion. And on the other hand, Theorem 1.1 does not follow from Lopez-Escobar's result even if we grant the Isomorphism Theorem (see below) because the complete $L_{\omega_1\omega}$ theory of a countable model for an uncountable type τ is not given by a single sentence of $L_{\omega_1\omega}$.

We remark that in Theorem 1.1 we could allow P to be any (finitary) relation or function on A. This is also true for the other results we will present in this paper. We have stated them just for unary relations solely for ease of exposition.

Theorem 1.1 is an easy consequence of the Countable Isomorphism Theorem of Scott [12], which is also the main tool which we shall require. In its simplest form it says that two countable models are isomorphic if and only if they are elementarily equivalent in $L_{\omega_1\omega}$. What we will actually use is the following theorem from which the usual formulation of the Isomorphism Theorem easily follows. (For a detailed account of Isomorphism theorems, see Chang's paper [4] in the present volume.)

THEOREM 1.2. (1) <u>If</u> $\mathfrak{A}$ <u>and</u> $\mathfrak{B}$ <u>are countable models and</u> $\mathfrak{A} \equiv_{\omega_1\omega} \mathfrak{B}$, <u>then for any</u> $a \in A$ <u>there is some</u> $b \in B$ <u>such that</u>

$$(\mathfrak{A},a) \equiv_{\omega_1\omega} (\mathfrak{B},b).$$

(2) <u>If</u> $\mathfrak{A}$ <u>is countable and</u> $a_0,\ldots,a_n \in A$ <u>then there is some formula</u> $\varphi(v_0,\ldots,v_n)$ <u>of</u> $L_{\omega_1\omega}$ <u>such that for any</u> $b_0,\ldots,b_n \in A$

$$\mathfrak{A} \models \varphi(b_0,\ldots,b_n) \text{ iff}$$

$$(\mathfrak{A},a_0,\ldots,a_n) \equiv_{\omega_1\omega} (\mathfrak{A},b_0,\ldots,b_n).$$

In the following section we present our main results, Theorems 2.1 and 2.2, Theorem 2.1 is a generalization of Theorem 1.1, and is an $L_{\omega_1\omega}$ analogue of the definability theorem of Chang [3] and Makkai [7]. Theorem 2.1 was obtained independently by G. E. Reyes and the author.

Theorem 2.2 gives an $L_{\omega_1\omega}$ definability characterization of those countable models with fewer than 2^ω automorphisms. From this we show that if a countable model $\mathfrak{A}$ is elementarily equivalent in $L_{\omega_1\omega}$ to an uncountable model, then $\mathfrak{A}$ has 2^ω automorphisms.

In §3 we consider briefly the existence and properties of uncountable models which are $L_{\omega_1\omega}$ elementarily equivalent to countable models. This topic is suggested by the corollary to Theorem 2.2.

Finally, in §4, we consider analogues of the results in §2 for uncountable models. Our results there are mainly negative, except for models of cardinality λ cofinal with ω.

2. <u>Definability in countable models</u>.

For ease in stating our main results, we introduce the following notation.

(1) If $\mathfrak{A}$ is any model and $P \subset A$, then

$$M(\mathfrak{A},P) = \{P' \subset A : (\mathfrak{A},P) \cong (\mathfrak{A},P')\}.$$

(2) For any model $\mathfrak{A}$, $F(\mathfrak{A})$ is the set of all automorphisms of $\mathfrak{A}$.

In this notation, condition (i) of Theorem 1.1 says that $M(\mathfrak{A},P)$ has exactly one element.

The main results of this paper are the following.

THEOREM 2.1. *Let $\mathfrak{A}$ be a countable model and let $P \subset A$. Then the following are equivalent:*

(i) $|M(\mathfrak{A},P)| \leq \omega$.

(ii) $|M(\mathfrak{A},P)| < 2^{\omega}$.

(iii) *There is a formula $\varphi(x,v_0,\ldots,v_k)$ of $L_{\omega_1\omega}$ such that*

$$(\mathfrak{A},P) \models \exists v_0,\ldots,v_k \; \forall x[\underline{P}(x) \leftrightarrow \varphi(x,v_0,\ldots,v_k)].$$

THEOREM 2.2. *Let $\mathfrak{A}$ be a countable model. Then the following are equivalent:*

(i) $|F(\mathfrak{A})| \leq \omega$.

(ii) $|F(\mathfrak{A})| < 2^{\omega}$.

(iii) *For every $P \subset A$, $|M(\mathfrak{A},P)| < 2^{\omega}$.*

(iv) *There are formulas $\varphi_n(x,v_0,\ldots,v_k)$ of $L_{\omega_1\omega}$, for $n \in \omega$, such that*

$$\mathfrak{A} \models \exists v_0,\ldots,v_k \; \forall x \bigvee_{n\in\omega} [\varphi_n(x,v_0,\ldots,v_k) \wedge \exists! \, z \, \varphi_n(z,v_0,\ldots,v_k)].$$

COROLLARY 2.3. *Let $\mathfrak{A}$ be countable. If $\mathfrak{A}$ is $L_{\omega_1\omega}$ elementarily equivalent to any uncountable model, then $\mathfrak{A}$ has 2^{ω} automorphisms.*

REMARKS. (1) Theorem 2.1 was obtained independently by G. E. Reyes. His proof is very different from the author's, and may be found in [11].

(2) Condition (iv) of Theorem 2.2 says that there are a finite number of elements $a_0,\ldots,a_k \in A$ in terms of which every element of A is definable (in $L_{\omega_1\omega}$). It follows that every subset of A is also definable in terms of $a_0,\ldots,a_k$. By Theorem 2.1, condition (iii) of Theorem 2.2 implies that every subset of A is definable in terms of some finite number of points of A, with different subsets perhaps requiring different individuals from A. Thus condition (iii) is apparently weaker than condition (iv).

(3) The equivalence, in Theorem 2.2, of (i), (ii), and (iii) is a purely model-theoretic fact of some interest in itself, and appears to be new.

Corollary 2.3 is an almost immediate consequence of Theorem 2.2, since if $\mathfrak{A}$ is countable and has fewer than 2^ω automorphisms, then $\mathfrak{A}$ satisfies condition (iv) of Theorem 2.2, and so any model which is $L_{\omega_1\omega}$ elementarily equivalent to $\mathfrak{A}$ also satisfies (iv). But any model satisfying (iv) must be countable, so $\mathfrak{A}$ cannot be $L_{\omega_1\omega}$ elementarily equivalent to any uncountable model.

It is easy to find examples of countable models with 2^ω automorphisms which are not $L_{\omega_1\omega}$ elementarily equivalent to any uncountable model, so the converse to the corollary fails.

The proof of Theorem 2.1 is similar to Chang's proof in [3] of the Chang-Makkai Theorem, by which it was suggested. In addition, we use the same general method in the proof of Theorem 2.2. Because of this, we give here only the proof of Theorem 2.2, which involves several points which do not arise in the other proof.

Proof of Theorem 2.2. The implications from (i) to (ii) and from (ii) to (iii) are immediate. To see that (iv) implies (i) simply note that (iv) says that there are elements $a_0,\ldots,a_k \in A$ in terms of which every element of A is definable by one of the φ_n. Thus every automorphism of $\mathfrak{A}$ is uniquely determined by its action on $a_0,\ldots,a_k$, and so $F(\mathfrak{A})$ is countable.

The remaining direction, from (iii) to (iv), will take more time. We will show that if (iv) fails then (iii) must also fail.

Let S be the set of all finite sequences of 0's and 1's, that is, all functions on some $n \in \omega$ into 2. If $s \in S$ has domain n, then we define

$$s^0 = s \cup \{\langle n,0\rangle\},$$

$$s^1 = s \cup \{\langle n,1\rangle\}.$$

Thus s^0 and s^1 are the immediate successors of s in the natural ordering of inclusion on S.

Let the sequence $\{a_k\}_{k\in\omega}$ enumerate the elements of A.

Assuming that (iv) fails, we construct functions G and H such that the following hold:

(1) domain of G = domain of $H = S$.

(2) for each $s \in S$, $G(s)$ and $H(s)$ are functions on the domain of s into A.

(3) if $s,t \in S$ and $s \subset t$ then $G(s) \subset G(t)$ and $H(s) \subset H(t)$.

(4) if $s \in S$ has fomain n, then

$$(\mathfrak{A},G(s)(i))_{i<n} \equiv_{\omega_1\omega} (\mathfrak{A},H(s)(i))_{i<n}.$$

(5) if $s \in S$ has domain $n = 3k$, then

$$G(s^0) = G(s^1) = G(s) \cup \{\langle n,a_k\rangle\}.$$

(6) if $s \in S$ has domain $n = 3k + 1$, then

$$H(s^0) = H(s^1) = H(s) \cup \{\langle n,a_k\rangle\}.$$

(7) if $s \in S$ has domain $n = 3k + 2$, then

(a) $H(s^0) = H(s^1)$

(b) $G(s^0)(n) \neq G(s^1)(n)$

(c) for any $t \in S$ of domain n,

$$G(s^0)(n) = G(t^0)(n), \text{ and}$$
$$G(s^1)(n) = G(t^1)(n)$$

(d) for any $t \in S$ of domain $\leq n$ and for any $j <$ domain of t,

$$G(s^0)(n) \neq G(t)(j) \text{ and}$$
$$G(s^1)(n) \neq G(t)(j).$$

$G(s)$ and $H(s)$ are defined by induction on the domain of s. So, let $n \in \omega$ and assume that G and H have been defined for all $s \in S$ with domain $\leq n$. We define G and H on those sequences in S with domain $n + 1$ by cases on n.

First, if $n = 3k$ or $n = 3k + 1$ then it is easy, using Theorem 1.2 (1), to define G and H on the sequences with domain $n + 1$. So we assume that $n = 3k + 2$.

Let $b_1,\ldots,b_m$ enumerate all the points of A of the form $G(t)(j)$ for $t \in S$ of domain $\leq n$ and $j <$ domain of t.

We first show

(A) There are points $c, c' \in A$ such that $c \neq c'$ and

$$(\mathfrak{A}, b_1, \ldots, b_m, c) \equiv_{\omega_1\omega} (\mathfrak{A}, b_1, \ldots, b_m, c').$$

Assume that no such points c, c' existed. For each $i \in \omega$ let $\varphi_i(x, v_1, \ldots, v_m)$ be a formula of $L_{\omega_1\omega}$ which determines the complete $L_{\omega_1\omega}$ type of the sequence $a_i, b_1, \ldots, b_m$ in $\mathfrak{A}$. Such formulas φ_i exist by Theorem 1.2 (2). In particular, then, for any $a' \in A$ we have

$$\mathfrak{A} \models \varphi_i(a', b_1, \ldots, b_m) \text{ iff}$$

$$(\mathfrak{A}, a_i, b_1, \ldots, b_m) \equiv_{\omega_1\omega} (\mathfrak{A}, a', b_1, \ldots, b_m);$$

and by our assumption that (A) fails, this happens if and only if $a' = a_i$. Hence

$$\mathfrak{A} \models \forall x \bigvee_{i \in \omega} [\varphi_i(x, b_1, \ldots, b_m) \wedge \exists! z\, \varphi_i(z, b_1, \ldots, b_m)].$$

But this implies that (iv) holds, contradicting our hypothesis that it fails.

Choosing points c, c' such that (A) holds, we define

$$G(s^0) = G(s) \cup \{\langle n, c\rangle\} \text{ and}$$

$$G(s^1) = G(s) \cup \{\langle n, c'\rangle\},$$

for all $s \in S$ with domain n.

Then (7)(b) and (7)(c) hold by definition. Also, $c \neq b_i$ for each $i = 1, \ldots, m$ since if $c = b_i$ then, by (A), we would also have $c' = b_i$ and hence $c = c'$. Similarly $c' \neq b_i$ for each i. Thus (7)(d) holds.

Now, if $s \in S$ has domain n, then $G(s)$ is a sequence $(c_0, \ldots, c_{n-1})$ and $H(s)$ is a sequence $(d_0, \ldots, d_{n-1})$ of points of A such that

$$(\mathfrak{A}, c_0, \ldots, c_{n-1}) \equiv_{\omega_1\omega} (\mathfrak{A}, d_0, \ldots, d_{n-1}).$$

Notice that $c_0,\dots,c_{n-1}$ is a subsequence of $b_1,\dots,b_m$ and hence (A) implies that

$$(\mathfrak{A},c_0,\dots,c_{n-1},c) \equiv_{\omega_1\omega} (\mathfrak{A},c_0,\dots,c_{n-1},c').$$

By Theorem 1.2 (1) choose $d \in A$ such that

$$(\mathfrak{A},d_0,\dots,d_{n-1},d) \equiv_{\omega_1\omega} (\mathfrak{A},c_0,\dots,c_{n-1},c),$$

and define

$$H(s^0) = H(s^1) = H(s) \cup \{\langle n,d\rangle\}.$$

Then for this definition of $H(s^0)$ and $H(s^1)$ we have shown that (7)(a) and (4) hold. Since all the other conditions obviously hold, we have defined G and H on all sequences of domain $n + 1$. This completes our induction.

We now define

$$P = \{G(s^0)(n) : s \in S \text{ of domain } n = 3k + 2\}.$$

It is for this set P for which we will eventually show that $|M(\mathfrak{A},P)| = 2^\omega$. But first we show

(B) For every $s \in S$ of domain $n = 3k + 2$,

$$G(s^1)(n) \notin P.$$

If, on the contrary, $G(s^1)(n) \in P$, then $G(s^1)(n) = G(t^0)(n')$ for some $t \in S$ of domain $n' = 3k'+ 2$. But if $k = k'$ this contradicts (7)(b)(c), and if $k \neq k'$ then this contradicts (7)(d).

We now extend G and H to be defined on any function s on ω into 2 by defining

$$G(s) = \bigcup_{n\in\omega} G(s|n) \quad\text{and}$$

$$H(s) = \bigcup_{n\in\omega} H(s|n).$$

Recall that $s|n$ is the restriction of s to domain n. By (3) $G(s)$ and $H(s)$ are well-defined ω-termed sequences of elements of A, and they each enumerate A by (5) and (6).

Finally, if s is a function on ω into 2, we define f_s on A by

$$f_s(G(s)(n)) = H(s)(n), \quad \text{for all } n \in \omega.$$

Then f_s maps A onto A, and by (4) f_s is an isomorphism. Hence for any such s, f_s is an automorphism of $\mathfrak{A}$.

Let S' be the set of all functions s on ω into 2 such that $s(n) = 0$ if n is not of the form $3k + 2$. Notice that $|S'| = 2^\omega$.

For any $s \in S'$, let $P_s = f_s(P)$. Then $(\mathfrak{A},P) \cong (\mathfrak{A},P_s)$ under the isomorphism f_s, so $P_s \in M(\mathfrak{A},P)$. Thus, we will have shown that $|M(\mathfrak{A},P)| = 2^\omega$, and hence that (iii) fails, as soon as we have shown

(C) If $s,t \in S'$ and $s \neq t$, then $P_s \neq P_t$.

If $s,t \in S'$ and $s \neq t$, then there is some $n = 3k + 2$ such that $s(m) = t(m)$ for $m < n$ but $s(n) \neq t(n)$. Say that $s(n) = 0$, $t(n) = 1$. Then $s|n = t|n$ is some $r \in S$ with domain n, and $s|(n + 1) = r^0$, $t|(n + 1) = r^1$. Therefore $G(s)(n) \in P$ by the definition of P, and $G(t)(n) \notin P$ by (B). Hence

$$H(s)(n) = f_s(G(s)(n)) \in P_s \quad \text{and}$$

$$H(t)(n) = f_t(G(t)(n)) \notin P_t.$$

But $H(s)(n) = H(t)(n)$ by (7)(a), hence we have shown that $P_s \neq P_t$, which completes the proof.

3. Complete $L_{\omega_1\omega}$ sentences.

In this section we depart briefly from definability in order to consider some questions raised by Corollary 2.3.

The first question is whether or not the hypothesis of the corollary is satisfied by a significant class of models; in other words, are there many non-trivial examples of countable models which are

$L_{\omega_1\omega}$ elementarily equivalent to uncountable models? The answer to this is yes; in fact one can show the following.

THEOREM 3.1. Let T be any $L_{\omega\omega}$ theory for a countable type τ which has infinite models. Then T has a countable model $\mathfrak{A}$ which is $L_{\omega_1\omega}$ elementarily equivalent to models of arbitrarily large cardinality.

Theorem 3.1 was obtained independently by Chang, Makkai, and the author. A proof of some more general results, using the method of indiscernables, appears in Chang [2].

The second question is less precise. It asks what properties are possessed by uncountable models which are $L_{\omega_1\omega}$ elementarily equivalent to countable models. To make the discussion clearer we introduce the following definition.

DEFINITION. A sentence σ of $L_{\omega_1\omega}$ is complete for $L_{\omega_1\omega}$ (or simply complete) if σ has a model, and any two models of σ are $L_{\omega_1\omega}$ elementarily equivalent.

By the Isomorphism theorem and the downward Löwenheim-Skolem Theorem, a complete $L_{\omega_1\omega}$ sentence is simply the complete $L_{\omega_1\omega}$ theory of a countable model (for a countable type τ). So Corollary 2.3 implies that if σ is a complete $L_{\omega_1\omega}$ sentence with uncountable models, then the countable models of σ have 2^{ω} automorphisms.

Our second question, then, is what can one say about the uncountable models of a complete $L_{\omega_1\omega}$ sentence, particularly with regard to the existence of many automorphisms. The main positive result here is the following theorem, due to Malitz (unpublished).

THEOREM 3.2. Let σ be a sentence of $L_{\omega_1\omega}$ which has models of arbitrarily large cardinality. Then in every infinite power α, σ has a model with 2^{α} automorphisms.

Theorem 3.2 may be derived from the following two facts:

(1) As Lopez-Escobar [6] has shown, the class of models of a sentence of $L_{\omega_1\omega}$ is precisely the class of reducts of ω-models of a theory in ω-logic with at most a countable number of new predicates.

(2) Morley's proof in [9] that the Hanf number of ω-logic is $\beth_{\omega_1}$ actually shows that the Ehrenfeucht-Mostowski results on indiscernables hold for theories of ω-logic with ω-models of

arbitrarily large cardinality; in particular, such theories have ω-models with 2^{α} automorphisms in each infinite power α.

As Malitz shows in his paper [8] in this volume, there are complete $L_{\omega_1\omega}$ sentences which have uncountable models but not models of arbitrarily large cardinality. Thus, Theorem 3.2 does not apply to every complete $L_{\omega_1\omega}$ sentence with uncountable models (in particular, it does not imply Corollary 2.3, even for models $\mathfrak{A}$ for a countable type τ). In fact, using Malitz's examples one can show the following (assuming the G.C.H.):

(A) For every ordinal $\xi < \omega_1$ there is a complete $L_{\omega_1\omega}$ sentence σ which has models in all infinite powers $\leq \beth_\xi$, but such that none of its models has more than 2^{ω} automorphisms.

Let σ be the complete sentence that Malitz defines which has models in all infinite powers $\leq \beth_\xi$, but in no larger powers. Note that if $\mathfrak{A}$ is a model of σ, then any automorphism of $\mathfrak{A}$ is uniquely determined by what it does to the tree part of $\mathfrak{A}$. Since the tree is countable, $\mathfrak{A}$ can then have at most 2^{ω} automorphisms.

Finally, we have the following example which indicates that the conclusion of Theorem 3.2 cannot be strengthened even if we assume that the sentence σ is complete.

(B) There is a complete $L_{\omega_1\omega}$ sentence σ which has models in arbitrarily large powers which admit only the trivial automorphism.

Let σ be the sentence of $L_{\omega_1\omega}$ which characterizes dense linear orderings without end points (in fact, σ is a sentence of $L_{\omega\omega}$). Then σ is complete for $L_{\omega_1\omega}$ since any two countable models of σ are isomorphic. But it is well-known (cf. [10]) that there are dense linear orders of arbitrarily large powers which admit only the trivial automorphism.

4. Generalizations to higher powers.

In this section we consider the question of generalizing Theorem 1.1 and the results of §2 to uncountable powers. Any such generalization must depend on first finding a suitable generalization of the Isomorphism Theorem, Theorem 1.2. The result one would want would be the following:

(A) If $\mathfrak{A}$ and $\mathfrak{B}$ have power λ and $\mathfrak{A} \equiv_{\infty\lambda} \mathfrak{B}$ then $\mathfrak{A} \cong \mathfrak{B}$.
($L_{\infty\lambda}$ is the union of the languages $L_{\kappa\lambda}$ over all κ).

As Chang shows in his paper [4] in this volume, (A) is true if λ is cofinal with ω. But Morley has found an example (unpublished) which shows that (A) is false if λ is regular and uncountable.

One can also use Morley's example to show that the natural generalizations of Theorems 1.1 and 2.1 fail for uncountable regular cardinals. In fact, if λ is regular and uncountable, then there is a model $\mathfrak{A}$ of power λ and a set $P \subset A$ such that $M(\mathfrak{A},P)$ has exactly one element, but P is not definable by any formula of $L_{\infty\lambda}$ in terms of fewer than λ elements of A.

In the remainder of this section we give the positive results which may be obtained when λ is cofinal with ω. First define $\lambda^* = [\lambda^{\Delta}]^+$, where λ^{Δ} is the sum of all λ^{μ} for $\mu < \lambda$. Then Chang actually shows that (A) is true, for λ cofinal with ω, assuming only elementary equivalence in $L_{\lambda^*\lambda}$.

The first result is a straightforward generalization of Theorem 1.1.

THEOREM 4.1. Let $\mathfrak{A}$ be a model of cardinality λ, where λ is cofinal with ω. Let $P \subset A$. Then the following are equivalent:

(i) $M(\mathfrak{A},P)$ has exactly one element.

(ii) There is some formula $\varphi(x)$ of $L_{\lambda^*\lambda}$ such that

$$(\mathfrak{A},P) \models \forall x\,[\underline{P}(x) \longleftrightarrow \varphi(x)].$$

Proof. It is sufficient to show that if (ii) fails then (i) fails. If (ii) fails then one can show (using the fact that the complete $L_{\lambda^*\lambda}$ type of any $a \in A$ is determined by a single formula of $L_{\lambda^*\lambda}$) that there are elements $a,b \in A$ such that

$$a \in P,\quad b \notin P,\quad \text{but}\quad (\mathfrak{A},a) \equiv_{\lambda^*\lambda} (\mathfrak{B},b).$$

Then $(\mathfrak{A},a) \cong (\mathfrak{A},b)$ under some isomorphism f. Let $P' = f(P)$. Then $P' \in M(\mathfrak{A},P)$ and $b \in P'$ (since $a \in P$ and $f(a) = b$). Therefore $P \neq P'$, and so (i) fails.

The situation as regards generalizations of Theorems 2.1 and 2.2 is not as pleasant. We can establish the following "one-directional" generalization of Theorem 2.1.

THEOREM 4.2. Let $\mathfrak{A}$ be a model of cardinality λ, where λ is cofinal with ω. Let $P \subset A$. Assume (i) $|M(\mathfrak{A},P)| < \lambda^{\omega}$.

Then (ii) There is a formula $\varphi(x,v)$ of $L_{\lambda^*\lambda}$, where v is a sequence of fewer than λ variables, such that

$$(\mathfrak{A},P) \models \exists v \, \forall x \, [\underline{P}(x) \leftrightarrow \varphi(x,v)].$$

The "easy" direction, from (ii) to (i), no longer holds for $\lambda > \omega$. In fact it is easy to find examples of $\mathfrak{A}$ and P such that (ii) holds for a countable sequence v of variables, but $|M(\mathfrak{A},P)| = \lambda^\omega$. And we certainly cannot allow $|M(\mathfrak{A},P)| = \lambda^\omega$ in (i), since it is consistent to have to have $\lambda^\omega = 2^\lambda$, but not every subset of A need be definable.

On the other hand, one cannot get an equivalence by restricting v in (ii) to be a finite sequence of variables, since one can find $\mathfrak{A}$ and P such that $|M(\mathfrak{A},P)| < \lambda$ but (ii) does not hold for a finite sequence v. These remarks seem to indicate that the definability of a subset P in terms of individual parameters is no longer solely dependent on the cardinality of $M(\mathfrak{A},P)$.

If we add to (ii) the condition that

$$|\{P' \subset A : (\mathfrak{A},P') \models \exists v \, \forall x \, [\underline{P}(x) \leftrightarrow \varphi(x,v)]\}| < \lambda^\omega,$$

then we do obtain an equivalence, in a trivial fashion. But this added condition seems rather artificial.

A similar partial generalization of Theorem 2.2 can also be obtained. From it we derive the following generalization of Corollary 2.3.

THEOREM 4.3. Let $\mathfrak{A}$ have cardinality λ, where λ is cofinal with ω. If $\mathfrak{A}$ is $L_{\lambda^*\lambda}$ elementary equivalent to a model of cardinality $> \lambda$, then $\mathfrak{A}$ has (at least) λ^ω automorphisms.

We omit the proofs of these last results, since they are similar in outline to the proofs of the corresponding results of §2. We indicate here the main differences between them. Let $\lambda = \Sigma \lambda_n$ $(n \in \omega)$ where $\lambda_n < \lambda_{n+1}$ for all $n \in \omega$. Then instead of using the tree S in which every branch has length ω and which branches twice at every node (as in the proof of Theorem 2.2), we use the tree in which every branch has length ω and which branches λ_n times at every node on the nth level of the tree. An additional variation is needed to ensure that the resulting sequences each enumerate A. And finally, at the end of the proof, we use the fact that $\lambda^\omega = \prod \lambda_n$ $(n \in \omega)$.

REFERENCES

[1] Beth, E. W., On Padoa's method in the theory of definition, Indag. Math., 15 (1953), pp. 330-339.

[2] Chang, C. C., Infinitary properties of models generated from indiscernables, to appear in Proc. of Int'l. Cong. for Logic, Method., and Phil. of Science, Amsterdam, 1967.

[3] ___________, Some new results in definability, Bull. A.M.S., 70 (1964), pp. 808-813.

[4] ___________, Some remarks on the model theory of infinitary languages, in this volume.

[5] Lopez-Escobar, E. G. K., An interpolation theorem for denumerably long sentences, Fund. Math., 57 (1965), pp. 253-272.

[6] ___________, On defining well-orderings, Fund. Math., 59 (1966), pp. 13-21.

[7] Makkai, M., A generalization of a theorem of Beth, Acta Math. Acad. Sci. Hungar., 15 (1964), pp. 227-235.

[8] Malitz, J. I., The Hanf number of complete sentences of $L_{\omega_1\omega}$, in this volume.

[9] Morley, M., Omitting classes of elements, The theory of models, Amsterdam, 1965, pp. 265-273.

[10] Ohkuma, T., Sur quelques ensembles ordonnés linéairement, Fund. Math., 43 (1965), pp. 326-337.

[11] Reyes, G. E., Typical and generic relations in a Baire space for models, Doctoral dissertation, University of California, Berkeley, 1967.

[12] Scott, D. Logic with denumerably long formulas and finite strings of quantifiers, The theory of models, Amsterdam, 1965, pp. 329-341.

[13] Svenonius, L., A theorem on permutations in models, Theoria (Lund), 25 (1959), pp. 173-178.

UNIVERSITY OF CALIFORNIA, LOS ANGELES

THE HANF NUMBER FOR COMPLETE $L_{\omega_1,\omega}$ SENTENCES

JEROME MALITZ

In [2], Lopez-Escobar proves that any sentence σ in $L_{\omega_1,\omega}$ having models in all infinite powers less than $\beth_{\omega_1}$ has models in all infinite powers. Furthermore, $\beth_{\omega_1}$ is the least cardinal with this property, i.e., $\beth_{\omega_1}$ is the Hanf number of $L_{\omega_1,\omega}$. Our main result, proved using the generalized continuum hypothpses in §2, is that the set of complete $L_{\omega_1,\omega}$ sentences has the same Hanf number. (A sentence is $L_{\omega_1,\omega}$-complete if any two of its models have the same true $L_{\omega_1,\omega}$ sentences.) In §3 we construct non-isomorphic structures $\mathfrak{A}$ and $\mathfrak{B}$, both of power $\beth_1$, and both models of the same complete $L_{\omega_1,\omega}$ sentence σ, such that no model of σ properly extends either $\mathfrak{A}$ or $\mathfrak{B}$. We also construct countable structures, $\mathfrak{A}$ and $\mathfrak{B}$, whose intersection is an elementary substructure in the sense of $L_{\omega_1,\omega}$ of $\mathfrak{A}$ and of $\mathfrak{B}$, yet, for no extension $\mathfrak{C}$ of $\mathfrak{A}$ is there an isomorphism on $\mathfrak{B}$ into $\mathfrak{C}$ which is the identity on $\mathfrak{A} \cap \mathfrak{B}$. At the end of §3 we state an open problem relating complete $L_{\omega_1,\omega}$-sentences to finitary theories omitting complete types.

1. Preliminaries.

We use $\alpha, \beta, \gamma, \delta, \xi$ to denote ordinals, and κ, λ, μ to denote infinite cardinals. ω is the first infinite cardinal and ω_α the αth infinite cardinal. cA is the cardinality of A.

We assume that the ordinals are so defined that the order relation and the membership relation coincide, and that cardinals are initial ordinals.

If A and B are disjoint sets ($\{A_i : i \in I\}$ a family of disjoint sets) then we write $A + B$ ($\Sigma\{A_i : i \in I\}$) instead of $A \cup B$ ($\bigcup\{A_i : i \in I\}$).

The domain and range of a function f will be denoted by $\mathrm{Dom}\, f$ and $\mathrm{Rng}\, f$ respectively. $f \upharpoonright A$ is the restriction of the function f to A. $\Pi\{A_i : i \in I\}$ is the set of all functions f such that $\mathrm{Dom}\, f = I$ and $f(i) \in A_i$ for all $i \in I$. We write $A_1 \times A_2$ for $\Pi\{A_i : i = 1$ or $i = 2\}$, and ${}^A B$ for $\Pi\{B_a : a \in A\}$ when $B_a = B$ for all $a \in A$. We write ν^λ for $c^\lambda \nu$ when ν is a cardinal. The beth numbers are defined inductively as follows: $\beth_0 = \omega$, $\beth_\alpha = \bigcup\{2^\beta : \beta \in \alpha\}$. The generalized continuum hypothesis (G.C.H.) asserts that $\beth_\alpha = \omega_\alpha$ for all α.

An n-placed relation R will be identified with $\{\langle x_1,\ldots,x_n\rangle : Rx_1,\ldots,x_n\}$.

The language $L_{\omega_\kappa,\omega}$ is the least class Γ containing the atomic formulas (of finite length) that satisfies the following conditions:

(1) $\varphi \in \Gamma$ implies $\neg\varphi \in \Gamma$

(2) $\Gamma' \subseteq \Gamma$ and $c\Gamma' \in \kappa$ implies $\bigwedge \Gamma' \in \Gamma$

(3) $\varphi \in \Gamma$ and v a variable implies $\exists v\, \varphi \in \Gamma$.

For finite conjunctions $\bigwedge\{\varphi_1,\ldots,\varphi_n\}$ we write $\varphi_1 \wedge \cdots \wedge \varphi_n$. The symbols $\bigvee$, $\vee$, $\exists$, $\rightarrow$ $\leftrightarrow$, are introduced in the usual way. Occurrence, free occurrence, substitution, etc. are defined in analogy with the finitary calculus ($L_{\omega,\omega}$). A sencence is a formula having no free variables. The type $\tau\varphi$ of a formula φ is the set of non-logical constants occurring in φ. The type of a set of formulas Γ,$\tau\Gamma$, is $\bigcup\{\tau\varphi : \varphi \in \Gamma\}$.

If φ is a formula and A a unary relation symbol then φ^A, the relativization of φ to A, is defined as follows:

(1) $\varphi^A = \varphi$ if φ is atomic

(2) $\bigwedge\{\psi^A : \psi \in \Gamma\}$ if $\varphi = \bigwedge\Gamma$

(3) $\exists v\,(Av \wedge \psi^A)$ if $\varphi = \exists v\psi$.

A structure $\mathfrak{A}$ of type t is a set A, denoted by $|\mathfrak{A}|$, and a function on the type t such that an n-placed relation symbol R in t is mapped onto a n-placed relation $R^{\mathfrak{A}}$ on $\mathfrak{A}$. If $t = \{R_\xi : \xi \in \alpha\}$ we write $\mathfrak{A} = \langle A, R_\xi^{\mathfrak{A}}\rangle_{\xi\in\alpha}$, or, for $\alpha = n$, $\mathfrak{A} = \langle A, R_1^{\mathfrak{A}},\ldots,R_n^{\mathfrak{A}}\rangle$. If S_ξ, $\xi \in \alpha$, is a sequence of relations on A, then $\langle A, S_\xi\rangle_{\xi\in\alpha}$ is a structure for some suitable type t and some mapping on t onto $\{S_\xi : \xi \in \alpha\}$, but unless an argument requires otherwise, mention of the type and mapping will be omitted. If $\beta \in \alpha$ and $t' = \{R_\xi : \xi \in \beta\}$, then $\mathfrak{B} = \langle A, R_\xi^{\mathfrak{A}} \rangle_{\xi \in \beta}$ is the t' reduct of $\mathfrak{A}$, written $\mathfrak{A} \upharpoonright t'$, and $\mathfrak{A}$ is an expansion of $\mathfrak{B}$, $\mathfrak{A} = (\mathfrak{B}, R_\xi^{\mathfrak{A}})_{\beta \leq \xi \in \alpha}$. Where confusion is unlikely, we will use the same symbol for a relation symbol and its denotation in $\mathfrak{A}$, i.e., we will write R for $R^{\mathfrak{A}}$, or, we may use a letter different from R for the denotation of R in $\mathfrak{A}$.

Let $\mathfrak{A}_\gamma = \langle A_\gamma, R_\xi^{\mathfrak{A}_\gamma} \rangle_{\xi \in \alpha}$ for $\gamma \in \delta$ and suppose $A_{\gamma_1} \cap A_{\gamma_2} = 0$ for $\gamma_1 \neq \gamma_2$. By $\sum_{\gamma \in \delta} \mathfrak{A}_\gamma$, the strong cardinal sum of the $\mathfrak{A}_\gamma$, $\gamma \in \delta$, we mean

$$\left(\left\langle \sum_{\gamma \in \delta} A_\gamma, \sum_{\gamma \in \delta} R_\xi^{\mathfrak{A}_\gamma} \right\rangle_{\xi \in \alpha}, A_\gamma \right)_{\gamma \in \delta} .$$

By a tree, we mean a structure $\langle T, < \rangle$ where for all $x, y, z \in T$

(1) $x \not< x$,

(2) $x < y$ and $y < z$ implies $x < z$,

(3) $x < z$ and $y < z$ implies $x < y$ or $y < x$ or $x = y$.

$B \subseteq T$ is a branch of T if for all x and y in B, $x < y$ or $y < x$ or $x = y$. A branch is maximal if it is not properly contained in any branch of T. In all the trees we consider, maximal branches have the same order type as $\langle \omega, \epsilon \rangle$. The height of $x \in T$ is n if x has precisely n predecessors.

A structure $\langle A, < \rangle$ is linearly ordered if (1) and (2) above are satisfied and if for all $x, y \in A$ either $x = y$ or $x < y$ or $y < x$. If $<$ linearly orders A and if B and C are subsets of A then B is said to be dense in C if whenever $x, y \in C$ and $x < y$ then there is some $z \in B$ such that $x < z < y$. The linearly ordered structure $\langle A, < \rangle$ is densely ordered if A is dense in A.

If z is a function from the variables free in φ into $|\mathfrak{A}|$, then, as usual, we write $\mathfrak{A} \models \varphi[z]$ if z satisfies φ in $\mathfrak{A}$. For example $\mathfrak{A} \models \bigwedge\Gamma[z]$ iff $\mathfrak{A} \models \gamma[z]$ for all $\gamma \in \Gamma$.

We say that $\mathfrak{A}$ is an $L_{\kappa,\omega}$-substructure of $\mathfrak{B}$ (written $\mathfrak{A} \prec_{\kappa,\omega} \mathfrak{B}$) if $\mathfrak{A}$ is a substructure of $\mathfrak{B}$ and for all formulas φ of $L_{\kappa,\omega}$ with finitely many free variables, and all assignments z to $\mathfrak{B}$, $\mathfrak{A} \models \varphi[z]$ iff $\mathfrak{B} \models \varphi[z]$.

A sentence σ is complete for a set of sentences Γ if whenever $\rho \in \Gamma$ and $\tau\rho \subseteq \tau\sigma$, then either $\sigma \to \rho$ or $\sigma \to \neg \rho$ is valid, but not both. Let $L^c_{\omega_1,\omega}$ be the set of $L_{\omega_1,\omega}$ sentences that are complete for $L_{\omega_1,\omega}$. Much of the interest in $L^c_{\omega_1,\omega}$ is due to the following fundamental result of Scott [5].

THEOREM. <u>For every countable structure</u> $\mathfrak{A}$, <u>there is a sentence</u> $\sigma \in L_{\omega_1,\omega}$ <u>such that a countable structure</u> $\mathfrak{B}$ <u>having the same type as</u> $\mathfrak{A}$ <u>is isomorphic to</u> $\mathfrak{A}$ <u>iff</u> $\mathfrak{B} \models \sigma$.

Chang and Kueker have independently shown that a sentence which is $L_{\omega_1,\omega}$-complete is $L_{\kappa,\omega}$-complete for all κ. Hence the sentence σ in Scott's theorem is $L_{\kappa,\omega}$-complete for every κ.

We will make frequent use of the downward Löwenheim-Skolem theorem for sentences of $L_{\omega_1,\omega}$.

THEOREM. <u>A sentence</u> $\sigma \in L_{\omega_1,\omega}$ <u>that has a model, has a countable model</u>. (Cf. 1 and 5).

These two theorems immediately give

COROLLARY. <u>A sentence</u> $\sigma \in L_{\omega_1,\omega}$ <u>is complete iff any two countable models of</u> σ <u>are isomorphic</u>.

How tightly does an $L^c_{\omega_1,\omega}$ sentence tie down the cardinality of its models? Say that κ is the Hanf number of a set Σ of sentences if κ is the least cardinal such that any $\sigma \in \Sigma$ that has for each $\mu \in \kappa$ a model of power $> \mu$, has models of arbitrarily high power.

THEOREM. (Lopez-Escobar [2]). <u>The Hanf number of</u> $L_{\omega_1,\omega}$ <u>is</u> $\beth_{\omega_1}$.

Our main result is

THEOREM. 1. <u>Assuming the</u> G.C.H., <u>the Hanf number of</u> $L^c_{\omega_1,\omega}$ <u>is</u> $\beth_{\omega_1}$.

In view of Lopez-Escobar's theorem, it suffices to show that for each $\alpha \in \omega_1$, there is a sentence $\sigma \in L^c_{\omega_1,\omega}$ that has models in every infinite power $\leq \beth_\alpha$, but in no higher power. This is done in §2.

2. <u>Proof of Theorem 1</u>.

We say that a one-placed relation H on $\mathfrak{A}$ is completely homogenous if every permutation of the members of H can be extended to an automorphism on $\mathfrak{A}$. We obtain Theorem 1 as an immediate consequence of Lopez-Escobar's theorem and the following.

THEOREM. 2. <u>Assume the</u> G.C.H. <u>For each successor ordinal</u> $\alpha \in \omega_1$, <u>there is a</u> $\sigma_\alpha \in L^c_{\omega_1,\omega}$ <u>such that</u>

(i) σ_α <u>has no models of power</u> $> \beth_\alpha$.

(ii) <u>If $\mathfrak{A}$ is a countable model of σ_α then $A_\alpha^{\mathfrak{A}}$ is completely homogeneous of power</u> ω.

(iii) <u>There is a model $\mathfrak{B}$ of σ_α with</u> $cA_\alpha^{\mathfrak{B}} = \beth_\alpha$.

<u>Hence, for each</u> $\alpha \in \omega_1$, <u>there is a sentence</u> σ_α <u>having models of power</u> $\beth_\alpha$, <u>but no models of higher power</u>.

It is easy to see that the second clause follows from the first: For σ_0 we can take a sentence whose interpretation is "$<$ is a linear ordering with no last element, such that every element has finitely may predecessors." (For example, the last clause is the interpretation of $\forall x \bigvee_{n\in\omega} \psi^{n+1}$ where $\psi^{n+1} = \neg \exists y_0 y_1 \cdots y_n \left(\bigwedge_{i\neq j \in n+1} y_i \neq y_j \wedge \bigwedge_{i\in n+1} y_i < x \right)$). For α a limit ordinal > 0, let $\alpha = \bigcup_{i\in\omega} \alpha_i + 1$, and let $\sigma_\alpha = \bigwedge_{i\in\omega} \sigma_{\gamma_i+1}^{A_i} \wedge \forall x \bigvee_{i\in\omega} A_i x \wedge \neg\exists x \bigwedge_{i\neq j\in\omega} (A_i x \wedge A_j x)$, where the σ_{γ_i+1} satisfy the conclusion of Theorem 2. $\sigma_\alpha \in L^c_{\omega_1,\omega}$, since if $\sum_{i\in\omega} \mathfrak{A}_{\gamma_i+1}$ and $\sum_{i<\omega} \mathfrak{B}_{\gamma_i+1}$ are countable models of σ_α with $\mathfrak{A}_{\gamma_i+1}$ and $\mathfrak{B}_{\gamma_i+1}$ models of σ_{γ_i+1}, then $\mathfrak{A}_{\gamma_i+1}$ and $\mathfrak{B}_{\gamma_i+1}$ are isomorphic for all $i \in \omega$. Combining these isomorphism gives an isomorphism on $\sum \mathfrak{A}_{\gamma_i+1}$ onto $\sum_{i<\omega} \mathfrak{B}_{\gamma_i+1}$. The completeness of σ_α now follows from the downward Löwenbeim-Skolem Theorem.

The proof of Theorem 2 is by induction on α.

Basis $\alpha = 1$: Let $\mathcal{K}$ be the class of structures of the form

$$\mathfrak{A} = \langle T \cup A \cup A_1, <, R, E \rangle$$

where

(1) T, A, A_1 are pairwise disjoint.

(2) $\langle T,< \rangle$ is a full binary tree, i.e., there are two minimal elements and every element has exactly two immediate successors.

(3) $R \subseteq T \times A$, and for each $a \in A$ the set $R_a = \{t \in T : tRa\}$ is a maximal branch of $\langle T,< \rangle$.

(4) $E \subseteq A \times A_1$, and the sets $E_d = \{R_a : aEd\}$ partition A such that in each member of the partition there are infinitely many branches passing through each point of T.

For each $i = 1,2,3,4$ it is easy to find a $\rho_i \in L_{\omega_1,\omega}$ such that $\mathfrak{A} \models \rho_i$ iff $\mathfrak{A}$ satisfies condition (i). For example, for ρ_2 we can take the relativization to T of the conjunction of the universal closures of the following formulas:

$$x < y \rightarrow \neg (y < x)$$

$$(x < y \wedge y < z) \rightarrow x < z$$

$$\exists y,z \forall w ((x < w \leftrightarrow y \leq w \vee z \leq w) \wedge (y \neq z))$$

$$\exists y_0, y_1 \forall z ((y_0 < z \vee y_1 < z) \wedge (y_0 \neq y_1))$$

$$\bigvee_{n\in\omega} \exists y_0,\ldots,y_n \forall z (z \leq x \leftrightarrow z \approx y_1 \vee \cdots \vee z \approx y_n).$$

Let σ_1 be a conjunction of such ρ_i's $i = 1,2,3,4$. Then $\sigma_1 \in L_{\omega_1,\omega}$. Clearly σ_1 has no models of power $> \beth_1$, since if $\mathfrak{A} \in K$, the number of elements in $\mathfrak{A}$ is $c(T \cup A \cup A_1) \leq \omega + 2^\omega + 2^\omega = 2^\omega = \beth_1$.

We next describe a model $\mathfrak{A}$ of σ_1 such that any countable model of σ_1 is isomorphic to $\mathfrak{A}$. This and the Löwenheim-Skolem theorem enables us to conclude that $\sigma_1 \in L^c_{\omega_1,\omega}$.

Let $T = \bigcup_{n\in\omega} {}^n\{0,1\}$

$t_1 < t_2$ iff $t_1, t_2 \in T$ and t_2 extends t_1.

$A = \sum_{0\in i\in\omega} A^i$ where

$A^i = \{f \in {}^\omega\{0,1\}$: for some n and all $n \geq m$, $f(n) = 0$ iff $i/n\}$.

tRg iff $t \in T$, $g \in A$ and g extends t

$A_1 = \omega - \{0\}$.

gEn iff $g \in A^n$, $n \in A_1$.

Clearly, $\mathfrak{A} = \langle T \cup A \cup A_1, <, R, E\rangle$ is a model of σ_1.

Let $\mathfrak{B}$ be a countable model of σ_1, say

$$\mathfrak{B} = \langle U \cup B \cup B_1, <, S, F\rangle.$$

Notice that $\langle T,<\rangle$ is the only structure, up to isomorphism, satisfying condition 2. Hence we may suppose that $\langle U,<\rangle = \langle T,<\rangle$.

Map B into ${}^{\omega}\{0,1\}$ as follows:

For each $b \in B$ and $n \in \omega$ let $f_b(n) = 0$ iff for some $t \in T$, tSb and $t(n) = 0$. Then let $B' = \{f_b : b \in B\}$, and define $tS'f_b$ iff tSb.

Let $\{b_1, b_2, \ldots\}$ be an enumeration of B_1. We define $B^i = \{f_b : bFb_i\}$ and $f_b F' B^i$ iff bFb_i. Let $B_1' = \{B^i : i = 1,2,\ldots\}$. Clearly, the structure $\mathfrak{B}' = \langle T \cup B' \cup B_1', <, S', F'\rangle$ is isomorphic to $\mathfrak{B}$. To show that $\mathfrak{B}'$ is isomorphic to $\mathfrak{A}$ we need a lemma, which we state with enough generality to be of use later on.

LEMMA. 3. Let $N = \omega$ or a finite segment $\{0,1,\ldots,r\}$ of ω. Let $A = \sum_{0 \in i \in \omega} A^i$ where $A^i = \{f : f \in {}^{\omega}N$ and for some n and all $m \geq n$, $f(m) = 0$ iff i/m and $f(m) = 1$ if not$\}$. $B = \sum_{0 \in i \in \omega} B^i$ where $B \subseteq {}^{\omega}N$ and $cB = \omega$. Suppose that for all $i \in \omega$, $n \in \omega$, and $t \in {}^{n}\omega$ there are infinitely many $g \in B^i$ such that $g \mid n = t$. Let $\alpha(f_1,f_2)$ be the least n such that $(f_1(n) \neq f_2(n))$. Then there is a function $H \in {}^{B}A$ such that

(1) $\alpha(g_1,g_2) = \alpha(Hg_1,Hg_2)$ for $g_1,g_2 \in B$.

(2) H maps B^i onto A^i.

Proof of Lemma. Since similar proofs can be given for N finite and $N = \omega$, we consider only the case where $N = \omega$.

$$\text{Let } A^i = \{f_1^i, f_2^i, \ldots\}$$

$$B^i = \{g_1^i, g_2^i, \ldots\}.$$

Enumerate $\omega \times \omega$, denoting the nth pair by (n_1,n_2). We define a nest of partial functions $H_1 \subseteq H_2 \subseteq \cdots$ as follows.

Let $\mathrm{Dom}\, H_1 = \{g_1^1\}$ and let $Hg_1^1 = f_1^1$. At step $n+1$ we extend H_n to H_{n+1} by adding at most two functions in $B^{(n+1)_1}$ to $\mathrm{Dom}\, H_n$. Let $z = (n+1)_1$. Let g be the first function in B^z not in $\mathrm{Dom}\, H_n$. Let $k = \max\{\alpha(g,h) : h \in \mathrm{Dom}\, H_n\}$ and let h^* be such that $\alpha(g,h^*) = k$ and $h^* \in \mathrm{Dom}\, H_n$.

Define $H_{n+1}g(j) = H_n h^*(1)$ for $j < k$

$H_{n+1}g(k) =$ the least $r \notin \{(H_n h)(k) : h \in \text{Dom } H_n$ and $\alpha(g,h) = k\}$

$H_{n+1}g(j) = 0$ if $j > k$ and $z \mid j$

$H_{n+1}g(j) = 1$ if $j > k$ and $z \nmid j$.

The definition of $H_{n+1}g(k)$ depends on the fact that for each $i, n < \omega$, and $t \in {}^n\omega$ there are infinitely many $f \in A^i$ such that f extends t. Clearly, if H_n satisfies condition (1) so does H_{n+1}. Next let f be the least function in A^z not in $\text{Rng } H_n \cup \{H_{n+1}g\}$.
Let $f^{\#} \in \text{Rng } H_n \cup \{H_{n+1}g\}$, with $\alpha(f,f^{\#}) = k$ maximal. Let $g^{\#}$ be the inverse image of $f^{\#}$ under H_{n+1}. Let $\hat{g}$ be the first element of $B^z - (\text{Dom } H_n \cup \{g\})$ such that $\hat{g} \upharpoonright k = g^{\#} \upharpoonright k$, $\hat{g}(k) \neq g^{\#}(k)$ (there are infinitely many such $\hat{g}$'s by the hypothesis of the lemma). The definition of H_{n+1} is completed by letting $H_{n+1}\hat{g} = f$. As before, H_{n+1} satisfies (1). Finally take $H = \bigcup_{n \in \omega} H_n$. Clearly, H satisfies (1) and (2).

Returning to the proof of the theorem, we take H as in the lemma and extend H to an isomorphism on $\mathfrak{B}'$ onto $\mathfrak{A}$, as follows: Let $t' \in t$ and let f' be an extension of t' in B'. Define $Ht' = Hf' \upharpoonright \text{Dom } t'$. The lemma assures us that this extension of H is well defined and an isomorphism. Hence every countable model of σ_1 is isomorphic to $\mathfrak{A}$ and so $\sigma_1 \in L^c_{\omega_1,\omega}$.

Notice that the above argument also shows that A_1 is completely homogeneous.

Next we construct another model of σ_1, again calling it $\mathfrak{B} = \langle T \cup B \cup B_1, <, S, F\rangle$, such that $cA_1 = \beth_{\omega_1}$.

Let $\langle T,<\rangle$ be as above and let

$B = {}^{\omega}\{0,1\}$,

tSf iff $t \in T$ and $f \in B$ and f extends t,

$B_1 = \{\hat{f} : f \in B\}$, where $\hat{f} = \{g :$ for some m and all $n \geq m$, $g(n) = f(m)\}$,

$gF\hat{f}$ iff $g \in B$ and $\hat{f} \in B_1$ and $g \in \hat{f}$.

It is easy to see that $\mathfrak{B} \models \sigma_1$. Also, $cB_1 = \beth_1$ since B_1 is a partition of B and each number of B_1 has cardinality ω.

To conclude that σ_1 is as required by Theorem 2, we need only notice that σ_1 has no models of power $> \beth_1$, since if $\langle T \cup B \cup B_1, <, S, F\rangle \models \sigma_1$, then $cT = \omega$, $cB \leq c({}^{\omega}T) = \beth_1$, and

$cB_1 \leq cB = \beth_1$.

For the induction step we consider two cases: α the successor of a limit ordinal, and α the successor of a successor ordinal.

Case 1. $\alpha = \beta + 1$, β a limit ordinal: Since $\beta \in \omega_1$, $\beta = \bigcup_{i\in\omega} \beta_i$ for some sequence of successor ordinals β_i, each less than β. Our induction hypothesis is that for each successor ordinal $\gamma \in \beta$, there is a σ_γ satisfying the requirements of Theorem 2. Let K be the class of all structures of the form $\mathfrak{B} = \left(\left(\sum_{i\in\omega} \mathfrak{B}_i\right) + \langle B\rangle + \langle B_\alpha\rangle, <_\alpha, S_\alpha, F_\alpha\right)$ where

(1) $\mathfrak{B}_i \models \sigma_{\beta_i}$

(2) $<_\alpha$ is a binary relation on $\sum_{i\in\omega} B_i$, where $B_i = A_{\beta_i}^{\mathfrak{B}_i}$, and $\left\langle \sum_{i\in\omega} B_i, <_\alpha \right\rangle$ is a tree such that B_i is the set of elements of height $i - 1$ and each element has infinitely many $<_\alpha$-successors.

(3) $S_\alpha \subseteq \bigcup_{i\in\omega} B_i \times B$, and the map sending b to $\{x : x\, S_\alpha b\}$ is a 1 - 1 function on B into the set of maximal $<_\alpha$ branches.

(4) $F_\alpha \subseteq B \times B_\alpha$, and the family of sets of the form $\{y : y\, F_\alpha\, z\}$ is a partition of B such that for each $x \in \sum_{i\in\omega} B_i$ and each $z \in B_\alpha$, there are infinitely many $y \in B$, such that $x\, S_\alpha y$ and $y\, F_\alpha z$.

It is not difficult to see that there is a sentence $\sigma_\alpha \in L_{\omega_1,\omega}$ such that $\mathfrak{A} \models \sigma_\alpha$ iff $\mathfrak{A} \in K$. Also, no model of σ_α has power $> \beth_\alpha$ since if $\mathfrak{B}$ is as above then

$$\begin{aligned} c|\mathfrak{B}| &= \Sigma\, c\mathfrak{B}_i + cB + cB_\alpha \\ &\leq \sum_{i\in\omega} \beth_{\beta_i} + c^{\omega}\sum_{i\in\omega} \beth_{\beta_i} + c^{\omega}\sum_{i\in\omega} \beth_{\beta_i} \\ &\leq \beth_\beta + \beth_\alpha + \beth_\alpha = \beth_\alpha . \end{aligned}$$

To get a model of σ_α of power $\beth_\alpha$, take $\mathfrak{B} \in K$ as above with $\mathfrak{B}_i \models \sigma_{\beta_i}$ such that

(1) $cB_i = \beth_{\beta_i}$, with $B_i = \{b_t : \text{dom } t = i + 1 \text{ and for } j \leq i,\ t(j) \in \beth_{\beta_j}\}$

(2) $b_t <_\alpha b_{t'}$ iff t' extends t.

(3) $B = \{f : \text{dom } f = \omega \;\&\; f(i) \in B_i\}$.

(4) $b_t\, S_\alpha\, f$ iff $b_t \in \sum_{i\in\omega} B_i$ and $f \in B$ and f extends t.

(5) For $f, f' \in B$ say $f \sim f'$ if for some $n \in \omega$ and all $m \geq n$, $f(m) = f'(m)$. Then '$\sim$' is an equivalence relation. Let $B_\alpha = \{\overline{f} : f \in B\}$ where $\overline{f} = \{f' : f' \sim f\}$.

(6) $gF_\alpha \overline{f}$ iff $g \in B$, $\overline{f} \in B_\alpha$ and $g \in \overline{f}$.

Clearly, $\mathfrak{B}$ is a model of σ_α. Also $cB = \beth_\alpha$: For let g map the power set of each B_i one to one into B_{i+1}. If $A \subseteq \sum_{i\in\omega} (A \cap B_i)$ then $A = \sum_{i\in\omega} (A \cap B_i)$. Map A into f_A where $f_A(0)$ is some arbitrary element of B_0, and $f_A(i + 1) = g(A \cap B_i)$. This sets up a one to one correspondence between the subsets of $\sum_{i\in\omega} B_i$ and B. Hence $cB \geq \beth^{\beta}2 = \beth_\alpha$. Since each $\overline{f}$ has power $\beth_\beta$, it follows that $cB_\alpha = \beth_\alpha$ as needed.

To conclude that σ_α satisfies the requirements of Theorem 2, it remains to show that whenever $\mathfrak{A}$ is a countable model of σ_α then A_α is completely homogeneous, and any countable model $\mathfrak{B}$ of σ_α is isomorphic to $\mathfrak{A}$.

Let $\mathfrak{A} = (\sum_{i\in\omega} \mathfrak{A}_i + \langle A\rangle + \langle A_\alpha\rangle, <_\alpha, R_\alpha, E_\alpha)$ where

(1) $\mathfrak{A}_i$ is a countable model of β_i, and $A_{\beta_i} = \{f : {}^{i+1}\omega\}$.

(2) $t <_\alpha t'$ iff $t, t' \in \sum_{i\in\omega} A_{\beta_i}$ and t' extends t.

(3) $A = \sum_{i\in\omega} A^i$ where $A^i = \{f : f \in {}^\omega\omega$ and for some n and an $n \geq m$, $f(m) = 1$ if $i | m$ and $f(m) = 0$ if $i \not| m\}$.

(4) $tR_\alpha f$ iff $t \in \sum_{i\in\omega} A_{\beta_i}$ and $f \in A$ and f extends t.

(5) $A_\alpha = \{A^i : i \in \omega\}$ where A^i is as defined in (3).

(6) $fE_\alpha A^i$ iff $f \in A$, $A^i \in A_\alpha$, and $f \in A^i$.

It is easy to see that $\mathfrak{A} \models \sigma_\alpha$.

Let $\mathfrak{B}$ be another countable model of σ_α. Since any two countable models of σ_{β_i} are isomorphic, we may suppose that $\mathfrak{B} = (\sum_{i\in\omega} \mathfrak{A}_i + \langle B\rangle + \langle B_\alpha\rangle, <_\alpha, S_\alpha, F_\alpha)$ where the $\mathfrak{A}_i$ and $<_\alpha$ are as above, and

(3') $B = \sum_{i\in\omega} B^i$ where $B^i \subseteq {}^{\omega}\omega$ such that for every $n \in \omega$ and every $t \in {}^{n}\omega$ there are infinitely many g's $\in B^i$ extending t.

(4') $tR_\alpha g$ iff $t \in \sum_{i\in\omega} A_{\beta_i}$ and $g \in B$ and g extends t.

(5') $B_\alpha = \{B^i : i \in \omega\}$.

(6') $g \models_\alpha B^i$ iff $g \in A$, $B^i \in B_\alpha$ and $g \in B^i$.

Now let H be as in the conclusion of Lemma 3, and extend H to an isomorphism on $\mathfrak{B}$ onto $\mathfrak{A}$ as follows:

For $t \in B_{\beta_i} = A_{\beta_i}$ we let $H(t) = H(g) \upharpoonright i + 1$ where g is any element of B extending t. The conditions on H imposed by Lemma 3 assure us that $H(t)$ is well defined, and that $tS_\alpha g$ iff $(Ht)\, R_\alpha(Hg)$, and that $t <_\alpha t'$ iff $(Ht) <_\alpha (Ht')$. H is onto A_{β_i}, for if $t \in A_{\beta_i}$ and if f extends t, $f \in A_\alpha$, then $t = H(H^{-1}f \,|\, \mathrm{Dom}\, t)$. Since each A_{β_i} is completely homogeneous in the structure $\mathfrak{A}_i$, H can be extended to $\mathfrak{A}_{\beta_i}$ giving the necessary isomorphism.

Hence $\sigma_\alpha \in L^c_{\omega_1,\omega}$.

The same argument, taking $\mathfrak{A}$ for $\mathfrak{B}$, shows that A_α is completely homogeneous.

This finishes the induction step for α a successor of a limit ordinal.

Case 2. α is the successor of a successor ordinal $\beta = \gamma + 1$.

We consider the class K of structures $\mathfrak{B} = (\mathfrak{A}'_\beta + \langle B\rangle + \langle B_\alpha\rangle, <_\alpha, E_\alpha)$ where

(1) $\mathfrak{A}_\beta \models \sigma_\beta$

(2) $<_\alpha$ is a dense linear ordering without endpoints on $A_\beta + B$, with A_β dense in B.

(3) $xE_\alpha y$ implies $x \in B$ and $y \in B_\alpha$, and if we define $B^d = \{b \in B : bE_\alpha d\}$, then $\{B^d : d \in B_\alpha\}$ is a partition of B into infinitely many equivalence classes, each dense in the other.

As before, it is easy to see that K is the class of models of some sentence of $L_{\omega_1,\omega}$. This will be our sentence σ_α.

No structure of power $> \beth_\alpha$ belongs to K, since, if $\mathfrak{B}$ is as above then

$$c|\mathfrak{B}| = c|\mathfrak{A}_\beta| + cB + cB_\alpha$$

$c|\mathfrak{A}_\beta| \leq \beth_\beta$. Because A_β is dense in B, $cB \leq c^{A_\beta}2 \leq \beth_\alpha$. Also, it is clear from (3) that $cB_\alpha \leq cB$. Hence $c|\mathfrak{B}| \leq \beth_\alpha$.

In order to show that σ_α has models of power $\beth_\alpha$ we need the following lemmas:

LEMMA. 4. Assume the G.C.H. For each infinite cardinal κ, there is a structure $\langle D \cup D', < \rangle$ such that

(1) $D \cap D' = 0$

(2) $cD = \kappa$, $cD' = 2^\kappa$

(3) $<$ is a dense ordering without endpoints on $D \cup D'$, such that between any two elements of D there are 2^κ elements of D'.

Proof. Let $D = \{f \in {}^\kappa\kappa : \text{for some } \gamma \in \kappa \text{ and all } \delta, \delta' \geq \gamma, f(\delta) = f(\delta')\}$, let $D' = {}^\kappa\kappa - D$, and let $<$ be the lexicographic ordering on ${}^\kappa\kappa$.

LEMMA. 5. Let $\langle D \cup D', < \rangle$ be a densely ordered structure satisfying the conclusion of Lemma 4. Then D' can be partitioned into $c({}^D2)$ classes each dense in every other and in D.

Proof. Let a_α, $0 < \alpha < c({}^D2)$ enumerate D'. We define a sequence D_α, $\alpha \in c({}^D2)$ such that

(1) $D = D_0$

(2) $\gamma < \delta \Rightarrow D_\gamma \cap D_\delta = 0$

(3) D_γ is dense in D_β for all $\gamma, \beta \in c({}^A2)$

(4) $a_\delta \in \bigcup_{\gamma \in \delta} D_\gamma$ for $0 \in \delta \in c({}^A2)$

(5) $cD_\gamma \leq c(\gamma \times D)$

(6) Between any two points of $\bigcup_{\gamma \in \delta} D_\gamma$ there are $c({}^D2)$ points of $D' - \bigcup_{\gamma \in \delta} D_\gamma$.

Suppose D_γ has been defined for each $\gamma \in \delta$ where $\delta \in c({}^D2)$, and the sequence satisfies (1) - (5). For each $x, y \in \bigcup_{\gamma \in \delta} D_\gamma$ such that $x < y$, let $f(x,y)$ be some point of $D' - \bigcup_{\gamma \in \delta} D_\gamma$ between x and y. Take $D_\delta = \text{Rng } f \cup \{a_\delta\}$ if $a_\delta \notin \bigcup_{\alpha \in \delta} D_\gamma$, $D_\gamma = \text{Rng } f$ otherwise. Clearly, (1), (2), (3), (4) are satisfied by $\{D_\alpha : \alpha \leq \delta\}$. $cD_\delta = c\left(\bigcup_{\gamma \in \delta} D_\delta\right) \leq \sum_{\gamma \in \delta} c(\gamma \times D) = cD \cdot \sum_{\gamma \in \delta} c\gamma \leq cD \cdot c(\delta \times \delta) \leq c(\delta \times D)$. Hence (5) holds for $\{D_\alpha : \alpha \in \delta + B\}$. Since $\bigcup_{\gamma \in \delta+1} D_\gamma \supseteq D$, $\bigcup_{\gamma \in \delta+1} D_\gamma$ is dense in D' and hence in $D' - \bigcup_{\gamma \in \delta+1} D_\gamma$. Since $<$ densely orders $D \cup D'$, between any two points of $D \cup D'$ there are infinitely many, and hence, at least 2 different points of D (D is dense in D'). Hence between any two points of $D \cup D'$ there are $c(^{D}2)$ distinct points of D'. Since $c\left(\bigcup_{\gamma\in\delta+1} D_\gamma\right) < c(^{D}2)$, between any two points of $\bigcup_{\gamma\in\delta+1} D_\gamma$ there are $c(^{D}2)$ distinct points of $D' - \bigcup_{\gamma\in\delta+1} D_\gamma$. Thus (1) - (6) holds for $\{D_\alpha : \alpha \in \delta + 1\}$. Then clearly, $\{D_\alpha : \alpha \in c(^{D}2)\}$ partitions D' as required by the theorem.

Now let $\mathfrak{A}_\beta$ be any model of σ_β with $cA_\beta = \beth_\beta$. By Lemma 4, there is a structure $\langle A_\alpha \cup B, <_\alpha\rangle$ satisfying the hypotheses of Lemma 5 with $D = A_\beta$ and $D' = B$, and $< = <_\alpha$. Let B_α be a partition on B that satisfies the conclusion of Lemma 5. Let $xE_\alpha y$ hold iff $x \in B$, $y \in B_\alpha$ and $x \in y$. Then $(\mathfrak{A}_\beta + \langle B\rangle + \langle B_\alpha\rangle, <_\alpha, E_\alpha)$ is a model of σ_α of power $\beth_\alpha$.

It remains to show that any two countable models of σ_α are isomorphic and that $A_\alpha^{\mathfrak{A}}$ is completely homogeneous for any countable model $\mathfrak{A}$ of σ_α. For this we need the following trivial extension of a well known theorem of Cantor.

LEMMA. 6. Let $<$ $(<')$ be a countable dense ordering without endpoints on $\sum_{i\in\omega} B^i \left(\sum_{i\in\omega} C^i\right)$, where B^i is dense in B^j (C^i is dense in C^j) for all $i,j \in \omega$. Then there is an isomorphism on $\left\langle \sum_{i\in\omega} B^i, <\right\rangle$ onto $\left\langle \sum_{i\in\omega} C^i, <'\right\rangle$ that maps B^i onto C^i for all $i \in \omega$.

Now let $\mathfrak{B} = (\mathfrak{A}_\beta + \langle B\rangle + \langle B_\alpha\rangle, <_\alpha, E_\alpha)$ and $\mathfrak{C} = (\mathfrak{A}'_\beta + \langle C\rangle + \langle C_\alpha\rangle, <'_\alpha, F_\alpha)$ be two countable models of σ_α. Since $\mathfrak{A}_\beta \models \sigma_\beta$ and $\mathfrak{A}'_\beta \models \sigma_\beta$ we may assume that $\mathfrak{A}_\beta = \mathfrak{A}'_\beta$. Let H be a one to one map on B_α onto C_α, say $H(b_i) = c_i$ where $B_\alpha = \{b_0, b_1, \ldots\}$, and $C_\alpha = \{c_0, c_1, \ldots\}$. Let $B^0 = A_\beta = C^0$, and for $i = 1,2,\ldots$ let $B^i = \{x \in B : xE_\alpha b_i\}$ and $C^i = \{x \in C : xF_\alpha c_i\}$. Using Lemma 6, extend H to an order isomorphism on $\left\langle \sum_{i\in\omega} B^i, <_\alpha\right\rangle$ onto $\left\langle \sum_{i\in\omega} C^i, <'_\alpha\right\rangle$ such that B^i is mapped onto C^i. In particular, $B^0 = A_\beta$ is mapped onto $C^0 = A_\beta$ and so H can be extended to an isomorphism on $\mathfrak{A}_\beta$ onto $\mathfrak{A}_\beta$, and so on $\mathfrak{B}$ onto $\mathfrak{C}$. Hence $\sigma_\alpha \in L^c_{\omega_1,\omega}$ and if $(\mathfrak{A}_\beta + \langle A^*\rangle + \langle A_\alpha\rangle, <_\alpha, E_\alpha)$ is any countable model of σ_α then A_α is completley homogeneous.

This concludes the induction and the proof of Theorem 2.

3. In this section we give some examples illustrating the differences between the model theory of $L^c_{\omega_1,\omega}$ and $L_{\omega_1,\omega}$. We conclude the section with a few problems.

DEFINITION. $\mathfrak{C}$ is an amalgamation of $\mathfrak{A}$ and $\mathfrak{B}$ if there are isomorphisms f on $\mathfrak{A}$ into $\mathfrak{C}$

and g on $\mathfrak{B}$ into $\mathfrak{C}$ such that $f \upharpoonright |A| \cap |B| = g \upharpoonright |A| \cap |B|$.

Recall that in the finitarly calculus, if $\mathfrak{A} \cap \mathfrak{B}$ is an elementary substructure of both $\mathfrak{A}$ and $\mathfrak{B}$, then $\mathfrak{A}$ and $\mathfrak{B}$ have an amalgamation $\mathfrak{C}$ such that the isomorphisms f and g are elementary embeddings. This property is central to a great many constructions in the model theory of $L_{\omega,\omega}$. However, as the following theorem shows, $L^c_{\omega_1,\omega}$ does not enjoy this property.

THEOREM. <u>There are countable structures</u> $\mathfrak{A}$ <u>and</u> $\mathfrak{B}$ <u>such that</u> $\mathfrak{A} \cap \mathfrak{B} \prec_{\omega_1,\omega} \mathfrak{A}$ <u>and</u> $\mathfrak{A} \cap \mathfrak{B}\ \alpha_{\omega_1,\omega} \mathfrak{B}$ <u>but no structure</u> $L_{\omega_1,\omega}$<u>-equivalent to</u> $\mathfrak{A}$ <u>is an amalgamation of</u> $\mathfrak{A}$ <u>and</u> $\mathfrak{B}$.

<u>Proof</u>. Let $<$ be the extension relation on $T = \bigcup_{n<\omega} {}^n\{0,1\}$, and let R be the extension relation between T and $A = \{f \in {}^\omega\{0,1\} :$ for some n and all m, $n \in m$ implies $f(m) = 0\}$. We take $\mathfrak{A}$ to be $\langle T + A, <, R, S\rangle$ where S is a dense ordering without endpoints on A such that: Given $t \in T$ and $x,y \in A$ with $x < y$ there are infinitely many extensions of t between x and y. Let $a \in A$ and let $b \notin T + A$. Extend R by defining tRb iff tRa, and extend S to a dense ordering without endpoints on $A + \{b\}$. For $\mathfrak{B}$ we take $\langle T + (A - \{a\}) + \{b\}, <, R, S\rangle$. (Restrictions and extensions of R and S are being denoted by R and S also). An argument like that used in Lemma 3 shows that for each finite subset X of $\mathfrak{A} \cap \mathfrak{B}$, there is an isomorphism of $\mathfrak{A} \cap \mathfrak{B}$ onto $\mathfrak{A}$ which is the identity on X. Hence, $\mathfrak{A} \cap \mathfrak{B} \prec_{\omega_1,\omega} \mathfrak{A}$. Similarly, $\mathfrak{A} \cap \mathfrak{B} \prec_{\omega_1,\omega} \mathfrak{B}$.

Now suppose that $\mathfrak{C}$ is $L_{\omega_1,\omega}$ equivalent to $\mathfrak{A}$ and that f and g map $\mathfrak{A}$ and $\mathfrak{B}$ respectively into $\mathfrak{C}$ with $f \upharpoonright |\mathfrak{A} \cap \mathfrak{B}| = g \upharpoonright |\mathfrak{A} \cap \mathfrak{B}|$. Then the sentence

$$\forall xy(Tx \vee Ty \vee x \approx y \vee \exists z\,(zRx \wedge \neg\, zRy))$$

is true in $\mathfrak{A}$, $\mathfrak{B}$, and $\mathfrak{C}$. Since f maps $T^{\mathfrak{A}}$ onto $T^{\mathfrak{C}}$ and $f \upharpoonright T = g \upharpoonright T$, and since tRa iff tRb for all $t \in T$, we have $f(a) = g(b)$. On the other hand, some $d \in A - \{a\}$ is between a and b, say $aS^{\mathfrak{A}}d$ and $dS^{\mathfrak{B}}b$. Hence $f(a)\ S^{\mathfrak{C}} f(d)$ and $g(d)\ S^{\mathfrak{C}} g(b)$, i.e., $f(a)\ S^{\mathfrak{C}} d$ and $d\ S^{\mathfrak{C}} f(b)$, so $f(a) \neq g(b)$-contradiction. Hence, no such amalgamation $\mathfrak{C}$ exists.

The authors first example of the failure of the amalgamation property used uncountable structures, although we were able to show that countable examples exist. Subsequently, Kueker found the first countable example of which the above is a variant.

If Σ is the $L_{\omega,\omega}$ theory of an infinite structure whose type is of cardinality $\leq \kappa$, then Σ

has a κ^+-universal (indeed, κ^+-saturated) model of power 2^κ [4]. The next theorem shows that the situation for $L^c_{\omega_1,\omega}$ is much different.

THEOREM. There is a sentence $\sigma \in L^c_{\omega_1,\omega}$ having models $\mathfrak{B}$ and $\mathfrak{C}$ such that

(1) $c|\mathfrak{B}| = c|\mathfrak{C}| = \beth_1$.

(2) No model of σ properly contains $\mathfrak{B}$ or properly contains $\mathfrak{C}$.

(3) $\mathfrak{B}$ and $\mathfrak{C}$ are not isomorphic.

Proof. Let $\mathfrak{A}$ be the model of power $\beth_1$ that we constructed for σ_1. Expand $\mathfrak{A}$ to $\mathfrak{A}'$ by adding unary relations P_i $(i \in \omega)$ where $P_i t$ iff $t \in T$ and t has exactly $i-1$ predecessors. Suppose that $\mathfrak{A}'$ is a substructure of $\mathfrak{D}$ and that $\mathfrak{D}$ is $L_{\omega_1,\omega}$ equivalent to $\mathfrak{A}'$. Since there are exactly 2^{i-1} elements in $P_i^{\mathfrak{A}'}$ and in $P_i^{\mathfrak{D}}$, we have $P_i^{\mathfrak{A}'} = P_i^{\mathfrak{D}}$, and so $T^{\mathfrak{A}'} = T^{\mathfrak{D}}$. Since elements of $A^{\mathfrak{A}'}$ and $A^{\mathfrak{D}}$ are determined by the elements of T that they extend, and since $A^{\mathfrak{A}'}$ is the set of all maximal branches of T, we have $A^{\mathfrak{A}'} = A^{\mathfrak{D}}$. $A_1^{\mathfrak{A}'}$ and $A_1^{\mathfrak{D}}$ are partitions on A, and each member of $A_1^{\mathfrak{A}'}$ is a subset of some member of $A_1^{\mathfrak{D}}$. Hence $A_1^{\mathfrak{A}'} = A_1^{\mathfrak{D}}$. Hence $\mathfrak{D} = \mathfrak{A}'$. This shows that $\mathfrak{A}'$ has no proper extensions. So if $<'$ and $<''$ are two non-isomorphic, dense ordering, without endpoints, on A_1, then clearly $\mathfrak{B} = (\mathfrak{A}', <')$ and $\mathfrak{C} = (\mathfrak{A}', <'')$ satisfy the conclusion of the theorem.

A type is a set of finitary (i.e. $L_{\omega,\omega}$) formulas, each having at most the variable v_0 free. A structure $\mathfrak{A}$ realizes the type T if some assignment satisfies $\bigwedge\Gamma$ in $\mathfrak{A}$, otherwise $\mathfrak{A}$ is said to omit Γ. Γ is complete with respect to a set of finitary sentences Σ if whenever $\mathfrak{A} \models \bigwedge\Sigma \wedge \bigwedge\Gamma\,[x]$ and $\mathfrak{B} \models \bigwedge\Sigma \wedge \bigwedge\Gamma\,[y]$. Then $(\mathfrak{A}, x(v_0)) \equiv_{\omega,\omega} (\mathfrak{B}, y(v_0))$.

It is well known that for every sentence $\sigma \in L_{\omega_1,\omega}$, there is a set Σ of finitary sentences and a type Γ such that every model of σ can be expanded to a model of Σ that omits the type Γ, and every model of Σ omitting Γ is a model of σ.

Question. Is this true if we replace '$\sigma \in L_{\omega_1,\omega}$' by '$\sigma \in L^c_{\omega_1,\omega}$' and 'a type Γ' by 'a complete type Γ'?

In 3, Morley asks which cardinals are characterized by complete types, i.e., for which cardinals κ are there countable theories Σ and complete types Γ such that Σ has models omitting Γ in all powers up to but not including κ. A positive answer to our question and Theorem

2 would show that every $\beth_\alpha$, $\alpha < \omega_1$, can be so characterized if we assume the G.C.H.

We don't know if Theorem 2 can be strengthened by dropping the condition 'α is a successor ordinal' from the hypothesis.

Nor do we know what the Hanf number for $L^c_{\omega_1,\omega}$ is when the G.C.H. is not assumed. In particular, regarding Lemma 4, for what cardinals κ is there a densely ordered set of power 2^κ having a dense subset of power κ (trivially, $\kappa = \sum_{\lambda\epsilon\kappa} 2^\lambda$ is sufficient).

REFERENCES

[1] Karp, C., Languages with expressions of infinite length, North Holland, 1964.

[2] Lopez-Escobar, E. G. K., On defining well-orderings, Fund. Math. (1966), pp. 13-21.

[3] Morley, M., Omitting classes of elements, The Theory of Models, eds. J. Addison, L. Henkin, and A. Tarski, Amsterdam, 1965, pp. 265-273.

[4] Morley, M. and Vaught, R., Homogeneous universal models, Math. Scand., 2 (1962), pp. 37-57.

[5] Scott, D., Logic with denumerably long formulas and finite strings of quantifiers, The Theory of Models, eds. J. Addison, L. Henkin and A. Tarski, Amsterdam, 1965, pp. 329-341.

UNIVERSITY OF CALIFORNIA, LOS ANGELES

QUANTIFIED ALGEBRAS

A. PRELLER*

0. Introduction.

This paper has two aspects: First we define quantified algebras, particular cases of which are the Lindenbaum-Tarski-algebras associated with infinitary languages. Some properties of the category of quantified algebras are investigated. Then these algebraic methods are used to propose a solution to a problem formulated in [2]: For which (infinitary) languages $L_{\alpha\beta}$ can we provide a notion of probability with the property that exactly the valid formulas are provable? To make the question precise: A set of formulas of the language is to be singled out to serve as axioms, closure conditions have to be specified to serve as rules of inference. A formula is then provable if it is the last element of a sequence of fewer than α formulas each of which is either an axiom or the result of applying a rule of inference to formulas appearing earlier in the sequence. Denote the set of provable formulas by $\mathfrak{P}_{\alpha\beta}$. The problem is to define $\mathfrak{P}_{\alpha\beta}$ in the way described above such that the provable formulas are exactly the valid formulas. The answer proposed here is: for all $L_{\alpha\beta}$ languages such a set $\mathfrak{P}_{\alpha\beta}$ can be defined provided that the language has enough variables. (2^{α} are always enough, but often less are sufficient.)

1. Individualized Sets.

Let us fix the notation: $\alpha, \alpha_0, \alpha_1, \ldots$ are regular infinite cardinals. $I, I_0, I_1, \ldots$ are sets of power at least α, respectively α_0, α_1, etc. We denote the power of a set I by $|I|$, so we have $|I| \geq \alpha$. The elements of I are called individuals. Let X and C be subsets of I such that $|X| \geq \alpha$, $X \cap C = \emptyset$, $I = X \cup C$. The elements of X are called variables, the elements of C, constants

---

*I gratefully acknowledge the help of Mr. James Johnson in preparing this paper for publication. Work partially supported by National Science Foundation Grant GS-7578.

C may be the empty set. The set of "substitutions" I_X is defined by

$$I_X = \{\sigma \in I^I;\ \sigma x = x \ \text{ if } \ x \in I - X\}\ .$$

I_X is a semigroup with unity under the law of ordinary composition of functions and may be identified with a one-object-category. Consequently, a function f from a semi-group S_1 to a semi-group S_2 is called a functor if $f(1) = 1$ and $f(\sigma\tau) = f(\sigma)f(\tau)$.

1.1 DEFINITION. A $(X \cup C,\alpha)$-structure on a set E is given by a functor f from I_X to E^E such that

(1) For all $a \in E$ there exists $J \subset X$ such that $|J| < \alpha$ and for all $\sigma,\tau \in I_X$, if $\sigma x = \tau x$ for $x \in J$, then $f(\sigma)(a) = f(\tau)(a)$.

1.2 REMARKS. (i) We always write σa instead of $f(\sigma)(a)$, if the context permits.

(ii) If J has the property (1) for a fixed element a of E, then we say that "J supports a" or that "J is a α-support of a."

If $K \subset X$, $|K| < \alpha$, J supports a, and $J \subset K$, then K supports a.

If J and K support a, then $J \cap K$ supports a.

(iii) A $(X \cup \emptyset,\alpha)$-structure on E is called a (I,α)-structure, E an (I,α)-set or an "individualized set without constants." If $C \neq \emptyset$, then a set E with an $(X \cup C,\alpha)$-structure is called an "individualized set with constants."

1.3 Examples. (A) Let $(K_r)_{r\in R}$ be a family of sets, each of power less than α. For $a \in I^{K_r}$ and $\sigma \in I^I$ let us designate by σa the function $(\sigma a)(k) = \sigma(a(k))$ for $k \in K_r$. This defines a (I,α)-structure on I^{K_r} and on $\coprod_{r\in R} I^{K_r}$ ($\coprod$ stands here for the direct sum in the category of sets). For every couple of variables X and constants C in I, the (I,α)-structure introduces an obvious $(X \cup C,\alpha)$-structure on $\coprod_{r\in R} I^{K_r}$.

It is clear that the set of atomic formulas of a $L_{\alpha\beta}$ language is a $(X \cup C,\alpha)$-set of type described in (A).

(B) Let $\mathfrak{F}$ be a set of functions between sets, defined on sets of power less than α. If $a \in \mathfrak{F}$ is defined on K and if $\sigma \in I^I$ suppose that σa defined by

$$\sigma a(k) = \begin{cases} a(k), & \text{if } a(k) \notin I \\ \sigma(a(k)), & \text{if } a(k) \in I \end{cases}$$

is an element of $\mathfrak{J}$. Then this defines an (I,α)-structure on $\mathfrak{J}$ which introduces obvious $(X \cup C,\alpha)$-structures on $\mathfrak{J}$ for every couple of variables X and constants C in I.

The set of all formulas of a $L_{\alpha\beta}$ language is a $(X \cup C,\alpha)$-set of type (B).

1.4 DEFINITION. A morphism from the $(X_0 \cup C_0, \alpha_0)$-set E_0 to the $(X_1 \cup C_1, \alpha_1)$-set E_1 is a couple (j,h) where j is a functor from I_{0X_0} to I_{1X_1}, h a function from E_0 to E_1 such that

$$h(\sigma a) = j(\sigma)h(a) \quad \text{for } a \in E_0,\ \sigma \in I_{0X_0} .$$

1.5. <u>Example</u>. Let i be an injection from I_0 into I_1 such that $i(X_0) \subset X_1$ and $i(C_0) \subset C_1$. For $\sigma \in I_{0X_0}$ define $j(\sigma)$ by

$$j(\sigma)x = \begin{cases} x, & \text{if } x \notin i(I_0) \\ i\sigma i^{-1}(x), & \text{if } x \in i(I_0) , \end{cases}$$

j is a functor from I_{0X_0} to I_{1X_1}. If i is the identity, so is j. If we say that h, function from the $(X \cup C,\alpha)$-set E_0 to the $(X \cup C,\alpha)$-set E_1, is a morphism, then we mean that $(1,h)$ is a morphism.

Let us denote by $\mathbf{Ind}$ the category with individualized sets as objects and the morphisms as in 1.4. For fixed I, X, C, α we denote by $\mathbf{J}_{(X\cup C,\alpha)}$ the subcategory of $\mathbf{Ind}$ formed by the $(X \cup C,\alpha)$-sets and morphisms (j,h) with $j = 1$.

1.6 PROPOSITION. $\mathbf{J}_{(X\cup C,\alpha)}$ has direct sums and direct products.

1.7. Let i be an injection from I_0 into I, suppose that $\alpha_0 \leq \alpha$. We define a "compression" functor S_i from $\mathbf{J}_{(I,\alpha)}$ to $\mathbf{J}_{(I_0,\alpha_0)}$ by $S_iE = \{a \in E;$ there is a support I_a of a with $|I_a| < \alpha_0$ and $I_a \subset I_0\}$ where E is an (I,α)-set. For an (I,α)-morphism f from the (I,α)-set E, let $S_if = f/S_iE$. Then we have

1.8 THEOREM. <u>The functor</u> S_i <u>from</u> $\mathbf{J}_{(I,\alpha)}$ <u>to</u> $\mathbf{J}_{(I_0,\alpha_0)}$ <u>defined in</u> 1.7 <u>has a left adjoint</u>

which is faithful.

1.9. Let us introduce some notation:

For $\sigma,\tau \in I_X$, $K \subset I$ let $\sigma \underset{K}{\otimes} \tau$ be the element of I_X defined by

$$(\sigma \underset{K}{\otimes} \tau)(x) = \begin{cases} \tau x & \text{if } x \in K \\ \sigma x & \text{if } x \notin K . \end{cases}$$

For $M \subset X$, I_M is the set $\{\sigma \in I_X;\ \sigma x = x \text{ if } x \notin M\}$. If $\sigma \in I_M$ and $\tau \in I_N$ and $M \cap N = \emptyset$ then $\sigma \underset{N}{\otimes} \tau = \tau \underset{M}{\otimes} \sigma$ and we shall denote both of them by $\sigma \otimes \tau$. Generally, if $M_\ell \subset X$, $\sigma_\ell \in I_{M_\ell}$, $\ell \in L$, and $M_\ell \cap M_{\ell'} = \emptyset$ for $\ell \neq \ell'$, then $\sigma = \underset{\ell \in L}{\otimes} \sigma_\ell$ is defined by

$$\sigma x = \begin{cases} \sigma_\ell x & \text{if } x \in M_\ell \\ x & \text{if } x \notin \bigcup_{\ell \in L} M_\ell . \end{cases}$$

This law is clearly associative. For $K \subset I$, $\sigma \in I_X$, the element $\tau = \sigma \restriction K$ of I_X is defined by

$$\tau x = \begin{cases} \sigma x, & \text{if } x \in K \\ x, & \text{if } x \notin K . \end{cases}$$

For $M \subset I$, $\sigma \in I_M$, $\varphi \in I^I$ we write $\overline{\varphi\sigma}$ instead of $(\varphi\sigma) \restriction M$. Clearly, the notation $\overline{\varphi\sigma}$ is ambiguous and shall be used only if the context makes clear to which subset M the function $\varphi\sigma$ is restricted.

If $K, J \subset I$, $\sigma \in I^I$, then $K \overset{\sigma}{\rightarrowtail} J$ (resp. $K \overset{\sigma}{\twoheadrightarrow} J$) has the following meaning: "the restriction of σ to K is an injection from K into (resp. onto) J and σ equals the identity outside of K." $K \underset{\tau}{\overset{\sigma}{\rightleftarrows}} J$ stands for "$K \overset{\sigma}{\rightarrowtail} J$ and $\sigma(K) \overset{\tau}{\rightarrowtail} J$ and $\tau\sigma \restriction K = 1$."

2. Free Quantified Algebras.

Let $\alpha,\alpha_0,\alpha_1,\ldots,I,I_0,I_1,\ldots,X,X_0,X_1,\ldots,C,C_0,C_1,\ldots$ have the same meaning as in Section 1. Let $\beta,\beta_0,\beta_1,\ldots$ be infinite cardinals or equal to 1 such that $\beta \leq \alpha$, $\beta_0 \leq \alpha_0$, $\beta_1 \leq \alpha_1, \ldots$. The Boolean operations will be designated by $\vee, \wedge, '$ or $\cup, \cap, '$, if we want to emphasize that a Boolean algebra can always be identified with a subfield of sets.

2.1 DEFINITION. A $\nearrow\!\alpha$-algebra A (i.e. a Boolean algebra A such that $\bigvee_{p\in P} a_p$ exists in A whenever $|P| < \alpha$) is said to be a $(X \cup C,\alpha,\beta)$-algebra if there is a functor f from I_X to $\mathrm{Hom}_{\nearrow\alpha}(A,A)$ such that

(1) every $a \in A$ has an $\nearrow\!\alpha$-support

(2) for all $a \in A$, $K \subset X$, $|K| < \beta$, all $\tau \in I_X$ $\bigvee_{\sigma\in I_K} \tau \underset{K}{\otimes} \sigma a$ exists and is equal to $\tau \bigvee_{\sigma\in I_K} \sigma a$.

(Notice, letting $\tau = 1$, $\bigvee_{\sigma\in I_K} \sigma a$ exists.)

A $(X \cup \emptyset,\alpha,\beta)$-algebra is called an (I,α,β)-algebra. Instead of $\bigvee_{\sigma\in I_K} \sigma a$, we often write $\exists Ka$, instead of $\bigwedge_{\sigma\in I_K} \sigma a$ we may use $\forall Ka$.

2.2 Example. Let E be the $(X \cup C,\alpha)$-set described in 1.3. (A). Let $\mathfrak{F}_{\alpha\beta}$ be the smallest set such that

(I) $E \subset \mathfrak{F}_{\alpha\beta}$

(II) if $a_p \in \mathfrak{F}_{\alpha\beta}$ for $p \in P$ and $|P| < \alpha$, then $\bigvee_{p\in P} a_p$ and $\bigwedge_{p\in P} a_p$ is in $\mathfrak{F}_{\alpha\beta}$.

(III) if $a \in \mathfrak{F}_{\alpha\beta}$ and $K \subset X$, $|K| < \beta$, then $\exists Ka$ and $\forall Ka$ is in $\mathfrak{F}_{\alpha\beta}$.

This definition has only a intuitive meaning, for we have not made precise how the formula $\bigvee_{p\in P} a_p$, for example, is defined.

Now suppose that $\mathfrak{D}_{\alpha\beta} \subset \mathfrak{F}_{\alpha\beta}$ is the "basic formal system" described in [2], Chapter 11, of "provable" formulas. Then the set of equivalence classes $\mathfrak{L}_{\alpha\beta} = \mathfrak{F}_{\alpha\beta}/\mathfrak{D}_{\alpha\beta}$ of the relation "$a \longleftrightarrow b \in \mathfrak{D}_{\alpha\beta}$" between formulas a and b of $\mathfrak{F}_{\alpha\beta}$ has a natural (I,α,β)-structure. For each couple (X,C) where X is the set of variables and C the set of constants, the $(X \cup C,\alpha,\beta)$-structure on $\mathfrak{F}_{\alpha\beta}/\mathfrak{D}_{\alpha\beta}$ is introduced by the (I,α,β)-structure. The same is true, if we omit in the set of axioms of $\mathfrak{D}_{\alpha\beta}$ the equality axioms.

2.3 REMARK. If I_a supports the element a of the $(X \cup C,\alpha,\beta)$-algebra A, then $I_a \cap K'$ supports $\bigvee_{\sigma\in I_K} \sigma a$.

2.4 DEFINITION. A morphism between a $(X_0 \cup C_0,\alpha_0,\beta_0)$-algebra A_0 and a $(X_1 \cup C_1,\alpha_1,\beta_1)$-algebra A_1 is a triplet (k,j,h) where k maps the subsets of X_0 of power less than β_0 to subsets of X_1 of power less than β_1, j is a functor from I_{0X_0} to I_{1X_1} and $h \in \mathrm{Hom}_{\nearrow\alpha_0}(A_0,A_1)$ such that

(*) for all $K \subset X_0$, $|K| < \beta_0$, $j(\sigma) \in I_{1k(K)}$ when $\sigma \in I_{0K}$.

(**) $h(\sigma a) = j(\sigma)h(a)$ when $a \in A_0$, $\sigma \in I_{0X_0}$.

(***) $h\left(\bigvee_{\sigma \in I_{0K}} \sigma a\right) = \bigvee_{\tau \in I_{1k(K)}} \tau h(a)$.

An (I,α,β)-algebra is said to be a quantified algebra (without constants) and an $(X \cup C,\alpha,\beta)$-algebra a quantified algebra (with constants). Let us denote by $\mathbf{Q}$ the category of quantified algebras and the morphisms between them.

2.5. An injection i from I_0 to I_1 as described in 1.5 introduces a couple (k,j) such that (*) is true; $k(K) = i(K)$ and j is the functor described in 1.5.

For (X,C,α,β) fixed we denote by $\mathbf{Q}_{(X \cup C,\alpha,\beta)}$ the subcategory of $\mathbf{Q}$ of the $(X \cup C,\alpha,\beta)$-algebras and the morphisms (k,j,h) between them with $k = 1$, $j = 1$.

2.6 PROPOSITION. $\mathbf{Q}_{(X \cup C,\alpha,\beta)}$ has direct products.

2.7. There is an obvious forgetful functor $S_{(X \cup C,\alpha,\beta)}$ (or simply S, if the context permits) from $\mathbf{Q}_{(X \cup C,\alpha,\beta)}$ to $\mathbf{J}_{(X \cup C,\alpha)}$.

Let E, F be $(X \cup C,\alpha)$-sets. The meaning of "E is a $(X \cup C,\alpha)$-subset of F" is obvious. If A is a $(X \cup C,\alpha,\beta)$-algebra, then "E is a $(X \cup C,\alpha)$-subset of A" if it is a $(X \cup C,\alpha)$-subset of SA.

2.8 PROPOSITION. Let E be an $(X \cup C,\alpha)$-subset of an $(X \cup C,\alpha,\beta)$-algebra A. The $(X \cup C,\alpha,\beta)$-algebra B generated by E, i.e., the least $(X \cup C,\alpha,\beta)$-subalgebra of A containing E, is the union of the following ascending sequence of $(X \cup C,\alpha)$-subsets of A:

$B_0 = E$. For $\alpha > i > 0$, B_i has the following property: a is an element of B_i if and only if at least one of the following conditions is true:

(i) there exists $b \in \bigcup_{j<i} B_j$ such that $a = b'$.

(ii) there exists P, $|P| < \alpha$, $a_p \in \bigcup_{j<i} B_j$ for $p \in P$ such that $a = \bigvee_{p \in P} a_p$.

(iii) there exist $K \subset X$, $|K| < \beta$, $b \in \bigcup_{j<i} B_j$ such that $a = \bigvee_{\sigma \in I_K} \sigma b$.

COROLLARY. (I) B <u>has power less or equal to</u> $2^{\max}(|X|,|E|)$.

(II) If two $(X \cup C,\alpha,\beta)$-morphisms defined on B agree on E, then they agree on B.

2.9 THEOREM. The forgetful functor S from $\mathbb{Q}_{(X\cup C,\alpha,\beta)}$ to $\mathbb{J}_{(X\cup C,\alpha)}$ has a left adjoint T. If Φ_E is the natural morphism from the $(X \cup C,\alpha)$-set E to STE then TE is generated by a $(X \cup C,\alpha)$-set $\Phi_E(E)$. TE is called the free $(X \cup C,\alpha,\beta)$-algebra generated by E.

Proof. We use a criterion which is a transcription into categorical language of a proof used by Rieger to show the existence of $\aleph_\alpha$-complete free Boolean algebras. [Fund. Math. 38, 35-52 (1951)]:

(a) Let S be a functor from $\mathbb{C}$ to $\mathbb{E}$. We say that an object B of $\mathbb{C}$ is S-generated by $f \in Hom_{\mathbb{E}}(E,SA)$, if there are $u \in Hom_{\mathbb{C}}(B,A)$ and $j \in Hom_{\mathbb{E}}(E,SB)$ such that $f = (Su)j$ and if $(Sh)j = (Sg)j$ then $h = g$ for $h,g \in Hom_{\mathbb{C}}(B,D)$. A "class of S-generated objects" is a function G which associates to $f \in Hom_{\mathbb{E}}(E,SA)$ and to A an object B of $\mathbb{C}$, S-generated by f.

(b) Suppose that $\mathbb{C}$ has direct products. Then S has a left adjoint if and only if S commutes with direct products, $\mathbb{C}$ has a class G of S-generated objects and for every E of $\mathbb{E}$ there is a reflecting family, i.e., a family $(A_\ell)_{\ell \in L}$ of objects of $\mathbb{C}$ such that for all $f \in Hom_{\mathbb{E}}(E,SA)$ and all A of $\mathbb{C}$, there exist $\ell \in L$ and $f_\ell \in Hom_{\mathbb{C}}(A_\ell,A)$ such that Sf_ℓ factorizes f.

Then we have $TE = G\Phi_E$ where Φ_E is the natural morphism from E to STE.

In the case where $S = S_{(X\cup C,\alpha,\beta)}$ we can take the family of all $(X \cup C,\alpha,\beta)$-algebras with elements in a fixed set of power $2^{\max(|I|,|E|)}$ as the reflecting family associated to E.

2.10 Example. Suppose that $\mathfrak{F}_{\alpha\beta}$ is the set of all formulas of a $L_{\alpha\beta}$ language and $\mathfrak{P}_{\alpha\beta}$ the set of all formally provable formulas of $\mathfrak{F}_{\alpha\beta}$ not containing the equality axioms (see 2.2). Then $\mathfrak{F}_{\alpha\beta}/\mathfrak{P}_{\alpha\beta} = T\mathcal{E}_\alpha$ where $\mathcal{E}_\alpha$ is the set of atomic formulas of $\mathfrak{F}_{\alpha\beta}$.

2.11 DEFINITION. Suppose that A is a $(X \cup C,\alpha,\beta)$-algebra and C a complete Boolean algebra. A Boolean homomorphism h from A to C is said to be valid, if and only if $h(\bigvee_{p\in P} a_p) = \bigvee_{p\in P} h(a_p)$ for all P such that $|P| < \alpha$, $a_p \in A$ for $p \in P$ and $h(\bigvee_{\sigma\in I_K} \sigma a) = \bigvee_{\sigma\in I_K} h(\sigma a)$ for all $a \in A$ and $K \subset X$, $|K| < \beta$.

REMARK. If h is a valid homomorphism from B to C and if f is a $(X \cup C,\alpha,\beta)$-homomorphism from A to B then hf is a valid homomorphism from A to C.

2.12 THEOREM. Let C be a complete Boolean algebra and h a map from the $(X \cup C,\alpha)$-set E to C. Then for all $\beta \leq \alpha$ there exist a $(X \cup C,\alpha,\beta)$-algebra $B_{C,h}$, a $(X \cup C,\alpha)$-map $\Phi_{C,h}$ from E to $SB_{C,h}$ and a valid homomorphism p from $B_{C,h}$ to C such that $p\Phi_{C,h} = h$ satisfying the following property: For all $(X \cup C,\alpha,\beta)$-algebras A, all $(X \cup C,\alpha)$-maps f from E to SA and all valid homomorphisms g from A to C such that gf = h there exists a unique $(X \cup C,\alpha,\beta)$-homomorphism $\bar{g}$ from A to $B_{C,h}$ such that $\bar{g}f = \Phi_{C,h}$ and $p\bar{g} = g$.

COROLLARY 1. For every map f from E to the complete Boolean algebra C there is a unique valid homomorphism g from TE to C such that $g\Phi_E = f$.

COROLLARY 2. $\Phi_E : E \to STE$ in an injective map.

3. Complete Quantified Algebras.

3.1 DEFINITION. A $(X \cup C,\alpha,\beta)$-algebra A is said to be complete, if for every $a \neq 0$ in A, there is a valid homomorphism h from A to 2 (the two-element Boolean algebra) such that h(a) = 1.

3.2 THEOREM. Let A be a $(X \cup C,\alpha,\beta)$-algebra. Then A is complete iff A is isomorphic to a $(X \cup C,\alpha,\beta)$-field, i.e., iff there is a set $\mathcal{V}$ and an injective Boolean homomorphism r from A into $\mathcal{P}(\mathcal{V})$ such that $r(\bigvee_{p\in P} a_p) = \bigcup_{p\in P} r(a_p)$ for every family $(a_p)_{p\in P}$ of elements of A such that $|P| < \alpha$ or there is $K \subset X$, $|K| < \beta$, $a \in A$ with $P = I_K$, $a_p = pa$ for all $p \in I_K$.

Proof. Suppose A complete, let $\mathcal{V}$ be the set of valid homomorphisms from A to 2. Identify A with the dual algebra of the Stone space A* of A. Define $r : A \to \mathcal{P}(\mathcal{V})$ by $r(a) = a \cap \mathcal{V}$. The completeness of A simply says that r is injective. Clearly, r(A) is the required $(X \cup C,\alpha,\beta)$-field if the quantified structure on r(A) is that defined by the isomorphism r. If A is a $(X \cup C,\alpha,\beta)$-field, then for $a \neq 0$ in A there is $x \in a$, the point homomorphism h_x from A to 2 defined by $h_x(b) = 1$ iff $x \in b$ is a valid homomorphism such that $h_x(a) = 1$.

3.3 The map r from A to $\mathcal{P}(\mathcal{V})$ in proof 3.2 is defined and a Boolean homomorphism having the property $r(\bigvee_{p\in P} a_p) = \bigcup_{p\in P} r(a_p)$ for the families considered in 3.2, whether A is complete or not. We shall see later that in the case A = TE, r(A) can be given a natural $(X \cup C,\alpha,\beta)$-

structure such that r becomes a $(X \cup C,\alpha,\beta)$-homomorphism, even if TE is not complete. So there is, in a sense we shall make precise later, a "free complete" $(X \cup C,\alpha,\beta)$-algebra for every $(X \cup C,\alpha)$-set E, which is a $(X \cup C,\alpha)$-set E, which is a $(X \cup C,\alpha,\beta)$-field, in fact a subalgebra of $\mathcal{P}(2^E)$.

Translating this into properties of a $L_{\alpha\beta}$-language, we see that there is always a set of axioms such that the formal theorems of $L_{\alpha\beta}$ yield a complete language. The axioms will certainly express the fact that the associated $(X \cup C,\alpha,\beta)$-algebra has the distributive properties of a $(X \cup C,\alpha,\beta)$-field. In order to define a set of axioms as "the set of all formulas having the distributive property..." we shall introduce the following notion of "nice families" of elements of a $(X \cup C,\alpha)$-set and of "nice elements" of a $(X \cup C,\alpha,\beta)$-algebra.

3.4 DEFINITION. Let ξ be an ordinal less than α.

(A) A ξ-set is a couple of sets (K,L) where $K = \bigcup_{0\le i\le\xi} K_i$, $K_i \cap K_j = \emptyset$ if $i \ne j$, $K_i \subset X$, $|K_i| < \alpha$, $K_i = M_i \cup N_i$ with $M_i \cap N_i = \emptyset$ and $L = \prod_{0\le i\le\xi} L_i$, $L_i = P_i \times Q_i$ and P_i, Q_i are cardinals less than α. If (K,L) is a ξ-set, we denote by L^i the set $\prod_{i\le j\le\xi} L_j$. Elements of L^i will be denoted $\ell_0^i, \ell_1^i, \ell_2^i$ etc. $\ell_1^i \ge \ell_2^k$ is intended to mean $i \ge k$ and $\ell_{1j} = \ell_{2j}$ for $i \le j \le \xi$.

(B) Suppose E is a $(X \cup C,\alpha)$-set and (K,L) a ξ-set. A nice family of elements of E defined on (K,L) is a couple of functions (s,t) satisfying:

(I) s maps $\coprod_{0<i\le\xi} L^i$ into ξ such that

(0) $i > s\ell^i$ for all $i \le \xi$ and $\ell^i \in L^i$

(1) if $i = j + 1 \le \xi$, then $s\ell^i = j$

(2) if $i > j > s\ell^i$ and $\ell^i \ge \ell^j$, then $s\ell^j \ge s\ell^i$

(3) for all i, $\ell^i \in L^i$, all k, $0 \le k \le \xi$, all $\ell_1^k, \ell_2^k \in L^k$, if $\ell_{1j} = \ell_j = \ell_{2j}$ for $\xi \ge j \ge i$ and $\ell_{1j} = \ell_{2j}$ for $s\ell^i \ge j \ge k$ then $s\ell_1^k = s\ell_2^k$.

(II) t maps $I_X \times L$ to E such that

(4) for every $\ell \in L$ there is $a_\ell \in E$ with $t(\sigma,\ell) = \sigma a_\ell$ for all $\sigma \in I_X$.

(5) for all i, $0 < i \le \xi$, all $\ell^i \in L^i$, all $\ell_1, \ell_2 \in L$ if $\ell_{1j} = \ell_{2j} = \ell_j$ for $\xi \ge j \ge i$ and $s\ell^i \ge j \ge 0$, and if $\sigma \restriction J = \tau \restriction J$ where $J = \left(\bigcup_{i>j>s\ell^i} K_j \right)'$, then $t(\sigma,\ell_1) = t(\tau,\ell_2)$.

(6) for all i, ℓ^{i+1} there exists $J \subset X$, $|J| < \beta$ such that $t(\sigma,\ell_1) = t(\tau,\ell_1)$ if $\ell^{i+1} \ge \ell_1$ and $\sigma \restriction (J \cup K_i') = \tau \restriction (J \cup K_i')$.

(C) ξ is called the length of the nice family (s,t).

Nice families have the following properties:

(D) Induction Principle

Let (s,t) be a nice family of length ξ. For every $\ell^i \in L^i$ there is a unique nice family $({}^{\ell^i}s, {}^{\ell^i}t)$ of length $s(\ell^i)$ defined on $\left({}^{s\ell^i}L = \prod_{0 \leq j \leq s\ell^i} L_j,\ {}^{s\ell^i}K = \bigcup_{0 \leq j \leq s\ell^i} K_j\right)$ such that

(1) ${}^{\ell^i}s(\overline{\ell}^k) = s(\ell_1^k)$ whenever $k \leq s\ell^i$, $\ell_1^k \leq \ell^i$ and $\ell_{1j} = \overline{\ell}_j$ for $k \leq j \leq s\ell^i$.

(2) ${}^{\ell^i}t(\sigma,\overline{\ell}) = t(\sigma,\ell_1)$ whenever $\ell_1 \leq \ell^i$ and $\ell_{1j} = \overline{\ell}_j$ for $0 \leq j \leq s\ell^i$.

(E) For every nice family (s,t) the set $\{t(1,\ell);\ \ell \in L\}$ has power less than α and there is a subset S of X of power less than α which supports $t(1,\ell)$ for all $\ell \in L$.

(F) We now suppose that E is a $(X \cup C,\alpha)$-subset of the $(X \cup C,\alpha,\beta)$-algebra A. We are going to associate elements of A to the nice family (s,t) of length ξ and of elements in E in the following way. First notice that $\bigvee_{\sigma \in I_{J_1}} \sigma a$ exists and equals $\bigvee_{\sigma \in I_{J_2}} \sigma a$ in A, whenever $a \in A$, $J_2 \subset X$ and $|J_2| < \beta$, $J_1 \subset X$ and $|J_1| < \alpha$ such that $\sigma a = \tau a$ for all σ,τ with $\sigma \restriction (J_1 \cap J_2) \cup (J_2' \cap J_1') = \tau \restriction (J_1 \cap J_2) \cup (J_1' \cap J_2')$. In this case there is a support J of a such that $(J \cap J_2) = (J \cap J_1)$ which is equivalent to "$(J_1 \cap J_2) \cup (J_1' \cap J_2')$ supports a."

If (s,t) is a nice family of length 0 of elements of E. Let us define

$$J(0,s,t) = \bigvee_{\varphi \in I_M} \bigwedge_{\varphi \in I_N} \bigvee_{p \in P} \bigwedge_{q \in Q} \varphi \otimes \psi a_{pq} = \exists M\ \forall N \bigvee_{p \in P} \bigwedge_{q \in Q} a_{pq}$$

$$(\text{here } a_{pq} = t(1,(p,q))).$$

By axiom 2.1,2 and (B6) it is easily seen that $J(0,s,t)$ is defined and supported by $S \cap K'$ if only S supports all $t(1,\ell)$ for $\ell \in L$.

Let us suppose that for every nice family (s^+,t^+) of length ξ^+ less than ξ, $J(\xi^+,s^+,t^+)$ is defined and supported by $(K^+)' \cap S^+$, where S^+ supports all $t^+(1,\ell^+)$, $\ell^+ \in L^+$ and let (s,t) be a nice family of length ξ. Then

$$J(\xi,s,t) = \bigvee_{\varphi \in I_{M_\xi}} \bigwedge_{\psi \in I_{N_\xi}} \bigvee_{p \in P_\xi} \bigwedge_{q \in Q_\xi} \varphi \otimes \psi J(s(p,q), {}^{(p,q)}s, {}^{(p,q)}t) = \exists M_\xi\ \forall N_\xi \bigvee_{p \in P_\xi} \bigwedge_{q \in Q_\xi} J(\xi^+,s^+,t^+)$$

where ξ^+,s^+,t^+ stand for $s(p,q)$, ${}^{(p,q)}s$ and ${}^{(p,q)}t$ respectively.

We have to show that there is a support S_b of $b = \bigvee_{p\in P_\xi} \bigwedge_{q\in Q_\xi} J(\xi^+,s^+,t^+)$ such that $|S_b \cap K_\xi| < \beta$. That will assure us that $J(\xi,s,t)$ is defined. Next we have to establish that $S \cap K'$ supports $J(\xi,s,t)$ whenever S supports all $t(1,\ell)$, $\ell \in L$. So, suppose $|S| < \alpha$ and S supports all $t(1,\ell)$ for $\ell \in L$. Then S supports all $t^+(1,\ell^+)$ for $\ell^+ \in L^+$, all t^+, all L^+. Now fix ℓ_ξ and consider the nice family (s^+,t^+) associated to ℓ_ξ. By the induction hypothesis and by (C5), $S^+ = S \cap (\bigcup_{\xi>j\geq 0} K_j)'$ supports $J(\xi^+,s^+,t^+)$. Now, by (C6), there is $J \subset K_\xi$, $|J| < \beta$ such that $S \cap (J \cup K_\xi')$ supports all $t(1,\ell)$, $\ell \in L$. By replacing in S^+ the support S by $S \cap (J \cup K_\xi')$ we see that $S \cap (J \cup (\bigcup_{\xi\geq j\geq 0} K_j)')$ supports $J(\xi^+,s^+,t^+)$. We conclude that $S_b = S \cap (J \cup K')$ supports b, that $S_b \cap K_\xi \subset J$ and that $S \cap K'$ supports $\exists M_\xi \forall N_\xi b = J(\xi,s,t)$.

3.5. Let us call an element of the form $J(\xi,s,t)$ where (s,t) is a nice family of elements of E, nice relative to E or only nice if the context makes clear, what subset E we are referring to. It is easily seen by induction on the length ξ that the elements of A, nice relative to E, are all in the $(X \cup C,\alpha,\beta)$-subalgebra of A generated by E. The converse of this is true: if E is closed under negation, then every element of the sub-algebra generated by E is nice. To show this, we have to establish some lemmas.

LEMMA 1. <u>If</u> f <u>is a</u> $(X \cup C,\alpha)$-<u>map from</u> E <u>to</u> F, (s,t) <u>a nice family of</u> E <u>of length</u> ξ, <u>then</u> (s,ft) <u>is a nice family of</u> F <u>of length</u> ξ. <u>If</u> f <u>is a</u> $(X \cup C,\alpha,\beta)$-<u>homomorphism from</u> A <u>to</u> B, <u>then</u> $f(J(\xi,s,t) = J(\xi,s,ft)$ <u>for every nice family</u> (s,t) <u>of</u> A.

LEMMA 2. <u>Let</u> (K,L), (K^*,L) <u>be</u> ξ-<u>sets</u>, (s,t) <u>a nice family defined on</u> (K,L) <u>and</u> $M_i \xrightarrow{\pi_i} M_i^*$, $N_i \xrightarrow{\rho_i} N_i^*$ $0 \leq i \leq \xi$, <u>let</u> $\chi = \bigotimes_{0\leq i\leq \xi} \pi_i \otimes \rho_i$. <u>Suppose that</u> $\tau \in I_\chi$ <u>and that for some</u> α-<u>support</u> S <u>of all</u> $t(1,\ell)$, $\ell \in L$, $K^* \cap \tau(S \cap K') = \emptyset$. <u>Then, if we define</u> t_τ^χ <u>by</u> $t_\tau^\chi(\sigma,\ell) = t(\sigma(\tau \otimes_K \chi),\ell)$ <u>for all</u> $\sigma \in I_\chi$, $\ell \in L$, (1) (s,t_τ^χ) <u>is a nice family of length</u> ξ <u>defined on</u> (K^*,L) <u>and</u> (2) $\tau J(\xi,st) = J(\xi,s,t_\tau^\chi)$.

<u>Proof</u>. (1) We only show that (II 5) holds. Suppose $\ell_{1j} = \ell_j = \ell_{2j}$ for $\xi \geq j \geq i$ and $\ell_{1j} = \ell_{2j}$ for $s\ell^i \geq j \geq 0$ and $\sigma_1 \upharpoonright J^* = \sigma_2 \upharpoonright J^*$ where $J^* = \left(\bigcup_{i>j>s\ell^i} K_j^*\right)'$. Then $t(\sigma_1(\tau \otimes_K \chi),\ell_1) = t(\sigma_1(\tau \otimes_K \chi) \upharpoonright S,\ell_1) = t(\sigma_2(\tau \otimes_K \chi) \upharpoonright S,\ell_2) = t(\sigma_2(\tau \otimes_K \chi),\ell_2)$, i.e. $t_\tau^\chi(\sigma_1,\ell_1) = t_\tau^\chi(\sigma_2,\ell_2)$.

(2) We use induction on the length ξ of the nice family (s,t).

If $\xi = 0$, then $\tau J(0,s,t) = \bigvee_{\varphi\in I_M} \bigwedge_{\psi\in I_N} \bigvee_{p\in P} \bigwedge_{q\in Q} \tau \underset{K}{\otimes} (\varphi \otimes \psi)a_{pq} =$

$\bigvee_{\varphi} \bigwedge_{\psi} \bigvee_{p} \bigwedge_{q} (\overline{\varphi\hat{\pi}} \otimes \overline{\psi\hat{\rho}})(\tau \underset{K}{\otimes} \chi)a_{pq}$ where $\hat{\pi}$ and $\hat{\rho}$ are defined by $M \underset{\hat{\pi}}{\overset{\pi}{\rightleftarrows}} M^*$ and $N \underset{\hat{\rho}}{\overset{\rho}{\rightleftarrows}} N^*$. Now, let $b = \bigvee_{p\in P} \bigwedge_{q\in Q} (\tau \underset{K}{\otimes} \chi)a_{pq}$. If S supports all $t(1,\ell)$, then $\tau \underset{K}{\otimes} \chi(S)$ supports b, and we conclude that $\chi(K) \cup \tau(S \cap K')$ supports b. So $J(0,s,t_\tau^\chi) = \exists M^* \forall N^* b = \exists \pi(M) \forall \rho(N) b = \bigvee_{\varphi\in I_M} \bigwedge_{\psi\in I_N} \overline{\varphi\hat{\pi}} \otimes \overline{\psi\hat{\rho}} b =$ $\tau J(0,s,t)$.

Suppose (2) true for every nice family of length $i < \xi$. We have $J(\xi,s,t_\tau^\chi) = \bigvee_{\varphi^*\in I_{M_\xi^*}} \bigwedge_{\psi^*\in I_{N_\xi^*}} \bigvee_{p\in P_\xi} \bigwedge_{q\in Q_\xi} \varphi^* \otimes \psi^* J(\xi^+,s^+,t_\tau^{\chi^+})$ where ξ^+, s^+, etc. are defined as in (F). Reasoning as in the case $\xi = 0$, we obtain $J(\xi,s,t_\tau^\chi) = \bigvee_{\varphi\in I_{M_\xi}} \bigwedge_{\psi\in I_{N_\xi}} \bigvee_p \bigwedge_q \overline{\varphi\hat{\pi}} \otimes \overline{\psi\hat{\rho}}\, J(\xi^+,s^+,t_\tau^{\chi^+})$.

So, all we have to show is: $\overline{\varphi\hat{\pi}} \otimes \overline{\psi\hat{\rho}}\, J(\xi^+,s^+,t_\tau^{\chi^+}) = \tau \underset{K_\xi}{\otimes} (\varphi \otimes \psi)\, J(\xi^+,s^+,t^+)$ for all φ, ψ, ℓ_ξ.

Let $\overline{\tau} = \tau \underset{K_\xi}{\otimes} (\varphi \otimes \psi)$ and $\overline{\tau}^* = \overline{\varphi\hat{\pi}} \otimes \overline{\psi\hat{\rho}}$. Choose $\overline{K}$ such that $(\overline{K},L^+)$ yields a ξ^+-set, that $[\overline{\tau}(S \cap (K^+)') \cup \overline{\tau}^*((\tau \underset{K}{\otimes} \chi)\, S \cap (K^{*+})')] \cap \overline{K} = \emptyset$ and that there are injections from M_i^* into $\overline{M}_i$ and from N_i^* into $\overline{N}_i$ for $0 \leq i \leq \xi^+$. Call μ the corresponding injection from K^{*+} into $\overline{K}$ and ν the injection $\mu\chi^+$ from K^+ into $\overline{K}$ (χ^+, of course, is $\underset{0\leq i\leq \xi^+}{\otimes} \pi_i \otimes \rho_i$). Then $\overline{\tau}\, J(\xi^+,s^+,t^+) = J(\xi^+,s^+,t^{+\nu}_{\overline{\tau}})$ and $\overline{\tau}^*\, J(\xi^+,s^+,t_\tau^{\chi^+}) = J\left(\xi^+,s^+,(t_\tau^{\chi^+})^{\mu}_{\overline{\tau}^*}\right)$. We only have to show that $t^{+\nu}_{\overline{\tau}}(\sigma,\ell^+) = (t_\tau^{\chi^+})^{\mu}_{\overline{\tau}^*}(\sigma,\ell^+)$ for all $\sigma \in I_X, \ell^+ \in L^+$. But $t^{+\nu}_{\overline{\tau}}(\sigma,\ell^+) = t(\sigma(\overline{\tau} \underset{K^+}{\otimes} \nu),\ell_1)$ and $(t_\tau^{\chi^+})^{\mu}_{\overline{\tau}^*}(\sigma,\ell^+) = t(\sigma(\overline{\tau}^* \underset{K^{*+}}{\otimes} \mu)(\tau \underset{K}{\otimes} \chi),\ell_1)$ where $\ell_{1\xi} = \ell_\xi$, $\ell_{1j} = \ell_j^+$, $0 \leq j \leq \xi^+$. It is easily shown that for $J = \left(\bigcup_{\xi > j > \xi^+} K_j\right)'$, $\sigma(\overline{\tau}^* \underset{K^{*+}}{\otimes} \mu)(\tau \underset{K}{\otimes} \chi) \restriction S \cap J = \sigma(\overline{\tau} \underset{K^+}{\otimes} \nu) \restriction S \cap J$, which finishes the proof.

LEMMA 3. _Let_ (K,L), (K,L^*) _be_ ξ-_sets such that there are surjections_ f_i _from_ P_i^* _onto_ P_i _and_ g_i _from_ Q_i^* _onto_ Q_i _for_ $0 \leq i \leq \xi$. _Let_ (s,t) _be a nice family on_ (K,L). _Define_ (s^*,t^*) _on_ (K,L^*) _by_ $s^*(\ell^{*i}) = s(h^i(\ell^{*i}))$ _and_ $t^*(\sigma,\ell^*) = t(\sigma,h(\ell^*))$ _where_ h^i _is the surjection from_ L^{*i} _onto_ L^i _induced by_ (f_j,g_j) $i \leq j \leq \xi$ _and_ h _stands for_ h^0. _Then_

(1) (s^*,t^*) _is a nice family on_ (K,L^*)

(2) $J(\xi,s,t) = J(\xi,s^*,t^*)$.

Proof. We only show (2), assuming (1) true for all nice families (s,t). Suppose the nice family (s,t) has length 0. All we have to prove is: $\bigvee_{p\in P} \bigwedge_{q\in Q} a_{pq} = \bigvee_{p^*\in P^*} \bigwedge_{q^*\in Q^*} a_{f(p^*)g(q^*)}$ where f,g are surjections from P^* onto P and Q^* onto Q respectively. But this equality is evident. We assume (2) true for all nice families of length less than ξ. Let $a_{p^*_\xi q^*_\xi} = J(\xi^{*+}, s^{*+}, t^{*+})$ where $\xi^* = s^*(p^*,q^*)$ and $a_{p_\xi q_\xi} = J(\xi^+, s^+, t^+)$ where $\xi^+ = s(p_\xi, q_\xi)$. Then, reasoning as for the case $\xi = 0$, it is sufficient to prove that $a_{p^*_\xi q^*_\xi} = a_{f_\xi(p^*_\xi)g_\xi(q^*_\xi)}$ for all p^*_ξ, q^*_ξ. First $\xi^{*+} = s^*(p^*_\xi, q^*_\xi) = s(f_\xi(p^*_\xi), g_\xi(q^*_\xi)) = \xi^+$. Suppose that (s^{*+}, t^{*+}) is the family of length ξ^{*+} associated by (D) to (p^*_ξ, q^*_ξ) and that (s^+, t^+) is the family of length ξ^+ associated by (D) to $(f_\xi(p^*_\xi), g_\xi(q^*_\xi))$. We want to show that $s^{*+} = s^{+*}$ and $t^{*+} = t^{+*}$ which is a matter of routine. So we have $a_{p^*_\xi q^*_\xi} = J(\xi^+, s^{+*}, t^{+*}) = J(\xi^+, s^+, t^+) = a_{f_\xi(p^*_\xi)g_\xi(q^*_\xi)}$.

LEMMA 4. *Let* (K,L) *be a* ξ*-set,* (s,t) *a nice family on* (K,L), ζ *an ordinal,* $\xi \leq \zeta < \alpha$. *Define* $\overline{K} = E_\zeta(K)$ *by* $\overline{K}_i = \emptyset$ *if* $\zeta \geq i > \xi$ *and* $\overline{K}_i = K_i$ *if* $\xi \geq i \geq 0$, *and* $\overline{L} = E_\zeta(L)$ *by* $\overline{P}_i = \overline{Q}_i = 1$ *if* $\zeta \geq i > \xi$ *and* $\overline{P}_i = P_i, \overline{Q}_i = Q_i$ *if* $\xi \geq i \geq 0$. *Denote* h^i *the obvious bijection from* $\overline{L}^i$ *to* L^i *for* $\xi \geq i \geq 0$. *Define* $(\overline{s},\overline{t})$ *on* $(\overline{K},\overline{L})$ *by* $\overline{t}(\sigma,\overline{\ell}) = t(\sigma, h^0(\overline{\ell}))$ *and*

$$\overline{s}(\overline{\ell}^i) = \begin{cases} i-1 & \text{if } \zeta \geq i > \xi \text{ and } i \text{ is not a limit ordinal} \\ \xi & \text{if } \zeta \geq i > \xi \text{ and } i \text{ is a limit ordinal} \\ s(h^i(\overline{\ell}^i)) & \text{if } \xi \geq i \geq 0 \,. \end{cases}$$

Then

(1) $(\overline{K},\overline{L})$ *is a* ζ*-set*

(2) $(\overline{s},\overline{t})$ *is a nice family on* $(\overline{K},\overline{L})$

(3) $J(\xi,s,t)' = J(\zeta,\overline{s},\overline{t})$.

LEMMA 5. *There are a strictly increasing function* $\xi \to \overline{\xi}$ *from* α *into* α, *a function* $(K,L) \to (\overline{K},\overline{L})$ *from the* ξ*-sets to the* $\overline{\xi}$*-sets, and a function* $(s,t) \to (\overline{s},\overline{t})$ *from the nice families on* (K,L) *to the nice families on* $(\overline{K},\overline{L})$ *such that*

(i) *if all* $t(\sigma,\ell)$ *are in some subset* E *of* A, *then all* $\overline{t}(\sigma,\overline{\ell})$ *are in* $'E = \{a';\ a \in E\}$.

(ii) $J(\xi,s,t)' = J(\overline{\xi},\overline{s},\overline{t})$.

Proof. (i) Every ordinal ξ has a unique decomposition $\xi = k_\xi + n_\xi$ where k_ξ is a limit ordinal and n_ξ an integer ≥ 0. Define strictly increasing functions m, n from α into α by $m(\xi) = k_\xi + 4n_\xi$ and $n(\xi) = m(\xi) + 3$. Then for all j, $0 \leq j \leq n(\xi)$, there are unique i and r such that $0 \leq i \leq \xi$, $r \in \{0,1,2,3\}$ and $j = m(i) + r$. Note that $n(i) < m(i + 1)$.

Let $\bar{\xi} = n(\xi)$. Define $\bar{K}$ by $\bar{K} = \bigcup_{0 \leq j \leq \bar{\xi}} \bar{M}_j \cup \bar{N}_j$ where for $0 \leq i \leq \xi$

$\bar{M}_{m(i)} = \bar{N}_{m(i)} = \bar{M}_{m(i)+1} = \bar{N}_{m(i)+1} = \emptyset$, $\bar{M}_{m(i)+2} = N_i$, $\bar{N}_{m(i)+2} = \emptyset = \bar{M}_{m(i)+3}$, $\bar{N}_{m(i)+3} = M_i$. Define $\bar{L}$ by $\bar{L} = \prod_{0 \leq j \leq \bar{\xi}} \bar{P}_j \times \bar{Q}_j$ where for $0 \leq i \leq \xi$ $\bar{P}_{m(i)} = Q_i$, $\bar{Q}_{m(i)} = 1 = \bar{P}_{m(i)+1}$, $\bar{Q}_{m(i)+1} = P_i$, $\bar{P}_{m(i)+2} = \bar{Q}_{m(i)+2} = \bar{P}_{m(i)+3} = \bar{Q}_{m(i)+3} = 1$. Clearly, $(\bar{K},\bar{L})$ is a $\bar{\xi}$-set.

Define h from $\bar{L}$ onto L by $h((\bar{p}_j,\bar{q}_j)_{0 \leq i \leq \xi} = (\bar{q}_{m(i)+1},\bar{p}_{m(i)})_{0 \leq i \leq \xi}$. h is a one-to-one function from $\bar{L}$ onto L and the "restriction" of h to $\bar{L}^{m(i)}$ is a one-to-one function from $\bar{L}^{m(i)}$ onto L^i which we also denote by h. The function $\ell \to \bar{\ell}$ will designate the inverse of h, for example, $h(\bar{\ell}^{m(i)}) = \ell^i$ and $\overline{\ell^i} = \bar{\ell}^{m(i)}$. Now define $\bar{s}$ by

$$\bar{s}(\bar{\ell}^{m(i)+r}) = \begin{cases} m(i) + r - 1 & \text{if} \quad r \in \{1,2,3\} \\ n(s(h(\bar{\ell}^{m(i)}))) = \overline{s\ell^i}, & \text{if} \quad r = 0 \end{cases}$$

for $0 \leq i \leq \xi$.

$\bar{s}$ has properties (B I) of 3.4:

(0) from $s\ell^i = s(h(\bar{\ell}^{mi})) < i$ we infer $\bar{s}(\bar{\ell}^{mi}) = n(s\ell^i) < m(i)$ and (I 0) holds for $\bar{s}$.

(1) if $m(i) = j + 1$, then i has a predecessor $i - 1$ and $\bar{s}(\bar{\ell}^{mi}) = n(i - 1) = m(i) - 1 = j$, therefore (I 1) is true for $\bar{s}$.

(2) if $m(i) > m(j) > \bar{s}(\bar{\ell}^{mi})$ and $\bar{\ell}^{mi} \geq \bar{\ell}^{mj}$, then $s\ell^i < j < i$ and $\ell^i \geq \ell^j$ which implies $\bar{s}(\bar{\ell}^{mj}) = \overline{s\ell^j} \geq \overline{s\ell^i} = \bar{s}(\bar{\ell}^{mi})$. We see that (I 2) is verified by $\bar{s}$.

(3) if $\bar{\ell}_{1j} = \bar{\ell}_{2j}$ for $m(k) \leq j \leq \bar{s}(\bar{\ell}_1^{mi})$ and $mi \leq j \leq \bar{\xi}$, then $(\bar{q}_{1m(j)+1},\bar{p}_{1m(j)}) = (\bar{q}_{2m(j)+1},\bar{p}_{2m(j)})$, i.e., $\ell_{1j} = \ell_{2j}$ for $k \leq j \leq s\ell_1^i$ and $i \leq j \leq \xi$, where of course ℓ_1^k, ℓ_2^k stands for $h(\bar{\ell}_1^{mk})$ and $h(\bar{\ell}_2^{mk})$ respectively. We infer $\bar{s}(\bar{\ell}_1^{mk}) = \overline{s\ell_1^k} = \overline{s\ell_2^k} = \bar{s}(\bar{\ell}_2^{mk})$. Therefore (I 3) holds for $\bar{s}$.

Define $\overline{t}$ by $\overline{t}(\sigma,\overline{\ell}) = t(\sigma,\ell)'$ for all $\sigma \in I_X$, $\ell \in L$. $\overline{t}$ has properties (B II):

(4) it is easily seen that (4) holds for $\overline{t}$.

(5) if $\overline{\ell}_{1j} = \overline{\ell}_{2j}$ for $0 \leq j \leq \overline{s}(\overline{\ell}_1^{mi})$ and $mi \leq j \leq \xi$, then $\ell_{1j} = \ell_{2j}$ for $0 \leq j \leq s\ell_1^i$ and $i \leq j \leq \xi$. In addition $\bigcup_{\overline{s}(\overline{\ell}_1^{mi}) < j < mi} \overline{K}_j = \bigcup_{s(\ell_1^i) < j < i} K_j$, therefore, if we denote the complement of this set by J, then $\sigma \restriction J = \tau \restriction J$ implies $\overline{t}(\sigma,\overline{\ell}_1) = t(\sigma,\ell_1)' = t(\tau,\ell_2)' = \overline{t}(\tau,\overline{\ell}_2)$. We see that (5) is true for $\overline{t}$.

(6) $\overline{\ell}^{m(i)+r+1} \geq \overline{\ell}_1^{m(i)+r+1}$ implies $\overline{\ell}^{m(i+1)} \geq \overline{\ell}_1^{m(i+1)}$ and $\overline{K}'_{m(i)+r} \subset K'_i$ for $r \in \{0,1,2,3\}$. So, if J has the property II, (6) for i, $\ell^{i+1} = h(\overline{\ell}^{m(i+1)})$ and (s,t), then $\overline{\ell}^{m(i)+r+1} \geq \overline{\ell}_1^{m(i)+r+1}$ and $\sigma \restriction J \cup \overline{K}'_{m(i)+r} = \tau \restriction J \cup \overline{K}'_{m(i)+r}$ imply $\ell^{i+1} \geq \ell_1^{i+1}$ and $\sigma \restriction J \cup K'_i = \tau \restriction J \cup K'_i$, therefore $\overline{t}(\sigma,\overline{\ell}_1) = t(\sigma,\ell_1)' = t(\tau,\ell_1)' = \overline{t}(\tau,\overline{\ell}_1)$ which shows that II 6 holds for $\overline{t}$.

The proof of (i) is complete.

(ii) Let us show (ii) by induction on the length ξ of the nice family (s,t).

If $\xi = 0$, we have to show that

$$J(0,s,t)' = \bigwedge_{\varphi\in I_M} \bigvee_{\psi\in I_N} \bigwedge_{p\in P} \bigvee_{q\in Q} (\varphi \otimes \psi\, a_{pq})' = J(3,\overline{s},\overline{t})$$

which is an easy consequence of the definition of $(\overline{s},\overline{t})$. We suppose (ii) true for all families of length less than ξ.

$$J(\xi,s,t)' = \bigwedge_{\varphi\in I_{M_\xi}} \bigvee_{\psi\in I_{N_\xi}} \bigwedge_{p\in P_\xi} \bigvee_{q\in Q_\xi} \varphi \otimes \psi\, J(\xi^+,s^+,t^+)'$$

$$= \bigwedge_{\varphi\in I_{M_\xi}} \bigvee_{\psi\in I_{N_\xi}} \bigwedge_{p\in P_\xi} \bigvee_{q\in Q_\xi} \varphi \otimes \psi\, J(\overline{s((p,q))}, \overline{(p,q)_s}, \overline{(p,q)_t})$$

by the induction hypothesis. (Remember that the imprecise notation ξ^+, s^+, t^+ stands for $s((p,q))$, $(p,p)_s$, $(p,q)_t$ respectively). It is easily seen that

$$J(\overline{\xi},\overline{s},\overline{t}) = \bigwedge_{\varphi\in I_{M_\xi}} \bigvee_{\psi\in I_{N_\xi}} \bigwedge_{p\in P_\xi} \bigvee_{q\in Q_\xi} \varphi \otimes \psi\, J(\overline{s}(\overline{(p,q)}), \overline{(p,q)}_{\overline{s}}, \overline{(p,q)}_{\overline{t}}) .$$

So all we have to prove is that the nice families $(\overline{(p,q)}_s, \overline{(p,q)}_t)$ and $(\overline{(p,q)}_{\overline{s}}, \overline{(p,q)}_{\overline{t}})$ are the same. By definition $\overline{s}(\overline{(p,q)}) = \overline{s((p,q))}$. Therefore these two families have the same length and it is easily seen that their sets of definition, $(\overline{(p,q)}_K, \overline{(p,q)}_L)$ and $(\overline{(p,q)}_{\overline{K}}, \overline{(p,q)}_{\overline{L}})$ are the same.

Now suppose $0 \leq k \leq \overline{s((p,q))}$, $u^k \in (\overline{(p,q)}_L)^k$. If k has a predecessor $k - 1$, then then $\overline{(p,q)}_s(u^k)$ and $\overline{(p,q)}_{\overline{s}}(u^k)$ are equal to $k - 1$. If $k = m(i)$, then $\overline{(p,q)}_s(u^{mi}) = \overline{s(\ell_1^i)}$ where $\ell_{1\xi} = (p,q)$ and $\ell_{1j} = h(u^{mi})_j$ for $i \leq j \leq s((p,q))$. On the other side $\overline{(p,q)}_{\overline{s}}(u^{mi}) = \overline{s}(\overline{\ell}_1^{mi})$ where $\overline{\ell}_1^{m\xi} = \overline{(p,q)}$ and $\ell_{1j} = u_j$ for $mi \leq j \leq \overline{s((p,q))}$. But by definition $s(\overline{\ell}_1^{mi}) = \overline{s(\ell_1^i)}$. This shows that $\overline{(p,q)}_s = \overline{(p,q)}_{\overline{s}}$.

Let $u \in \overline{(p,q)}_L$, $\sigma \in I_X$. Then $\overline{(p,q)}_t(\sigma,u) = t(\sigma,\ell_1)'$ where $\ell_{1\xi} = (p,q)$ and $h(u)_j = \ell_{1j}$ $0 \leq j \leq s(p,q)$. On the other side, $\overline{(p,q)}_{\overline{t}}(\sigma,u) = \overline{t}(\sigma,\overline{\ell}_1)$ where $\overline{\ell}_1^{m\xi} = \overline{(p,q)}$ and $\overline{\ell}_{1j} = u_j$, $0 \leq j \leq \overline{s((p,q))}$. By definition $\overline{t}(\sigma,\overline{\ell}_1) = t(\sigma,\ell_1)'$ which proves that $\overline{(p,q)}_t = \overline{(p,q)}_{\overline{t}}$. This completes the proof of (ii) and of Lemma 5.

Using Lemmas 2, 3, 4 we see that for every cardinal R less than α there is a function which associates to a family $(s_r,t_r)_{r\in R}$ of nice families a nice family $({}^Rs, {}^Rt)$ such that

(1) ${}^Rt(\sigma, {}^R\ell)$ is an element of $\bigcup_{\substack{r\in R, \ell_r\in L_r \\ \sigma\in I_X^r}} \{t_r(\sigma,\ell_r)\}$ for all $\sigma \in I_X$ and all ${}^R\ell \in {}^RL$.

(2) $\bigvee_{r\in R} J(\xi_r, s_r, t_r) = J({}^R\xi, {}^Rs, {}^Rt)$.

Further we see that for every $J \subset X$, $|J| < \beta$ there is a function which associates to a nice family (s,t) of length ξ a nice family $({}^Js, {}^Jt)$ of length $\xi + 1$ such that

(1) ${}^Jt(\sigma, {}^J\ell)$ is an element of $\bigcup_{\substack{\sigma\in I_X \\ \ell\in L}} \{t(\sigma,\ell)\}$ for all $\sigma \in I_X$, ${}^J\ell \in {}^JL$.

(2) $\exists J\, J(\xi,s,t) = J(\xi + 1, {}^Js, {}^Jt)$.

We resume the results of this section in the following

THEOREM. Let E be a $(X \cup C,\alpha)$-subset of the $(X \cup C,\alpha,\beta)$-algebra A. If E is closed under negation, then the $(X \cup C,\alpha,\beta)$-algebra of A generated by E has as elements exactly the elements of A which are nice with respect to E.

3.6. We will now establish the connection between the notion of completeness of a $(X \cup C,\alpha,\beta)$-algebra A and the notion of nice elements in A.

DEFINITION. Let (s,t) be a nice family on the ξ-set (K,L). Recall that $K = \bigcup_{0\leq i\leq \xi} M_i \cup N_i$, $L = \prod_{0\leq i\leq \xi} P_i \times Q_i$. Write N^i for $\bigcup_{i\leq j\leq \xi} N_j$, Q^i for $\prod_{i\leq j\leq \xi} Q_j$, $0 \leq i \leq \xi$ and N for N^0, Q for Q^0, and define M^i, P^i, M, P similarly. Identify L and $P \times Q$. A choice function c associated to (s,t) is a couple $c = (c_1,c_2) \in C(\xi,K,L)$ where

$$c_1 \in \prod_{0\leq i\leq \xi} I_{M_i}^{(I_{N^{i+1}}\times Q^{i+1})} = c_1(\xi,K,L)$$

and

$$c_2 \in \prod_{0\leq i\leq \xi} P_i^{(I_{N^i}\times Q^{i+1})} = c_2(\xi,K,L).$$

For $\psi \in I_N$, $q \in Q$ write $\psi^i = \psi \restriction N^i$, $q^i = q \restriction Q^i$ (where $(q_j)_{0\leq j\leq \xi} \restriction Q^i = (q_j)_{i\leq j\leq \xi}$) and define

$$\bar{c}_1(\psi,q) = \bigotimes_{0\leq i\leq \xi} c_{1i}(\psi^{i+1},q^{i+1}) \in I_M$$

$$\bar{c}_2(\psi,q) = (c_{2i}(\psi^i,q^{i+1}))_{0\leq i\leq \xi} \in P .$$

We call choice-set associated to t and c the following set

$$(tc) = \{t(\bar{c}_1(\psi,q) \otimes \psi, (\bar{c}_2(\psi,q),q));\ \psi \in I_N,\ q \in Q\} .$$

We say that the choice-set (tc) is contradictory, if there exist $\psi_1,\psi_2 \in I_N$ and $q_1,q_2 \in Q$ such that

$$t(\bar{c}_1(\psi_1,q_1) \otimes \psi_1,(\bar{c}_2(\psi_1,q_1),q_1)) = (t(\bar{c}_1(\psi_2,q_2) \otimes \psi_2,(\bar{c}_2(\psi_2,q_2),q_2)))' .$$

3.7 LEMMA. Suppose h is a valid homomorphism from A to 2, $a \in A$ and $a = J(\xi,s,t)$ for some nice family (s,t) defined on the ξ-set (K,L). Then $h(a) = 1$ if and only if there is a choice function $c \in C(\xi,K,L)$ such that $h((tc)) = \{1\}$.

Proof. First let us show by induction on the length ξ of the nice family (s,t) that $h(J(\xi,s,t)) = 1$ whenever there is a choice function $c \in C(\xi,K,L)$ such that $h((t,c)) = \{1\}$. It is easily seen for $\xi = 0$. Assume it true for all families of length less than ξ. We have

$$h(J(\xi,s,t)) = \bigvee_{\varphi_\xi} \bigwedge_{\psi_\xi} \bigvee_{p_\xi} \bigwedge_{q_\xi} h\left(\varphi_\xi \otimes \psi_\xi J(s(p_\xi,q_\xi), {}^{(p_\xi,q_\xi)}s, {}^{(p_\xi,q_\xi)}t)\right).$$

For simplifying, write ρ instead of $c_{1\xi}(\emptyset)$, $j(\psi_\xi,q_\xi)$ instead of $s(c_{2\xi}(\psi_\xi),q_\xi)$, ${}^{(\psi_\xi q_\xi)}s$ instead of ${}^{(c_{2\xi}(\psi_\xi),q_\xi)}s$ and ${}^{(\psi_\xi),q_\xi)}t$ instead of ${}^{(c_{2\xi}(\psi_\xi),q_\xi)}t$. Then it is sufficient to show that

$$h\left(\rho \otimes \psi_\xi J(j(\psi_\xi,q_\xi), {}^{(\psi_\xi,q_\xi)}s, {}^{(\psi_\xi,q_\xi)}t)\right) = 1$$

for all $\psi_\xi \in I_N$ and all $q_\xi \in Q_\xi$.

Define for each (ψ_ξ,q_ξ) a choice function ${}^{\psi_\xi q_\xi}c$ in $C\left(j(\psi_\xi,q_\xi), {}^{j(\psi_\xi,q_\xi)}K, {}^{j(\psi_\xi,q_\xi)}L\right)$ by

$${}^{\psi_\xi q_\xi}c_{1i}(\mu^{i+1},r^{i+1}) = c_{1i}(\psi_1^{i+1},q_1^{i+1})$$

for $0 \leq i \leq j(\psi_\xi,q_\xi)$, $\mu^{i+1} = \bigotimes_{i<j\leq j(\psi_\xi,q_\xi)} \mu_j$ with $\mu_j \in I_{N_j}$ and $r^{i+1} \in \prod_{i<j\leq j(\psi_\xi,q_\xi)} Q_j$ where $\psi_{1\xi} = \psi_\xi$, $q_{1\xi} = q_\xi$, ψ_{1j} and q_{1j} are arbitrary but fixed for $\xi > j > j(\psi_\xi,q_\xi)$ and $\psi_{1j} = \mu_j$, $q_{1j} = r_j$ for $j(\psi_\xi,q_\xi) \geq j \geq i + 1$. Define ${}^{\psi_\xi q_\xi}c_{2i}$ similarly.

Then ${}^{(\psi_\xi,q_\xi)}t\left(\rho \otimes \psi_\xi \bigotimes_{K^+} ({}^{\psi_\xi q_\xi}\bar{c}_1(\mu,r) \otimes \mu), ({}^{\psi_\xi q_\xi}\bar{c}_2(\mu,r),r)\right) = 1$ for all μ, r (K^+ stands for ${}^{j(\psi_\xi,q_\xi)}K$). Using the induction hypothesis we show that $h\left(\rho \otimes \psi_\xi J(j(\psi_\xi,q_\xi), {}^{(\psi_\xi,q_\xi)}s, {}^{(\psi_\xi,q_\xi)}t)\right) = 1$.

Now let us show, also by induction on the length ξ of the nice family (s,t) that $h(J(\xi,s,t)) = 1$ implies that there is a choice function $c \in C(\xi,K,L)$ such that $h((t,c)) = \{1\}$. This is easily seen for $\xi = 0$. Suppose it is true for all families of length less than ξ. We have

$$1 = h(J(\xi,s,t)) = \bigvee_{\varphi_\xi} \bigwedge_{\psi_\xi} \bigvee_{p_\xi} \bigwedge_{q_\xi} h\Big(\varphi_\xi \otimes \psi_\xi\ J(s(p_\xi,q_\xi),\ {}^{(p_\xi,q_\xi)}s,\ {}^{(p_\xi,q_\xi)}t)\Big).$$

Therefore there is ρ in I_{M_ξ} such that for all $\psi_\xi \in I_{N_\xi}$ there is a $p_\xi = p_\xi(\psi_\xi)$ such that for all q_ξ we have

$$h\Big(\rho \otimes \psi_\xi\ J(s(p_\xi(\psi_\xi),q_\xi),\ {}^{(p_\xi(\psi_\xi),q_\xi)}s,\ {}^{(p_\xi(\psi_\xi),q_\xi)}t)\Big) = 1 .$$

Using the induction hypothesis we show that for every (ψ_ξ,q_ψ) there is a choice function

${}^{\psi_\xi q_\xi}c \in C\Big(j(\psi_\xi,q_\xi),\ {}^{j(\psi_\xi,q_\xi)}K,\ {}^{j(\psi_\xi,q_\xi)}L\Big)$ such that

$$h\Big({}^{(\psi_\xi,q_\xi)}t((\rho \otimes \psi_\xi) \underset{K^+}{\otimes} ({}^{\psi_\xi q_\xi}\overline{c}_1(\mu,r) \otimes \mu),({}^{\psi_\xi q_\xi}\overline{c}_2(\mu,r),r))\Big) = 1$$

for all μ, r. Here $j(\psi_\xi,q_\xi)$ stands for $s(p_\xi(\psi_\xi),q_\xi)$, the other notations are used as above. This permits us to define a choice function $c \in C(\xi,K,L)$ such that $h\Big(t(\overline{c}_1(\psi,q) \otimes \psi,\ (\overline{c}_2(\psi,q),q)\Big) = 1$ for all ψ, q.

Notice that in both parts of the proof, when we say "Using the induction hypothesis we show that...," this induction hypothesis is not directly applicable and we have to make use of a lemma which is not explicitly formulated here.

COROLLARY. <u>If</u> $A = TE$, <u>if</u> (s,t) <u>is nice with respect to</u> $E \cup {}'E$ (<u>all</u> $t(\sigma,\ell)$ <u>are elements or negations of elements of</u> E), <u>then there is a valid homomorphism</u> h <u>from</u> TE <u>to</u> 2 <u>such that</u> $h(J(\xi,s,t)) = 1$ <u>if and only if there is a choice function</u> $c \in C(\xi,K,L)$ <u>such that</u> (tc) <u>is not contradictory</u>.

3.8 THEOREM. <u>Let</u> E <u>be a</u> $(X \cup C,\alpha)$<u>-set. Then</u> TE <u>is complete if and only if for every nice family</u> (s,t) <u>of</u> TE (<u>nice with respect to</u> TE) <u>defined on a</u> ξ<u>-set</u> (K,L) $J(\xi,s,t) = 0$ <u>whenever</u>

(tc) *is contradictory for every choice function* $c \in C(\xi,K,L)$.

Proof. If TE is complete, then for $a \neq 0$ in TE there is a valid h such that $h(a) = 1$. Therefore if $a = J(\xi,s,t)$ there is by 3.7 $c \in C(\xi,K,L)$ such that $h((tc)) = \{1\}$, therefore (tc) cannot be contradictory. Now suppose that $a \in TE$ and $a \neq 0$. We want to show that there is a valid h such that $h(a) = 1$. E generates TE, *a fortiori*, $E \cup {}'E$ generates TE. Then there is a nice family (s,t) with respect to $E \cup {}'E$ such that $a = J(\xi,s,t)$ by Theorem 3.5. As $a \neq 0$, (tc) is not contradictory for some choice function c of $C(\xi,K,L)$ and by 3.7 there exists a valid h such that $h(a) = 1$.

4. *Nice Quantified Algebras*.

4.1 DEFINITION. We say that a quantified algebra A is nice if for every nice family (s,t) of A, defined on a ξ-set (K,L), the element $J(\xi,s,t) = 0$ whenever (tc) is contradictory for all choice functions c of $C(\xi,K,L)$. It is clear that every complete quantified algebra is nice.

4.2 LEMMA. *Let* E *be an* $(X \cup C,\alpha)$*-set, let* r *be the* $'\alpha$*-homomorphism from* TE *into* $\mathcal{P}(\mathcal{U})$ *defined in* 3.2 *and* 3.3. *Then* $r(TE)$ *has a natural* $(X \cup C,\alpha,\beta)$*-structure such that* r *is a* $(X \cup C,\alpha,\beta)$*-homomorphism*.

Proof. By showing that if (tc) is contradictory for every choice function c in $C(\xi,K,L)$, then $(t^X_\tau c^*)$ is contradictory for every c^* in $C(\xi,K^*,L)$ and all $\tau \in I_X$, all $X : K \mapsto K^*$.

4.3 THEOREM. (1) *For every* $(X \cup C,\alpha)$*-set* E *and every cardinal* $\beta \leq \alpha$, *there is a nice* $(X \cup C,\alpha,\beta)$*-algebra* JE *which contains* E *as a* $(X \cup C,\alpha)$*-subset and is generated by* E *such that every* $(X \cup C,\alpha)$*-mapping* f *from* E *to a nice* $(X \cup C,\alpha,\beta)$*-algebra* B *can be extended in a unique way to a* $(X \cup C,\alpha,\beta)$*-homomorphism from* JE *to* B.

(2) JE *is isomorphic to* $r(TE)$ *and to a* $(X \cup C,\alpha,\beta)$*-subfield of* $\mathcal{P}(2^E)$ *generated by the elementary clopen subsets of the Stone Space* 2^E (*an elementary clopen set* U_a *has the form* $U_a = \{h \in 2^E;\ ha = \mathcal{E}\}$ *where* $\mathcal{E} \in \{0,1\}$ *and* $a \in E$).

(3) JE *is complete*.

4.4 COROLLARY. TE *is complete if and only if* TE = JE.

4.5. Let us consider Example 2.2. Assume first that we omitted the equality axioms in the set of axioms of $\mathfrak{L}_{\alpha\beta}$. Let X be the set of variables and C the set of constants of the language.

Valid homomorphisms of $\mathcal{L}_{\alpha\beta}$ are in one-to-one correspondence with models of universe $X \cup C$. Let us call these models $(X \cup C)$-models and let us say that the formula a of the language is $(X \cup C)$-valid, if a is true in every $(X \cup C)$-model. When the power of the set of variables X is "big enough," for example, when $|X| = 2^{\alpha}$, then a formula is $(X \cup C)$-valid if and only if it is universally valid.

(1) Suppose that X is big enough. Then the language $L_{\alpha\beta}$ with formulas $\mathfrak{F}_{\alpha\beta}$ and provable formulas $\mathfrak{D}_{\alpha\beta}$ is complete if and only if $\mathcal{L}_{\alpha\beta} = \mathfrak{F}_{\alpha\beta}/\mathfrak{D}_{\alpha\beta}$ is complete. Let us translate: Denote by E the set of atomic formulas, by $\neg E$ the set of negations of atomic formulas of the language. To a nice family (s,t) of elements of $E \cup \neg E$ we associate a nice formula $J(\xi,s,t)$ similarly to 3.4. Now say that a nice formula $J(\xi,s,t)$ is super-nice if (tc) is contradictory for every choice function c. Then $L_{\alpha\beta}$ is complete if and only if $\mathfrak{D}_{\alpha\beta}$ contains the negations of all super-nice formulas.

(2) Let us define the complete language $C_{\alpha\beta}$ with conjunctions and disjunctions of less than α formulas and quantification over less than β variables as follows: Take as the set of axioms the negations of super-nice formulas plus the axioms of [2], Chapter 11 (at this moment without the equality axioms). Close the set under the three inference-rules of [2], Chapter 11. Call the obtained set $\mathfrak{M}_{\alpha\beta}$ the set of all provable formulas. Then $C_{\alpha\beta} = \mathfrak{F}_{\alpha\beta}|\mathfrak{M}_{\alpha\beta}$ is complete and equal to JE. The language $C_{\alpha\beta}$ with $\mathfrak{M}_{\alpha\beta}$ as provable formulas is complete. In the case $\alpha \leq \omega_1$, $\beta = \omega$ it is known that $\mathfrak{M}_{\alpha\beta}$ and $\mathfrak{D}_{\alpha\beta}$ are equal. Therefore the set of axioms in $\mathfrak{M}_{\alpha\beta}$ is unnecessarily big. But in most cases $\mathfrak{M}_{\alpha\beta}$ properly contains $\mathfrak{D}_{\alpha\beta}$, for example in the case $\alpha = \beta = \omega_1$, $L_{\alpha\beta}$ being not complete (see [2], Chapter 12).

(3) A filter F of a $(X \cup C,\alpha,\beta)$-algebra A is called a $(X \cup C,\alpha,\beta)$-filter, if

(0) $a_r \in F$ for $r \in R$ and $|R| < \alpha$, then $\bigwedge_{r\in R} a_r \in F$.

(1) if $a \in F$, $K \subset X$ and $|K < \beta$, then $\bigwedge_{\sigma\in I_K} \sigma a \in F$.

(2) if $a \in F$, $\sigma \in I_X$, then $\sigma a \in F$.

Of course, the set A/F of equivalence classes in A defined by the relation $(a + b)' \in F$ between elements a and b of A, has a natural $(X \cup C,\alpha,\beta)$-structure such that the natural surjection from A onto A/F is a $(X \cup C,\alpha,\beta)$-homomorphism. Now, let $\mathfrak{B}_{\alpha\beta}$ be the smallest subset of $\mathfrak{F}_{\alpha\beta}$ closend under the inference rules and containing $\mathfrak{M}_{\alpha\beta}$ and the equality axioms. Let $B_{\alpha\beta}$ be the $(X \cup C,\alpha,\beta)$-filter of $C_{\alpha\beta} = JE$ generated by the images of the equality axioms in JE. Then $\mathfrak{F}_{\alpha\beta}/\mathfrak{B}_{\alpha\beta} = JE/B_{\alpha\beta}$ is complete and so is the language with equality $E_{\alpha\beta}$ with formulas $\mathfrak{F}_{\alpha\beta}$ and provable formulas $\mathcal{B}_{\alpha\beta}$.

REFERENCES

[0] Fraissé, Roland, Cours de logique mathématique, Gauthiers-Villars, 1967.

[1] Halmos, Paul, Boolean Algebras, Princeton, New Jersey, Van Nostrand, 1963. Algebraic Logic, Chelsea Pub. Co. 1962.

[2] Karp, Carol, Languages with expressions of infinite length, North-Holland, 1964.

[3] Ponasse, Daniel, Problèmes d'Universalité s'introduisant dans l'Algébrisation de la Logique Mathématique, Nagoya, Mathematical Journal, 1962.

[4] Preller, Anne, Algèbres quantificés, Pub. Dept. Math. Lyon 1967, T-4, I.

[5] Rieger, L., On free $\aleph_\xi$-complete Boolean algebras, Fund. Math. 38, 35-52, 1951.

[6] Tarski, Alfred, A Representation Theorem for Cylindric Algebras, Bull. Amer. Math. Soc., vol. 58, 1.

UNIVERSITY OF CALIFORNIA, BERKELEY

NORMAL DERIVABILITY IN CLASSICAL LOGIC

W. W. TAIT*

The main result in this paper, the Elimination Theorem below, is a generalization of Gentzen's Hauptsatz [3] for finitary predicate logic to the logic of infinitary propositions, that is, propositions involving infinite disjunctions and conjunctions. For the sake of simplicity, I will discuss only classical logic; but the extension to intuitionistic logic is a routine matter. Of course, predicate logic already involves infinitary propositions in the form of quantifications. But these are of just one, particularly simple, kind of infinitary proposition; and there are others which also occur naturally. For example, arithmetical propositions and propositions of ramified type theory are built up from equations between numerical terms by means of countable disjunctions and conjunctions. Moreover, many of the results in the literature on these special kinds of infinitary proposition are simply instances of general theorems about infinitary logic—and are best understood in this way. Of course, this does not apply to those results about formal systems which depend on their combinatorial properties.

Although we will be dealing with infinitary propositions and infinite derivations of them, we will deal with these constructively—not, of course, in the narrow sense of Hilbert's finitism, but in the wider sense which admits (potential) infinities, in the form of rules of construction, as objects. As a consequence of this, we will obtain a constructive foundation for infinitary classical logic and for those parts of mathematics contained in it. Specifically, the Elimination Theorem will show constructively that the propositions with "cut-free" derivations are closed under the usual laws of classical logic. Also, the elementary propositions (atoms) with cut-free derivations are precisely

*This work has been partially supported by a NSF Grant GP-7640. The first draft of this paper was written while the author was an associate member of the Illinois Center for Advanced Study.

the axioms. Hence, by replacing the intended interpretation 'A is true' of the proposition 'A has a cut-free derivation,' a constructive foundation is obtained. A more technical point—of which I will not make anything in this paper—is that the constructive arguments are easily formalized in quantifier-free formal systems, the strength of which can readily be determined. This leads to a comparison of strength among those formal axiomatic theories which can be regarded as fragments of infinitary logic, and in particular, to consistency proofs for these theories relative to one another. We will discuss a less delicate comparison of these theories, which involves finding a bound for the "provable ordinals" (i.e., the ordinals of decidable well orderings for which induction can be proved) in them. This does not require formalizing the proof of the Elimination Theorem; but also, it does not serve to distinguish consistency problems: Of two theories with the same least bound on provable ordinals, the consistency of one may be a theorem of the other. This is because the provable ordinals are not changed when true purely universal propositions with decidable matrices (such as consistency statements) are added to the axioms. Nonconstructively, this applies to a much wider class of propositions, e.g., true arithmetical propositions.

The results on provable ordinals need a sharper form of the Elimination Theorem than Gentzen stated for finitary predicate logic. Gentzen's formulation is that every derivation can be transformed into a <u>cut-free</u> or <u>normal</u> derivation of the same result. We need, in addition, a bound on the (in general, transfinite) length of the normal derivation in terms of the length of the given derivation and the logical complexity (also transfinite, in general) of the cut formulae in it. Of course, in finitary logic, the derivations and propositions (even with quantifiers) are of finite length and complexity, resp.; and so the sharper formulation of the theorem does not add all that much. Besides the Elimination Theorem, we will also need the <u>Induction Theorem</u> below, which again is proved in the general setting of infinitary logic. It gives a bound on the ordinal of a decidable well-ordering in terms of the length of any normal derivation of the principle of induction on that ordering.

The Elimination and Induction Theorems have their origins in Gentzen's papers [4] and [5] on number theory, and more directly, in Schütte's work [9], [11], and [13] on number theory and ramified type theory. A constructive proof of the Elimination Theorem without ordinal bounds was given by Lorenzen [8] for ramified type theory.

The results of this paper were reported in [14], with a slightly different formulation of the system of infinitary propositional logic.

§1. The <u>propositional formulae</u>, or <u>pf</u> for short, are built up from <u>atoms</u> by means of the possibly infinitary operations

$$\bigvee_{i\in I} A_i \qquad \bigwedge_{i\in I} A_i$$

of disjunction and conjunction, resp. Here I is a constructive species, and for each i in I, A_i is a pf already obtained. The definition should be understood constructively: $\bigvee A_i$ is given by $\bigvee$, I and a function (i.e., rule) which associates the disjunct A_i with each i in I. Similarly for conjunction. We could take the species of pf to be the species which is inductively defined by the above conditions. I leave open the question in this paper of the species I for which this inductive definition of the species of pf can be constructively justifed. In the particular applications discussed here, I is always of the form $\{i : i \leq z\}$, where $z \leq \omega$; and in this case, the definition is no more problematical than Brouwer's definition of the second number class, since they are formally identical. However, it seems to me that undecidable species I can be introduced as well. But I will discuss this in another paper, in which such applications will be considered. In any case, for many applications, we want the pf to constitute, not the entire species which is inductively defined by the above conditions, but some subspecies of this. Such a subspecies is called <u>complete</u> if it contains each atom and, whenever it contains one of $\bigvee A_i$ and $\bigwedge A_i$, it contains both of them and the components A_i for each $i \in I$. Of course, the inductively defined species is itself complete.

Henceforth, the pf will all be assumed to belong to some fixed complete species. To call A a pf will mean that it belongs to this species.

Negation is not taken as a primitive operation, because for the purposes of this paper, it is more convenient to deal with it as follows: Assume that each atom p has associated with it an atom $\bar{p}$, called its <u>complement</u>; and that conversely, p is the complement of $\bar{p}$, i.e., $\bar{\bar{p}} = p$. The complement $\bar{A}$ of an arbitrary pf A is inductively defined by De Morgan's laws:

$$\overline{\bigvee A_i} = \bigwedge \bar{A}_i \qquad \overline{\bigwedge A_i} = \bigvee \bar{A}_i .$$

The negation of A will be identified with $\bar{A}$, with the advantage for us here that the classical law of double negation becomes the syntactical identity $\bar{\bar{A}} = A$. Note that, by completeness, $\bar{A}$ is a pf when A is.

Quantification is not introduced as primitive, either. This is because, without any loss, it can be dealt with in terms of infinite disjunctions and conjunctions. We will take this point up below.

In order to make the structure of derivations as simple as possible, the objects to be derived are taken to be finite sets Γ, Δ, etc., of pf, rather than single pf. These sets are interpreted <u>disjunctively</u>, so that $\{A_0,\ldots,A_{n-1}\}$ is valid just in case $\bigvee_{i<n} A_i$ is. $\Gamma + \Delta$ will denote the union of Γ and Δ, $\Gamma + A$ will denote $\Gamma + \{A\}$; and sometimes, A will denote $\{A\}$.

Let S be a collection of finite sets of atoms with the

INTERSECTION PROPERTY. If $\Gamma + p$ and $\Delta + \overline{p}$ are in S, then so is some subset of $\Gamma + \Delta$.

S will be called an <u>axiom system</u>, and its elements <u>axioms</u>. For example, the axiom system might consist of all true propositional constants (including the complements of false ones) together with the sets $p + \overline{p}$ for each propositional variable p. But, it will be useful, e.g., in treating predicate logic with identity below, to consider the more general kind of axiom system.

Relative to the choice of an axiom system, the normal rules of inference are the <u>rule of axioms</u>

$$\underset{\sim}{A} \qquad \Gamma + \Delta \qquad (\underline{\text{if}}\ \Delta\ \underline{\text{is an axiom}}),$$

the <u>rule of disjunction</u>

$$\underset{\sim}{\bigvee} \qquad \frac{\Gamma + A_j}{\Gamma + \bigvee A_i} \qquad (\underline{\text{some}}\ j\ \underline{\text{in}}\ I)$$

$$\underset{\sim}{\bigwedge} \qquad \frac{\Gamma + A_j}{\Gamma + \bigwedge A_i} \qquad (\underline{\text{all}}\ j\ \underline{\text{in}}\ I).$$

Besides these normal rules, there is the <u>cut rule</u>

$$\underset{\sim}{C} \qquad \frac{\Gamma + A \quad \Gamma + \overline{A}}{\Gamma}.$$

Rule $\underset{\sim}{A}$ has no premises, $\underset{\sim}{\bigvee}$ has one, $\underset{\sim}{C}$ has two, and $\underset{\sim}{\bigwedge}$ has one corresponding to each j in I—and so may have infinitely many premises. The atoms in Δ are called the <u>principle terms</u> (pt) of

A. $\bigvee A_i$ and $\bigwedge A_i$ are the pt of $\underset{\sim}{\bigvee}$ and $\underset{\sim}{\bigwedge}$, resp. $\underset{\sim}{C}$ has no pt. A_j is called the <u>minor term</u> (mt) of the premise of $\underset{\sim}{\bigvee}$; A_j is the mt of the premise $\Gamma + A_j$ of $\underset{\sim}{\bigwedge}$; and A and $\overline{A}$ are the mt of the premises $\Gamma + A$ and $\Gamma + \overline{A}$ of $\underset{\sim}{C}$, resp. Thus let the pf B_j, for j in J, be the mt of some inference and Δ the set of pt. Then for each Γ,

$$(*) \qquad \frac{\Gamma + B_j}{\Gamma + \Delta} \qquad (\underline{\text{all}}\ j\ \underline{\text{in}}\ J)$$

is an inference; and every inference is of this form. We will regard the inference (*) as given by the mt B_j (j in J), the pt Δ and the set Γ whose elements will be called the <u>side terms</u> (st) of the inference. Note that in every case but $\underset{\sim}{A}$ the st are determined by being the only pf occurring in all the premises and in the conclusion. In the case of $\underset{\sim}{A}$, any set Λ such that $\Gamma \subseteq \Lambda \subseteq \Gamma + \Delta$ can be the set of st of an instance of A with conclusion $\Gamma + \Delta$ (providing that Δ is an axiom).

<u>Derivations</u> are given in (possibly infinite) tree form. Thus, if (*) is an inference, and D_j is a derivation of the premise $\Gamma + B_j$ for each j in J, then

$$\frac{D_j}{\Gamma + \Delta} \qquad (\underline{\text{all}}\ j\ \underline{\text{in}}\ J)$$

is a derivation of $\Gamma + \Delta$. (*) is called the <u>last inference</u> of the derivation, and the D_j its <u>direct subderivations</u>. The instances of $\underset{\sim}{A}$ are derivations, and all other derivations are built up from these using the remaining rules of inference. $D \vdash \Delta$ will mean that D is a derivation of Δ, and $\vdash \Delta$ will mean that there is a derivation of Δ.

The species of derivations is thus inductively defined, relative to the species of pf. We have relativized the notion of pf from the full inductively defined species to some suitable (i.e. complete) subspecies. It is also possible to do this for the notion of a derivation. In this case, to see what a "suitable" subspecies would be, we would have to analyze the closure conditions on the notion of a derivation which suffice for the proof of the Elimination Theorem to go through. For a classical treatment of infinitary logic, this has recently been done by Barwise [1]: Both the pf and the derivations are relativized to some admissible set. (This includes the condition of completeness for the set of pf.) But, so far, no constructively meaningful treatment of this problem has been given.

A derivation is called normal if it involves only normal inferences, i.e., if it contains no cuts. In a normal inference, every pf which occurs in a premise is a part, or subformula, of a pf in the conclusion. It follows from this, for example, that if a set of atoms has a normal derivation, then some subset of it must be an axiom. The Elimination Theorem states in part that every derivable set has a normal derivation. An immediate consequence is the

CONSISTENCY THEOREM. Every derivable set of atoms includes an axiom.

Consistency in the usual sense means that not every set is derivable. But by the Consistency Theorem, this is equivalent to the condition that the null set is not an axiom. The statement of the Consistency Theorem is not itself significant, of course. Nonconstructively, it is a triviality. The significance lies in the fact that it is proved constructively.

A diagression. By a valuation, I will mean a set of atoms which contains at least one element of each axiom and at most one of p and $\overline{p}$ for each atom p. Each atom in a valuation will be called true for it. $\bigvee A_i$ ($\bigwedge A_i$) will be called true for a valuation if A_j is true for some (all) j in I. Δ is valid if, for each valuation, some pf in Δ is true. A pf is called countable if it contains only countable disjunctions and conjunctions.

COMPLETENESS THEOREM. If a finite set of countable pf is valid, it has a normal derivation.

I will omit the proof of this. (See Lopez-Escobar [7].) It is in complete analogy with the proof of completeness of the cut-free rules for predicate logic (which it implies). The main lemma needed is this:

If M is a (possibly infinite) set of atoms which intersects each valuation, then it includes an axiom.

The Completeness Theorem is formulated in slightly greater generality than usual, because normally it is stated only for logically complete axiom systems, i.e., systems in which some subset of $p + \overline{p}$ is an axiom for each atom p. For logically complete systems, a valuation contains exactly one of p and $\overline{p}$ for each atom p; and so, the definition of truth is the usual one for classical logic. We have not assumed logical completeness in the definition of an axiom system, simply because it is not needed for the Elimination Theorem. (It also turns out that the Intersection Property is exactly what is needed to prove the above lemma.)

The restriction to countable pf in the Completeness Theorem is known to be essential. (E.g. see Karp [6].) A particularly simple proof of this is possible using the present formulation of infinitary logic. Let the atoms be $p_0^0, p_1^0, p_2^0, \ldots$ and their complements $p_0^1, p_1^1, p_2^1, \ldots$, resp. The axioms are just the sets $p_n^0 + p_n^1$ for $n \geq 0$. (This is the weakest logically complete system.) If f ranges over the uncountable set 2^N of numerical functions with values < 2, then

$$(+) \qquad \bigvee_f \bigwedge_n p_n^{f(n)}$$

is valid. Suppose that it had a normal derivation D. Because of the subformula property of normal derivations, the only conjunctions occurring in D must be of the form $\bigwedge p_n^{\theta(n)}$ where θ is some function in 2^N. Hence, in every instance of rule $\underset{\sim}{\bigwedge}$-and so, of any rule—there are only a countable number of premises. It follows that D can contain only a countable number of pf. In particular, there are only a countable number of θ, say $\theta_0, \theta_1, \ldots$, such that an inference of the form

$$\frac{\Gamma + \bigwedge p_n^{\theta(n)}}{\Gamma + \bigvee_f \bigwedge p_n^{f(n)}}$$

occurs in D. Consequently, D would remain a correct derivation if f in (+) were restricted to range over $\theta_0, \theta_1, \ldots$. But, so modified, (+) is invalid. Therefore, there is no normal derivation of (+), and so, by the Elimination Theorem, no derivation at all.

A similar argument demonstrates the (known) nonderivability of the <u>axiom of choice</u>

$$\bigvee_m \bigwedge_n \overline{p}_{mn} + \bigvee_g \bigwedge_m p_{mg(n)}$$

where g ranges over N^N and the axioms are $p_{mn} + \overline{p}_{mn}$ for each m and n. Again, the crucial point is that the conjunctions are all countable, while g cannot be restricted to a countable range.

§2. The strong form of the Elimination Theorem requires the introduction of a suitable constructive system of ordinals to measure the complexity of pf and the length of derivations. In classical terms, let $\chi^0(x) = 2^x$, let $\chi^z(x)$, for $z > 0$, be the xth simultaneous solution y of the equations $\chi^u(y) = y$ for all $u < z$. It is well-known that $\chi^z(x)$ exists for all countable ordinals x and z [10]. Let $x = 2^{x_0} + \cdots + 2^{x_m}$ and $y = 2^{y_0} + \cdots + 2^{y_n}$, where $x_0 > \cdots > x_m$

and $y_0 > \cdots > y_n$, and let $z_0 \geq z_1 \geq \cdots \geq z_{m+n}$ be all the x_i and y_j (counting multiple occurrences). Then the natural sum $x \circ y$ of x and y is defined by $x \circ y = 2^{z_0} + \cdots + 2^{z_{m+n}}$. The natural sum is commutative and strictly increasing. Various authors, e.g., Schütte [10], have given constructive systems of ordinals on which the functions $\chi^z(x)$ and $x \circ y$ are defined. We will assume that some such system has been chosen; and all discussion of ordinals should be referred to it.

The relation $A \preceq u$ (A is of rank $\leq u$) is to measure the complexity of the pf A. It is inductively defined as follows: $p \preceq u$ for every atom p and every ordinal u. If for each i in I there is a $u_i < u$ with $A_i \preceq u_i$, then $\bigvee A_i \preceq u$ and $\bigwedge A_k \preceq u$. For a given pf A, there need not be a u with $A \preceq u$. This will depend on the choice of the system of ordinals and on the species I admitted in the construction of pf. Also, even if $A \preceq u$, there need not be a least such u which we can effectively determine. For this reason, we cannot regard the rank of A itself as an ordinal. $A \prec u$ will mean that $A \preceq x$ for some $x < u$. Thus, $\bigvee A_i \preceq u$ just in case $A_i \prec u$ for each i in I; and similarly for conjunction. It is clear that A and $\bar{A}$ have the same rank, i.e., that $A \preceq u$ if and only if $\bar{A} \preceq u$.

A derivation D is said to be of cut degree $\leq u$ if $A \prec u$ for every mt A of a cut

$$\frac{\Gamma + A \quad \Gamma + \bar{A}}{\Gamma}$$

in D. Thus, D is of cut degree 0 (i.e., ≤ 0) just in case it is normal. The mt of cuts are usually called the cut formulae.

The relation $D \preceq u$ (D is of rank $\leq u$) is inductively defined by the condition that, for every direct subderivation D' of D, $D' \prec u$ (i.e., $D' \preceq u'$ for some $u' < u$). This gives a measure of the length of a derivation. Instances of $\underset{\sim}{A}$ are of rank 0 (i.e., ≤ 0). As in the case of pf, a derivation needn't have a bound on its rank, or in any case, a least one.

$$D \vdash \Delta \; [u,v]$$

will mean that D is a derivation of Δ of rank $\leq u$ and cut degree $\leq v$. We will write $[u]$ for $[u,0]$ so that $D \vdash \Delta \; [u]$ means that D is a normal derivation of Δ of rank $\leq u$.

Let $D + \Delta$ denote the result of adding Δ to the st of each inference in the derivation D.

Then by a straightforward induction on u:

WEAKENING LEMMA. <u>If</u> $D \vdash \Gamma$ [u,v], <u>then</u> $D + \Delta \vdash \Gamma + \Delta$[u,v].

A finite set Θ of pf is called a <u>reduction</u> of a pf A if Θ contains a mt of each inference of which A is a pt. There are three kinds of reduction: First, A is an atom which does not occur in any axiom, and Θ is arbitrary. Secondly, A is a finite disjunction, and Θ contains each disjunct. (Infinite disjunctions, i.e., with infinitely many distinct disjuncts, have no reductions.) Thirdly, A is a conjunction, and Θ contains at least one of the conjuncts.

REDUCTION LEMMA. <u>If</u> Θ <u>is a reduction of</u> A <u>and</u> $D \vdash \Gamma + A$ [u,v], <u>then</u> $\vdash \Gamma + \Theta$ [u,v].

The proof is by induction on u. We can assume that A is not in Γ, since otherwise the result follows by weakening, taking $D + \Theta$. We can also assume that A is a st of the last inference of D, replacing D by $D + A$ if necessary. So this inference has the form

$$\frac{\Lambda + A + B_j}{\Lambda + \Delta + A} \quad (j \text{ in } J)$$

where $\Gamma = \Lambda + \Delta$, $\Lambda + A$ is the set of st, and either Δ or $\Delta + A$ is the set of pt. For each j in J there is a $u_j < u$ with $\vdash \Lambda + A + B_j$ $[u_j,v]$; and so by the induction hypothesis (ind. hyp.), $\vdash \Lambda + \Theta + B_j$ $[u_j,v]$. If A is not a pt, then the result follows by the inference

$$\frac{\Lambda + \Theta + B_j}{\Lambda + \Theta + \Delta} \quad (j \text{ in } J) .$$

If A is a pt, then B_j is in Θ for some j, and so $\vdash \Lambda + \Theta$ $[u_j,v]$. The result then follows by weakening. q.e.d.

Recall that an axiom system is <u>logically complete</u> if some subset of $p + \overline{p}$ is an axiom for each atom p. If we assume logical completeness, then the usual laws of classical logic can be derived. The crucial case is the

LAW OF EXCLUDED MIDDLE. <u>If the system is logically complete and</u> $A \preceq u$, <u>then</u> $\vdash A + \overline{A}$ $[2 \cdot u]$.

The proof is by induction on u. If A is an atom, then $A + \overline{A}$ is an instance of rule $\underset{\sim}{A}$. If A is not an atom, we can assume it is a disjunction $\bigvee A_i$, since otherwise $\overline{A}$ is a disjunction and

$\bar{\bar{A}} = A$. For each j in I there is a $u_j < u$ with $A_j \leq u_j$; and so by the ind. hyp., $\vdash A_j + \bar{A}_j$ $[2 \cdot u_j]$. By $\underset{\sim}{\vee}$, $\vdash A + \bar{A}_j$ $[2 \cdot u_j + 1]$ for each j; and so by $\underset{\sim}{\wedge}$, $\vdash A + \bar{A}$ $[2 \cdot u]$, since $2u_j + 1 < 2 \cdot u$. q.e.d.

We turn now to the Elimination Theorem. Most of the proof theory involved is contained in the following lemma.

ELIMINATION LEMMA. <u>Let</u> $D \vdash \Gamma + A$ $[x,v]$ <u>and</u> $D' \vdash \Delta + \bar{A}$ $[y,v]$, <u>where</u> $A \leq v$; <u>and let</u> $u = x \circ y$. <u>Then</u> $\vdash \Gamma + \Delta$ $[u,v]$.

Of course, we obtain a derivation of $\Gamma + \Delta$ of cut degree $\leq v + 1$ by an application of the cut rule; but the object of the lemma is to show that we needn't raise the cut degree.

The proof is by induction on u. Since $\bar{\bar{A}} = A$, $x \circ y = y \circ x$, and A and $\bar{A}$ have the same rank, the lemma is symmetric with respect to the two given derivations.

<u>Case</u> 1. Either A is not a pt in the last inference of D or else $\bar{A}$ is not a pt in the last inference of D'. By symmetry, we can assume the former. Then the last inference of D is of the form

$$\frac{\Lambda + A + B_j}{\Lambda + A + \Theta} \quad (j \text{ in } J)$$

with mt B_j, pt Θ and st $\Lambda + A$. $\Gamma = \Lambda + \Theta$. For each j in J there is an x_j with $\vdash \Lambda + A + B_j [x_j, v]$. Since $u_j = x_j \circ y < u$, it follows by the ind. hyp. that $\vdash \Lambda + \Delta + B_j [u_j, v]$ for each j. The result then follows by the inference

$$\frac{\Lambda + \Delta + B_j}{\Lambda + \Delta + \Theta} \quad (j \text{ in } J)$$

since the u_j are $< u$ and, if this inference is a cut, its mt must be of rank $< v$.

<u>Case</u> 2. A is a pt of the last inference of D and $\bar{A}$ is a pt of the last inference of D'.

<u>Case</u> 2a. A (and $\bar{A}$) are atoms. Then the last (and only) inferences of D and D' are instances of rule $\underset{\sim}{A}$. It follows from the intersection Property that $\Gamma + \Delta$ is also an instance of $\underset{\sim}{A}$.

<u>Case</u> 2b. A and $\overline{A}$ are not atoms. By symmetry, we can assume that A is a disjunction $\bigvee A_i$, and so $\overline{A} = \bigwedge \overline{A}_i$. We can assume that A is a st of the last inference of D, replacing D by $D + A$ if necessary. So that inference is of the form

$$\frac{\Gamma + A + A_j}{\Gamma + A}$$

for some j in I, and $\vdash \Gamma + A + A_j\ [z,v]$ for some $z < x$. Since $z \circ y < u$, it follows from the ind. hyp. that $\vdash \Gamma + \Delta + A_j\ [z \circ y, v]$. Also, by the Reduction Lemma, $\vdash \Delta + \overline{A}_j\ [y,v]$. Since $x > 0$, $y < x \circ y = u$; and $z \circ y < u$. Also, $A_i \prec v$, since $A \preceq v$. So the result follows from the cut

$$\frac{\Gamma + \Delta + A_j \quad \Gamma + \Delta + \overline{A}_j}{\Gamma + \Delta}$$

q.e.d.

Now we can prove the

ELIMINATION THEOREM. <u>If</u> $D \vdash \Delta\ [u, v + \omega^z]$, <u>then</u> $\vdash \Delta\ [\chi^z(u), v]$.

In particular, taking $v = 0$, a derivation of Δ of cut degree $\leq \omega^z$ can be transformed into a normal one by going from rank $\leq u$ to rank $\leq \chi^z(u)$.

The proof is by induction on z, and within that, by induction on u. I.e., we assume the theorem for all $\overline{z}$ and $\overline{u}$ with (i) $\overline{z} < z$ and (ii) $\overline{z} = z$ and $\overline{u} < u$; and we prove it for z and u.

First, assume that the last inference of D is not a cut. Then it is of the form

$$\frac{\Delta_j}{\Delta} \quad (\underline{\text{all}}\ j\ \underline{\text{in}}\ J)$$

where for each j there is a $u_j < u$ with $\vdash \Delta_j\ [u_j, v + \omega^z]$. By the ind. hyp. (ii), $\vdash \Delta_j\ [\chi^z(u_j), v]$; and the result follows since $\chi^z(u_j) < \chi^z(u)$.

So, we can assume that the last inference of D is a cut

$$\frac{\Delta + A \quad \Delta + \bar{A}}{\Delta}$$

where for some x and $y < u$, $\vdash \Delta + A\ [x, v + \omega^z]$ and $\vdash \Delta + \bar{A}\ [y, v + \omega^z]$. By the ind. hyp. (ii), $\vdash \Delta + A\ [\chi^z(x), v]$ and $\vdash \Delta + \bar{A}\ [\chi^z(y), v]$.

Case 1. $z = 0$. Since D is of cut degree $\leq v + \omega^z = v + 1$, $A \leq v$. So, by the Elimination Lemma, $\vdash \Delta\ [\chi^0(x) \circ \chi^0(y), v]$. But $\chi^0(x) \circ \chi^0(y) = 2^x \circ 2^y \leq 2^u = \chi^0(u)$.

Case 2. $z > 0$. Since $A < v + \omega^z$, there is a $w < z$ and a k such that $A < v + \omega^w \cdot k$. Hence, $\vdash \Delta\ [\max(\chi^z(x), \chi^z(y)) + 1, v + \omega^w + \cdots + \omega^w]$, with k summands ω^w. Let $w_0 = \max(\chi^z(x), \chi^z(y)) + 1$, and for $i < k$, $w_{i+1} = \chi^w(w_i)$. Since $w_0 < \chi^z(u)$ and $w < z$, it is easy to prove by induction on i that $w_i < \chi^z(u)$. But, by iterating the ind. hyp. (i) k times, $\vdash \Delta\ [w_k, v]$; and so $\vdash \Delta\ [\chi^z(u), v]$. q.e.d.

Set $2^x_0 = x$ and $2^x_{n+1} = 2^{2^x_n}$. Then writing $n = \omega^0 + \cdots + \omega^0$, and applying the Elimination Theorem n times (with $z = 0$):

COROLLARY. __If__ $\vdash \Delta\ [u, v + n]$, __then__ $\vdash \Delta\ [2^u_n, v]$.

§3. The induction Theorem is concerned with partial orderings $m \subset n$ of the natural numbers. Actually, the restriction to natural numbers is not essential. m and n can be regarded as ranging over any species, but the applications here concern the natural numbers.

Let $m \subset n$ be fixed. The relation $n \preceq u$ (n is of __rank__ $\leq u$) is inductively defined by the condition that, for every $m \subset n$ there is a $u_m < u$ with $m \preceq u_m$. $n \prec u$ means that $n \preceq x$ for some $x < u$. Of course, even when the ordering $m \subset n$ is well-founded, n needn't have a rank u, i.e., with $n \preceq u$. E.g., $m \subset n$ could be a well-ordering of which the system of ordinals is a proper segment. When $\subset$ coincides with the ordering $<$ of the ordinals, then $n \preceq z$ is clearly equivalent to $n \leq z$. Under fairly general conditions—left for the reader to determine—$m \preceq z$ means that there is an ordinal valued function φ, defined for all $m \subseteq n$, such that $k \subset m \subseteq n$ implies $\varphi k < \varphi m \leq z$.

Suppose that we have specified a particular system of propositional logic (i.e., its complete species of pf and its axiom system). Assume that $p_0, p_1, \ldots, \bar{p}_0, \bar{p}_1, \ldots$ are all distinct atoms, and that, for each n, the only axiom containing either p_n or $\bar{p}_n$ is $p_n + \bar{p}_n$. Let

$$D = \bigwedge_n \left(\bigvee_{m \subset n} \bar{p}_m \vee p_n \right)$$

$$\Theta(n) = \bar{D} + \bigwedge_{m \subset n} p_m .$$

Thus, $\Theta(n)$ expresses induction on $\subset$ up to n for the "predicate variable" p.

INDUCTION THEOREM. <u>If the null set is not an axiom and</u> $\vdash \Theta(n)[u]$ <u>and</u> $u < \chi^1(z)$, <u>then</u> $n < \chi^1(z)$.

The results of Section 2 on propositional logic remain valid when we extend this system by adding the new rule of inference

$$\text{K.} \qquad \frac{\Gamma + p_m}{\Gamma + p_n} \qquad (\underline{\text{all}}\ m \subset n) .$$

We have only to amend Case 2a in the proof of the Elimination Lemma, to cover the new possibility: $A = p_n$, and A and $\bar{A}$ are pt in the last inferences of D and D', resp. (where the last inference of D is an instance of K). But the only inferences with pt $\bar{p}_n$ are the instances $\bigwedge + p_n + \bar{p}_n$ of A. Hence, Δ must contain p_n; and so by weakening, $D + \Delta \vdash \Gamma + \Delta\ [x,v]$.

In the extended system, the following is a normal derivation of D:

$$\frac{\bar{p}_k + p_k}{\bigvee_{m \subset n} \bar{p}_m + p_k} \qquad (\underline{\text{all}}\ k\ \underline{\text{and}}\ n\ \underline{\text{with}}\ k \subset n)$$

$$(\underline{\text{all}}\ k\ \underline{\text{and}}\ n\ \underline{\text{with}}\ k \subset n)$$

$$\frac{}{\bigvee_{m \subset n} \bar{p}_m + p_n} \qquad (\underline{\text{all}}\ n,\ \underline{\text{by}}\ \text{K})$$

$$\frac{}{\bigvee_{m \subset n} \bar{p}_m \vee p_n + p_n} \qquad (\underline{\text{all}}\ n,\ \underline{\text{by}}\ \vee)$$

$$\frac{}{\bigvee_{m \subset n} \bar{p}_m \vee p_n} \qquad (\underline{\text{all}}\ n,\ \underline{\text{by}}\ \vee)$$

$$\frac{}{D} \qquad (\text{by}\ \wedge)$$

Suppose now that $\vdash \Theta(n)[u]$ in the original system of propositional logic, and that $u < X^1(z)$. Since $\vdash D\,[5]$ in the extended system and $D \leq 3$, we have $\vdash \bigwedge_{m \subset n} p_m\,[\max(5,u)+1,4]$ in the extended system by applying rule $\underset{\sim}{C}$ with mt D and $\overline{D}$. So, by the Elimination Theorem for the extended system, $\vdash \bigwedge_{m \subset n} p_m\,[v]$, where $v = 2_4^{\max(5,u)+1}$. But if $u < X^1(z)$, then $v < X^1(z)$. So, by reduction, $\vdash p_m\,[v]$ in the extended system, for all $m \subset n$.

In a normal derivation of p_m, only the rule $\underset{\sim}{K}$ can be used, because of the sub-formula property (since the null set is not an axiom). But if only $\underset{\sim}{K}$ is involved (and more generally, providing instances $\Lambda + p_m + p_m$ of $\underset{\sim}{A}$ are not involved), then p_k is a reduction of p_m for all $k \subset m$. It easily follows from this that if $\vdash p_m\,[x]$ then there is a derivation of p_m of rank $\leq x$ using only instances of $\underset{\sim}{K}$ of the form

$$\frac{p_h}{p_k} \quad (\underline{\text{all}}\ h \subset k)$$

i.e., without st.. By induction on x,

$$\vdash p_m\,[x] \ \ \underline{\text{implies}}\ \ m \leq x\,.$$

Assume that this holds for all $x < y$, and that $\vdash p_m\,[y]$. Then for each $k \subset m$ there is an $x_k < y$ with $\vdash p_k\,[x_k]$. By the ind. hyp. then, $k \leq x_k$ for each $k \subset m$. Hence, $m \leq y$.

We have $\vdash p_m\,[v]$ for each $m \subset n$; and so, $m \leq v < X^1(z)$. Hence, $n \leq v + 1 < X^1(z)$. q.e.d.

Let $m \lessdot n$ be an atom for each m and n. Let $m \lessdot n$ be an axiom if $m \subset n$, and let $\overline{m \lessdot n}$ be an axiom if $m \not\subset n$. Induction on $\subset$ is usually encountered in the form

$$\Theta'(n) = \bigvee_k \left(\bigwedge_m (\overline{m \lessdot k} \vee p_m) \wedge \overline{p}_k \right) + \bigwedge_m (m \lessdot n \vee p_m)\,.$$

But it is easy to see that if $\vdash \Theta'(n)\,[u]$ with $u < X^1(z)$, then $\vdash \Theta(n)\,[v]$ for some $v < X^1(z)$. Simply notice that each inference of the form

$$\frac{\Gamma + \overline{m \lessdot n} \vee p_m}{\Gamma + \bigwedge_m (\overline{m \lessdot n} \vee p_m)} \quad (\underline{\text{all}}\ m)$$

can be transformed into

$$\cfrac{\cfrac{\Gamma + \overline{m \lessdot n} + p_m \qquad \Gamma + m \lessdot n + p_m \quad (\underline{\text{all}}\ m \subset n)}{\Gamma + p_m \quad (\underline{\text{all}},\ m \subset n)}}{\Gamma + \bigwedge_{m \subset n} p_m},$$

where the left-hand upper premise is obtained by reduction and the right-hand upper premise is an instance of $\underset{\sim}{A}$ (since $m \lessdot n$ is an axiom when $m \subset n$)).

§4. It remains to be seen how the Elimination Theorem applies when quantifiers are added to the system. In this case, the atoms are of the form $Ps_1 \cdots s_n$ $(n \geq 0)$, where P is a predicate symbol of n arguments and $s_1,\ldots,s_n$ are terms built up in the usual way from free variables a, b, etc., and function symbols (including individual constants). Each predicate symbol P has a complement $\overline{P}$, with $\overline{\overline{P}} = P$; and the complement $\overline{Ps_1 \cdots s_n}$ of an atom is defined to be $\overline{P}s_1,\ldots,s_n$. Besides disjunction and conjunction, formulae may be built up using existential and universal quantification

$$\bigvee xA(x) \qquad \bigwedge xA(x)$$

where x is a bound variable (notationally distinct from free variables) and $A(x)$ results from replacing a free variable b by x in a formula $A(b)$ which does not contain x. The formulae obtained in this way, and which contain only a finite number of free variables, will be called <u>quantificational formulae</u>, or <u>qf</u>. Complementation is extended to qf by

$$\overline{\bigvee xA(x)} = \bigwedge x\overline{A(x)} \qquad \overline{\bigwedge xA(x)} = \bigvee x\overline{A(x)},$$

and the rank relation by: If $A(b) \prec u$, then $\bigvee xA(x) \preceq u$ and $\bigwedge xA(x) \preceq u$. Again, we assume that we have fixed a <u>complete</u> species of qf, where in the definition of completeness, we must add the condition: If one of $\bigvee xA(x)$ and $\bigwedge xA(x)$ is in the species, then so are both of them and each $A(s)$ where s is a term.

An <u>axiom system</u> in quantification theory is a collection of finite sets of atoms with the Intersection Property and the

SUBSTITUTION PROPERTY. If $\Delta(b)$ is an axiom, then for each term s, some subset of $\Delta(s)$ is an axiom.

$\Delta(s)$ is, of course, the result of replacing b by the term s in each qf in $\Delta(b)$. The additional normal rules of inference for this system are the rule of existential quantification

$$\underset{\sim}{\exists} \qquad \frac{\Gamma + A(s)}{\Gamma + \bigvee xA(x)} \qquad (\underline{\text{some}}\ \underline{\text{term}}\ s)$$

and the rule of universal quantification

$$\underset{\sim}{\forall} \qquad \frac{\Gamma + A(b)}{\Gamma + \bigwedge xA(x)} \qquad (b\ \underline{\text{not}}\ \underline{\text{in}}\ \Gamma)\ .$$

Since the axioms are closed under substitutions, it is easy to see that $\vdash \Delta(b)\ [u,v]$ implies $\vdash \Delta(s)\ [u,v]$ for all terms s. In particular, if b is not in Γ, then $\vdash \Gamma + A(b)\ [u,v]$ implies $\vdash \Gamma + A(s)\ [u,v]$ for all s. This shows that the rule $\underset{\sim}{\forall}$ can be interchanged with the infinitary rule

$$\underset{\sim}{\forall}' \qquad \frac{\Gamma + A(s)}{\Gamma + \bigwedge xA(x)} \qquad (\underline{\text{all}}\ \underline{\text{terms}}\ s)$$

without affecting the rank or cut degree of the derivation. ($\underset{\sim}{\forall}$ is obtained from $\underset{\approx}{\forall}'$ by choosing s to be a variable which is not free in $\Gamma + A(x)$. Since there are only a finite number of free variables in these q.f., there will always be such an s.)

Let $t_0, t_1, \ldots,$ be an enumeration of all the terms. Each qf can be regarded as a pf by identifying

$$\bigvee xA(x) = \bigvee_n A(t_n),\quad \bigwedge xA(x) = \bigwedge_n A(t_n)\ .$$

It is evident that the relation $A \leq u$ is independent of whether A is regarded as a qf or as a pf. Moreover, if we regard qf as pf, then they form a complete species of pf, and the rules $\underset{\sim}{\exists}$ and $\underset{\approx}{\forall}'$ are simply special cases of $\underset{\approx}{\bigvee}$ and $\underset{\approx}{\bigwedge}$, resp. It therefore follows that

$\vdash \Delta$ [u,v] <u>holds in quantificational logic</u> (using the rules $\underset{\sim}{\exists}$ and $\underset{\sim}{\forall}$) <u>just in case it holds in propositional logic</u> (regarding Δ as a set of pf).

The Elimination Theorem thus holds for quantificational logic, too. This includes quantificational logic with identity, since we obtain the theory of identity by including the sets $s = s$ and $\overline{s = t} + \overline{A(s)} + A(t)$ among the axioms for all terms s and t, and all atoms $A(b)$.

A <u>finitary</u> qf is one which contains only binary disjunctions and conjunctions, $A \vee B$ and $A \wedge B$, resp. <u>Finitary predicate logic</u> refers to the systems obtained by restricting the rules of inference to finitary qf (which form a complete species). All finitary qf have finite rank, and all derivations in finitary predicate logic have finite rank, and so, have finite cut degree.

GENTZEN'S <u>HAUPTSATZ</u>. <u>If</u> $\vdash \Delta$ [m,n] <u>in a system of finitary predicate logic, then</u> $\vdash \Delta\ [2^m_n]$.

Gentzen proved this for pure logic, i.e., when the axioms are all of the form $Ps_1 \cdots s_n + \overline{P}s_1 \cdots s_n$.

Now, consider qf which contain no free individual variables, whose function symbols are computable function constants (including the numerals $\overline{0}, \overline{1}, \ldots$), and whose predicate symbols are decidable predicate constants and predicate variables. These qf will be called <u>numerical formulae</u> (nf). Each term of this system has a unique numerical value (on the intended interpretation) which we can compute. Besides the Intersection Property, we assume that the axiom system satisfies the condition that, if n is the value of the term s, then for every axiom $\Delta(s)$, some subset of $\Delta(\overline{n})$ is an axiom. ($\Delta(s)$ and $\Delta(\overline{n})$ are obtained from $\Delta(b)$ by replacing all occurrences of b by s and $\overline{n}$, resp.) <u>Number Theory</u> is the system whose formulae are the nf and whose rules of inference are those of propositional logic, $\underset{\sim}{\exists}$, and for universal quantifiers, the ω-<u>rule</u>

$$\underset{\sim}{N}. \qquad \frac{\Gamma + A(\overline{n})}{\Gamma + \bigwedge xA(x)} \quad (\underline{\text{all}}\ n).$$

It is evident that if n is the value of the term s and $\vdash \Delta(s)$ [u,v], then $\vdash \Delta(\overline{n})$ [u,v]. Thus, the rule $\underset{\sim}{\exists}$ can be replaced by its special case

$$\underset{\sim}{\exists}' \qquad \frac{\Gamma + A(\overline{n})}{\Gamma + \bigvee xA(x)} \quad (\underline{\text{some}}\ n)$$

without increasing the rank or cut degree of the derivation involved. If we identify

$$\bigvee xA(x) = \bigvee_n A(\bar{n}), \ \bigwedge xA(x) = \bigwedge_n A(\bar{n})$$

then every nf can be regarded as a pf of the same rank. Moreover, under this identification, $\exists'$ and $\underset{\sim}{N}$ are special cases of $\underset{\sim}{\bigvee}$ and $\underset{\sim}{\bigwedge}$, resp. Hence, the truth of $\vdash \Delta\,[u,v]$ for nf Δ is independent of whether we regard them as nf or as pf. So, the Elimination Theorem holds for number theory, too.

Peano arithmetic is a system of finitary predicate logic with the rule

$$\frac{\Gamma + A(\bar{0}) \quad \Gamma + \overline{A(b)} + A(b')}{\Gamma + A(s)}$$

of mathematical induction added to the rules of inference (where b is not free in Γ). (The numerals are $\bar{0}, \bar{1} = \bar{0}'$, etc.) The axioms are the axioms of identity mentioned above, $\overline{0 = s'}$, $\overline{s' = t'} + s = t$ and certain atoms which express defining equations for recursive functions. Suppose that Δ consists of formulae without quantifiers, and that there is a normal derivation of Δ in Peano arithmetic. Then it is clear that no quantifiers occur in the derivation. I.e., Δ is derivable in quantifier-free Peano arithmetic. Now, the consistency of quantifier-free Peano arithmetic is expressed by a quantifier-free formula (with a free variable), and is derivable in Peano arithmetic. Therefore, the cut rule $\underset{\sim}{C}$ is not redundant in Peano arithmetic. However, Schutte [6] has observed that the rule of mathematical induction can be eliminated in favor of the ω-rule $\underset{\sim}{N}$, using the schema

$$\cfrac{\cfrac{\Gamma + A(\bar{0}) \quad \Gamma + \overline{A(\bar{0})} + A(\bar{1})}{\Gamma + A(\bar{1})} \quad \Gamma + \overline{A(\bar{1})} + A(\bar{2})}{\Gamma + A(\bar{2})}$$

etc., applying the rule $\underset{\sim}{C}$ with cut formulae $A(\bar{n})$.

Since the derivation in Peano arithmetic is of finite rank and of finite cut degree, it transforms into a derivation of rank $< \omega^2$ and finite cut degree in number theory. So the theorems of Peano arithmetic all have normal derivations of rank $\leq 2_k^{\omega^2} < \varepsilon_0 (= \chi^1(1))$ for some k. So, ε_0 is a bound on the provable ordinals of Peano arithmetic. It is the least such bound, by Gentzen [5], in fact. Schutte [9] notes more generally that if mathematical induction is replaced in Peano arithmetic by induction on some decidable partial ordering $\subset$ of rank $\leq z$:

$$\frac{\Gamma + \bigwedge y(y \subset b \vee A(y)) \vee A(b)}{\Gamma + A(t)},$$

then each derivation is transformed into one in number theory of finite cut degree and rank $\leq z \cdot \omega$. Hence, the bound on the provable ordinals in this case is the least $\chi^1(u) > z$.

Suppose that we add to Peano arithmetic the above rule of induction on $\subset$ for some standard well-ordering of the natural numbers of type $2_k^{\omega+1}$ for each $k \geq 0$. It seems possible that the cut rule is redundant in this system; but I do not know a proof of this.

§5. Ramified systems can also be treated in the framework of infinitary propositional logic. Instead of ramified type theory, however, I will discuss ramified (or constructible) set theory, which is notationally simpler. But an analogous account of ramified type theory and ramified analysis can be given.

For each ordinal z, terms and formulae of level z, called z-terms and z-formulae, resp., are built up as follows: Each free z-variable a^z, b^z, etc., and z-constant is a z-term. If P is a predicate symbol of n arguments, s_i is a z_i-term for $i = 1,\ldots,n$, and $z = \max(z_1,\ldots,z_n)$, then $Ps_1,\ldots,s_n$ is a z-formula. If A and B are u- and v-formulae, resp., and $z = \max(u,v)$, then $A \vee B$ and $A \wedge B$ are z-formulae. If $A(b^v)$ is a u-formula which does not contain the bound v-variable x^v, and $z = \max(u, v + 1)$, then $\bigvee xA(x)$ and $\bigwedge xA(x)$ are z-formulae and $\lambda xA(x)$ is a z-term. (I am following the practice of dropping the level superscript on variables once the level has been established, or when the level is irrelevant.) Complementation is defined as usual, with $\overline{Ps_1,\ldots,s_n} = \bar{P}s_1,\ldots,s_n$ and $\bar{\bar{P}} = P$. The binary predicate symbol ε will play a special role, and we will write $s \varepsilon t$ for εst. The atoms are the formulae $Ps_1,\ldots,s_n$, except when P is ε or $\bar{\varepsilon}$. $s \varepsilon t$ or $s \bar{\varepsilon} t$ is an atom only when t is a variable or constant, i.e., when it is not of the form $\lambda xA(x)$.

The axiom system is to be a collection of finite sets of atoms with the Intersection Property and the Substitution Property, i.e., if $\Delta(b^z)$ is an axiom and s is a term of level $\leq z$, then some subset of $\Delta(s)$ is an axiom. The rules of inference are $\underset{\sim}{A}$, $\underset{\sim}{\bigvee}$ and $\underset{\sim}{\bigwedge}$ (restricted to binary disjunctions and conjunctions, of course), $\underset{\sim}{C}$, the rules of quantification

$\underset{\sim}{\exists}^z$ $$\frac{\Gamma + A(s)}{\Gamma + \bigvee x^z A(x^z)}$$ (s <u>is a term of level</u> $\leq z$)

$\underset{\sim}{\forall}^z$ $$\frac{\Gamma + A(s)}{\Gamma + \bigwedge x^z A(x^z)}$$ (<u>all</u> s <u>of level</u> $\leq z$)

and the rules of abstraction

$\underset{\sim}{\lambda}$ $$\frac{\Gamma + A(s)}{\Gamma + s \; \varepsilon \; \lambda x^z A(x^z)}$$ (s <u>of level</u> $\leq z$)

and

$\underset{\sim}{\overline{\lambda}}$ $$\frac{\Gamma + \overline{A}(s)}{\Gamma + s \; \overline{\varepsilon} \; \lambda x^z A(x^z)}$$ (s <u>is of level</u> $\leq z$)

The intended interpretation of the terms is this: The objects of level 0 are individuals. The $x + 1$ objects are attributes of objects of levels $\leq x$. If z is a limit ordinal, the z objects are precisely the objects of level $< z$. In keeping with this interpetation, we should consider only terms and formulae which contain no free variables or constants of limit levels. Also, instead of considering an arbitrary domain of individuals, we could take the 0 objects to be the natural numbers. In this case, as in number theory, we would consider only terms and formulae without free 0-variables. However, for those levels z (e.g., $z = x + 1$) for which the terms and formulae contain free z-variables, the rule $\underset{\sim}{\forall}^z$ can be interchanged with

$$\frac{\Gamma + A(b^z)}{\Gamma + \bigwedge x^z A(x^z)} \qquad (b \text{ } \underline{\text{not in}} \text{ } \Gamma)$$

without affecting the ranks of the derivations involved—as in the case of quantificational logic.

The z-degree A^* of a formula A is defined for each ordinal z as follows: If A is of level $< z$, then $A^* = 0$. Assume now that A is of level $\geq z$. If A is an atom, then $A^* = 0$. If A is $B \vee C$ or $B \wedge C$, then $A^* = \max(B^*, C^*) + 1$. If A is $s \; \varepsilon \; \lambda x B(x)$, $s \; \overline{\varepsilon} \; \lambda x B(x)$, $\bigvee x B(x)$ or $\bigwedge x B(x)$, then $A^* = B(b)^* + 1$. It is easy to show that:

If $A(b^x)$ *is of level* z, $z' < z$, *and* s *is of level* $\leq z'$, *then* $A(s)^* = A(b)^*$.

To each z-formula A we assign a pf A' of rank $\leq \omega \cdot z + A^*$, as follows: If $z = 0$, then $A' = A$. Assume that $z > 0$ and that B' is defined for all B of level $< z$ and for all z-formulae B of z-degree $< k$. Let A be a z-formulae of z-degree k. If A is an atom, then $A' = A$. If A is $s \in \lambda x^{z'} B(x^{z'})$ (where $z' < z$), then A is $B(s)'$ or the empty disjunction $\perp$, (i.e. $\bigvee_{i \in I} A_i$ with I empty), depending on whether s is of level $\leq z'$ or not. In either case, A' is defined, since if s is of level $\leq z'$, then $B(s)^* = B(b)^* < k$. If A is $s \bar{\varepsilon} \lambda x^{z'} B(x^{z'})$, then A' is $\overline{B(s)'}$ or $\perp$, depending on whether s is of level $\leq z'$ or not. If $A = B \vee C$ then $A' = B' \vee C'$, and if $A = B \wedge C$ then $A' = B' \wedge C'$. Let A be $\bigvee x^{z'} B(x^{z'})$, and let $t_0^{z'}, t_1^{z'}, \ldots$ be an enumeration of all terms of level $\leq z'$. Then A' is $\bigvee_n B(t_n^{z'})'$. Since each $B(t_n^{z'})$ is of z-degree $< k$, $B(t_n^{z'})' \prec \omega \cdot z + 1$. Hence, $A' \preceq \omega \cdot z + k$. If A is $\bigwedge x^{z'} B(x^{z'})$, then A' is $\bigwedge_n B(t_n^{z'})'$. It is clear that the species of A', for A a formula of ramified set theory, is complete.

Let Δ' denote the set of A' with A in Δ. Under this translation, $\underset{\sim}{\exists}^z$ and $\underset{\sim}{\forall}^z$ become special cases of $\underset{\sim}{\bigvee}$ and $\underset{\sim}{\bigwedge}$, resp., and $\underset{\sim}{\lambda}$ and $\underset{\sim}{\bar{\lambda}}$ become redundant, since the premise and conclusion translate into the same set. Thus:

Every derivation of rank $\leq u$ *of* Δ *in ramified set theory, which involves only formulae of level* $< z$, *transforms into a derivation of* Δ' *in propositional logic of rank* $\leq$ *and cut degree* $< \omega \cdot z$.

It immediately follows that the provable ordinals of ramified set theory of level $\leq \omega^z$, using derivations of rank $\leq u$, are all of rank $\leq \chi^{1+z}(u)$. (Cf Schutte [13] for the analogous result for ramified type theory.) Of course, we have not proved an elimination theorem for ramified set theory, itself; but it would not be difficult to do so, by showing that $\vdash \Delta'[u,v]$ in propositional logic implies that there is a derivation of Δ is ramified set theory of rank $\leq 2 \cdot u$ involving only formulae of level $\leq z$.

§6. In the remainder of this paper, I will discuss certain subsystems of classical analysis (i.e. second order number theory).

The <u>analytic formulae</u> (af) are built up from atoms $Rs_1 \cdots s_n$, where R is a predicate constant or, if $n = 1$, a free predicate variable and $s_1,\ldots,s_n$ are constant terms of number theory, by means of countable disjunctions and conjunctions and the quantifications

$$\bigvee Z\,A(Z) \qquad \bigwedge Z\,A(Z)$$

over predicates of one argument. $A(Z)$ is obtained from an af $A(P)$ which does not contain Z by replacing each part Ps or $\overline{P}s$ by Zs or $\overline{Z}s$, resp. P will always denote a prepredicate variable. Bound variables — X,Y,Z,X_1, etc. — are distinct from free ones. We will require of an af that it contain only a finite number of free variables. The complement $\overline{A}$ of an af A and the rank relation $A \le z$ are defined in the usual way.

The definition of an af is an extension of the usual one, which refers to <u>finitary</u> af. In these, the only infinite disjunctions and conjunctions are the numerical quantifications $\bigvee_n A(\overline{n})$ and $\bigwedge_n A(\overline{n})$; and so, finitary af are always of finite rank.

A <u>predicate</u> $F = F(b)$ is obtained from an af which contains no quantifiers by replacing zero or more constant terms by the individual variable b. $A(F_1,\ldots,F_n)$ denotes the result of replacing $P_i s$ and $\overline{P}_i s$ in $A(P_1,\ldots,P_n)$ by $F_i(s)$ and $F_i(s)$ for $i = 1,\ldots,n$. If $F_i \le z$ for $i = 1,\ldots,n$ and $A(P_1,\ldots,P_n) \le v$, then $A(F_1,\ldots,F_n) \le z + v$.

The axiom system is logically complete and contains the axioms for $=$: $s = t$ and $\overline{A(s)} + A(t)$ if s and t are equal numerical terms (i.e., denote the same number), and $\overline{s = t}$ if s and t are unequal. The only axioms containing the variable P are to be the sets $\overline{Ps} + Pt$ when s and t are equal. The rules of inference for <u>elementary analysis</u> are those of propositional logic and the quantification rules

$\underset{\sim}{\exists}$
$$\frac{\Gamma + A(F)}{\Gamma + \bigvee Z\,A(Z)}$$

$\underset{\sim}{\forall}$
$$\frac{\Gamma + A(P)}{\Gamma + \bigwedge Z\,A(Z)}$$

where the variable P does not occur in the formulae in Γ or in $A(Z)$. F and P are called the <u>predicate</u> and <u>variable of quantification</u> in $\underset{\sim}{\exists}$ and $\underset{\sim}{\forall}$, resp.

Let $\mathfrak{F}$ denote the list $F_0, F_1, \ldots$ of predicates $F_n = F_n(b)$. $\mathfrak{F}$ is of <u>rank</u> $\leq z$, written $\mathfrak{F} \leq z$, if $F_n \leq z$ for all n. $\mathfrak{F}$ is <u>closed</u> if it contains all the atomic predicates (i.e. containing no logical constants) and if, whenever $G(P)$ is in the list $\mathfrak{F}$, then so is $G(F_n)$ for each n. Suppose that $\mathfrak{F}$ is not closed. Let $\mathfrak{F}_1$ be an enumeration of $\mathfrak{F}$ together with all the atomic predicates; and let $\mathfrak{F}_{k+1}$ (for $k > 0$) be an enumeration of $\mathfrak{F}_k$ together with all predicates $G(G_1, \ldots, G_n)$ where $G(P_1, \ldots, P_n)$ is in $\mathfrak{F}_1$, $P_1, \ldots, P_n$ are variables, and $G_1, \ldots, G_n$ are in $\mathfrak{F}_k$. Let $\mathfrak{F}'$ be an inumeration of $\bigcup_k \mathfrak{F}_k$. Then $\mathfrak{F}'$ is closed, and is called the <u>closure</u> of $\mathfrak{F}$. If $\mathfrak{F} \leq z$, then clearly, $\mathfrak{F}_k \leq z \cdot k$; and so $\mathfrak{F}' \leq z \cdot w$.

Let $\mathfrak{F}$ be a closed list of predicates. A species of af is called <u>$\mathfrak{F}$-complete</u> if it contains all the atoms, if it contains $\overline{A}$ whenever it contains A, if it contains each A_n whenever it contains $\bigvee A_n$, and if it contains $A(F_n)$ whenever it contains $\bigvee ZA(Z)$. $(\mathfrak{F})$ will denote a system of elementary analysis whose af constitute an $\mathfrak{F}$-complete species and in which the predicates of quantification in instances of $\underset{\sim}{\exists}$ are restricted to predicates in $\mathfrak{F}$.

To each af A of $(\mathfrak{F})$ we assign a pf $A(\mathfrak{F}) = A'$ by: $A' = A$ if A is atomic. $(\bigvee A_n)' = \bigvee A_n'$, $(\bigvee ZA(Z))' = \bigvee_n A(F_n)'$, and $(\overline{A})' = \overline{A'}$. Then the A', for A an af of $(\mathfrak{F})$, form a complete species of pf. If $\mathfrak{F} \leq z$ and $A(P_1, \ldots, P_k) \leq u$, then $A(F_{i_1}, \ldots, F_{i_k})' \leq z + u$. Let $\mathfrak{F} \leq z$.

I. <u>If</u> $\vdash \Delta(P_1, \ldots, P_n)\ [u,v]$ <u>in</u> $(\mathfrak{F})$, <u>then</u> $\vdash \Delta(F_{i_1}, \ldots, F_{i_n})'\ [2 \cdot z + u,\ z \oplus v]$ <u>in propositional logic, where</u> $z \oplus 0 = 0$ <u>and for</u> $v > 0$, $z \oplus v = z + v$.

Let $\Delta = \Delta(P_1, \ldots, P_n)$ and $\Delta^+ = \Delta(F_{i_1}, \ldots, F_{i_n})$. The proof is by induction on u. If Δ is an instance of $\underset{\sim}{A}$, then either Δ^+ is an instance of $\underset{\sim}{A}$ or else it includes a set $\overline{F_{i_j}(s)} + F_{i_j}(t)$ where s and t are equal terms (and $\overline{P}_j s + P_j t$ is in Δ). But $\overline{F_{i_j}(s)} + F_{i_j}(t)$ has a normal derivation of rank $2 \cdot z$, since $\mathfrak{F}$ is of rank $\leq z$. (This is the law of excluded middle, when $s = t$. The proof is exactly the same when s and t are equal terms.) If the last inference in the derivation of Δ is an instance of $\underset{\sim}{\bigvee}$ or $\underset{\sim}{\bigwedge}$, there is no difficulty, since these inferences are preserved under the translation of A into A^+. A cut with cut formula B is transformed into a cut with cut formula B^+. Since $B < v$, we have $B^+ < z + v$. The inference

$$\frac{\Gamma + A(F)}{\Gamma + \bigvee ZA(Z)}$$

transforms into

$$\frac{\Gamma^+ + A(F^+)^+}{\Gamma^+ + \bigvee_n A(F_n)^+}$$

which is valid, since F^+ is in the closed list $\mathfrak{F}$ if F is. Finally, let the last inference be

$$\frac{\Gamma + A(P_1,\ldots,P_n,P)}{\Gamma + \bigwedge ZA(P_1,\ldots,P_n,Z)} .$$

By the induction hypothesis $\vdash \Gamma^+ + A(F_{k_1},\ldots,F_{i_n},F_k)^+ \; [2\cdot z + u', z \oplus v]$ for all k, where $u' < u$. So, $\vdash \Gamma^+ + \bigwedge_k A(F_{i_1},\ldots,F_{i_n},F_k)^+ \; [2\cdot z + u, z \oplus v]$. q.e.d.

II. <u>If</u> $\vdash \Delta'[u,v]$ <u>in the system of propositional logic which is associated with</u> ($\mathfrak{F}$), <u>then</u> $\vdash \Delta[u,v]$ <u>in</u> ($\mathfrak{F}$).

The proof is by induction on u, and is just like the proof in the case of predicate logic.

From I and II we immediately obtain the

ELIMINATION THEOREM FOR ELEMENTARY ANALYSIS. <u>If</u> $\mathfrak{F} \le z$, $z \oplus v \le x + w^{x_0} + \cdots + w^{x_n}$ <u>and</u> $\vdash \Delta[u,v]$ <u>in</u> ($\mathfrak{F}$), <u>then</u> $\vdash \Delta[x^{x_0}(\ldots x^{x_n}(2z + u)\ldots),x]$ <u>in</u> ($\mathfrak{F}$).

For example, let $\mathfrak{F}$ consist of the atomic predicates (i.e. containing no logical constants). Then $\mathfrak{F} \le 0$; and so $\vdash \Delta[u,v + w^x]$ in ($\mathfrak{F}$) implies $\vdash \Delta[x^x(u),v]$ in ($\mathfrak{F}$). <u>Pure second order Peano arithmetic</u> refers to the formal system obtained by adding predicate quantifiers to Peano arithmetic with the rules $\underset{\sim}{\exists}$ and $\underset{\sim}{\forall}$, but with the predicates of quantification restricted to atomic predicates. (Of course, the rule of mathematical induction is to apply to arbitrary formulae of this system, and not just those of Peano arithmetic.) By the above result, ε_0 is still the bound on the provable ordinals of this system. Kreisel has pointed out that this bound remains the same even if we add arbitrary true axioms of the from $\bigvee ZA(Z)$ to the system, where $A(P)$ contains no predicate quantifiers—although the proof of this is nonconstructive: Since $\bigvee ZA(Z)$ is true, we can introduce a solution R with true axioms $R\bar{n}$ or $\overline{R\bar{n}}$ for each n, so that there is a normal derivation of $A(R)$, and so of $\bigvee ZA(Z)$, of finite rank. So (nonconstructivity), every derivation of the extended system still transforms into a normal derivation of rank $< \varepsilon_0$. This remains so if we add arbitrary true Σ^1_1 formulae (i.e. all of whose predicate quantifiers are existential) as axioms by the same argument.

§7. Hyperarithmetical analysis. An af is called Σ^1_1 (Π^1_1) if all of its quantifiers are existential (universal).

Let $\mathfrak{G}$ be a list $G_0, G_1, \ldots$ of predicates such that some finite set of variables contains all the variables in any of the G_n. Then $A(\mathfrak{G})$ is an af for each af A. $\mathfrak{G} \subseteq \mathfrak{F}$ means that each G_i is an F_j.

LEMMA. *Let* $\mathfrak{F}$ *be closed and* $\mathfrak{G} \subseteq \mathfrak{F}$. *If* A *is* Σ^1_1 *and* $A(\mathfrak{G}) \leq u$, *then* $\vdash A + \overline{A(\mathfrak{G})}\ [2 \cdot u]$ *in* $(\mathfrak{F})$.

Let $A' = A(\mathfrak{G})$. If A is atomic, then $A + \overline{A'}$ is an instance of $\underset{\sim}{A}$. If $A = \bigvee A_i$ and $\vdash A_i + \overline{A}_i'[2 \cdot u]$ where $A_i' \leq u_i < u$, then $\vdash A + \overline{A}_i'\ [2 \cdot u_i + 1]$ by $\underset{\sim}{\bigvee}$, and so $\vdash A + \overline{A'}\ [2 \cdot u]$ by $\underset{\sim}{\bigwedge}$. Similarly for $A = \bigwedge A_i$. Let $A = \bigvee Z B(Z)$, so that $A' = \bigvee_n B(G_n)'$. If $\vdash B(G_n) + \overline{B(G_n)'}[2 \cdot u_n]$, where $B(G_n)' \leq u_n < u$, then $\vdash A + \overline{B(G_n)'}[2 \cdot u_n + 1]$ by $\underset{\sim}{\exists}$ (with the predicate of quantification G_n in $\mathfrak{F}$), and so $\vdash A + \overline{A'}[2 \cdot u]$ by $\underset{\sim}{\bigwedge}$. q.e.d.

Hyperarithmetical analysis refers to a system $(\mathfrak{F})$ with the added Δ^1_1 comprehension rule: If $A(\overline{n})$ is Σ^1_1 and $B(\overline{n})$ is Π^1_1, then

$$\Delta^1_1\ \underset{\sim}{\mathrm{CR}}. \qquad \frac{A(\overline{n}) + \overline{B(\overline{n})} \qquad \overline{A(\overline{n})} + B(\overline{n})}{\bigvee Z \bigwedge_n (Z\overline{n} \leftrightarrow A(\overline{n}))} \qquad (\text{all } n)\ .$$

Here, $A \leftrightarrow B$ is an abbreviation for $A \vee \overline{B}.\wedge.\overline{A} \vee B$. Let $A(\overline{n})$ and $B(\overline{n})$ be Σ^1_1 and Π^1_1, resp., and let them both be of rank $\leq x$. Suppose that $\vdash A(\overline{n}) + \overline{B(\overline{n})}[u,v]$ and $\vdash \overline{A(\overline{n})} + B(\overline{n})[u,v]$, for all n, in $(\mathfrak{F})$, where $\mathfrak{F}$ is a closed list of predicates of rank $\leq z$. Let $C' = C(\mathfrak{F})$. By I,

$$\vdash A(\overline{n})' + \overline{B(\overline{n})'}[2 \cdot z + u,\ z \oplus v]\ ,$$

in propositional logic. $A(\overline{n})'$ and $\overline{B(\overline{n})'}$ are not generally af, however, since they may contain infinitely many variables. But, let $P_1, \ldots, P_k$ be all the variables in $A(\overline{n})$ and $B(\overline{n})$, and let Q be a fixed predicate constant of one argument. Let G_n be the result of replacing each variable $P \neq P_1, \ldots, P_k$ in F_n by Q, and let $\mathfrak{G}$ be the list $G_0, G_1, \ldots$. Set $C^* = C(\mathfrak{G})$. Then C^* is an af when C is. We obtain

$$\vdash A(\bar{n})^* + \overline{B(\bar{n})^*}\ [2 \cdot z + u,\ z \oplus v]$$

from the given derivation of $A(\bar{n})' + \overline{B(\bar{n})'}$ by replacing each variable $\neq P_1, \ldots, P_k$ in it by Q. Since $\overline{B(\bar{n})}$ is Σ_1^1, the Lemma yields

$$\vdash \overline{B(\bar{n})} + B(\bar{n})^*\ [2 \cdot (z + x)]\ .$$

So by rule $\underset{\sim}{C}$,

$$\vdash A(\bar{n})^* + \overline{B(\bar{n})}\ [\bar{u}, \bar{v}]\ ,$$

where $\bar{u} = \max(2z + u,\ 2 \cdot (z + x)) + 1$ and $\bar{v} = \max(z \oplus v,\ z + x + 1)$, since the cut formula is $B(\bar{n})^*$ which is of rank $\leq z + x$. Hence, using the second of the original assumptions,

$$\vdash A(\bar{n})^* + \overline{A(\bar{n})}\ [\bar{u} + 1, \bar{v}]\ ,$$

by $\underset{\sim}{C}$ with cut formula $B(\bar{n})$. But, again by the Lemma,

$$\vdash A(\bar{n}) + \overline{A(\bar{n})^*}\ \ [2(z + x)]\ .$$

So, by applications of $\underset{\sim}{\vee}$ and $\underset{\sim}{\wedge}$,

$$\vdash A(\bar{n})^* \leftrightarrow A(\bar{n})\ [\bar{u} + 4, \bar{v}]\ .$$

I.e.,

$$\vdash \ \bigwedge_n\ (A(\bar{n})^* \leftrightarrow A(\bar{n}))\ [\bar{u} + 5, \bar{v}]$$

in $(\mathfrak{F})$. Of course, the predicate $A(b)^*$ is not generally in the list $\mathfrak{F}$; but let $\mathfrak{F}'$ be the closed list generated by $A(b)^*, F_0, F_1, \ldots$. Since each of these latter predicates is of rank $\leq z + x$, $\mathfrak{F}' \leq (z + x) \cdot \omega$. In $(\mathfrak{F}')$, we can then apply $\underset{\sim}{\exists}$ to obtain

$$\vdash \bigvee Z \bigwedge_n (Z\bar{n} \leftrightarrow A(\bar{n}))\ [\bar{u} + 6, \bar{v}]\ .$$

EXAMPLE. Let $\mathfrak{H}$ be the system obtained from pure second order Peano arithmetic by adding the rule: If A(b) and B(b) are Σ^1_1 and Π^1_1, resp., then

$$(*) \qquad \frac{A(b) + \overline{B(b)} \qquad \overline{A(b)} + B(b)}{\Gamma + \bigvee Z \ \bigwedge x(Zx \leftrightarrow A(x))}\ .$$

$\mathfrak{H}$ is the union of a hierarchy $(\mathfrak{H}^k)$ of systems, $k \geq 0$, where $\mathfrak{H}^0$ does not contain the rule (*), and, for each instance of (*) in a derivation in $\mathfrak{H}^{k+1}$, the subderivations of the premises $A(b) + \overline{B(b)}$ and $\overline{A(b)} + B(b)$ are derivations in $\mathfrak{H}^k$. The formulae of $\mathfrak{H}$ can be regarded as finitary af by reading $\bigvee x$ and $\bigwedge x$ as disjunction and conjunction, resp. Let $\mathfrak{F}^0$ be a list of all atomic predicates, and let $\mathfrak{F}^{k+1}$ be a closed list generated by $\mathfrak{F}^k$ and all the predicates $A(\mathfrak{Q})$ when $\mathfrak{Q}$ is obtained by replacing all but a finite number of predicate variables in the predicate is $\mathfrak{F}^k$ by the fixed constant Q and $A = A(b)$ is a finitary Σ^1_1 af. Thus, $\mathfrak{F}^k \leq \omega^k$. Let $\mathfrak{F}'$ be an enumeration of the union of the $\mathfrak{F}^k (k \geq 0)$. So $\mathfrak{F}'$ is closed and of rank $\leq \omega^\omega$. Let $\varphi(0) = \omega^2$ and $\varphi(k+1) = \omega^k + \varphi(k)$, and let $\psi(0) = \omega$ and $\psi(k+1) = \omega^k + \psi(k)$.

<u>If</u> $\vdash \Delta$ <u>in</u> $\mathfrak{H}^k$, <u>then</u> $\vdash \Delta\ [u,v]$ <u>in</u> $(\mathfrak{F}^k)$ <u>for some</u> $u < \varphi(k)$ <u>and</u> $v < \psi(k)$. <u>So,</u> $\vdash \Delta$ <u>in</u> $\mathfrak{H}$ <u>implies</u> $\vdash \Delta\ [u,v]$ <u>in</u> $(\mathfrak{F})$ <u>for some</u> u <u>and</u> $v < \omega^\omega$.

For $k = 0$, this is clear, since $\mathfrak{H}^0$ is pure second order Peano arithmetic. Assume that it is true for k. Each derivation of Δ in $\mathfrak{H}^{k+1}$ yields

$$\vdash \Delta + \left\{ \overline{\bigwedge \mathfrak{X} \bigwedge_{\mathfrak{m}} \bigvee Z \bigwedge_n (Z\bar{n} \leftrightarrow A_i(\mathfrak{X}, \bar{\mathfrak{m}}, \bar{n}))}:\ i = 1, \dots, p \right\} [u,v]$$

in $(\mathfrak{F}^0)$ with $u < \omega^2$ and $v < \omega$, and

$$\vdash A_i(\mathfrak{P}, \bar{\mathfrak{m}}, \bar{n}) + \overline{B_i(\mathfrak{P}, \bar{\mathfrak{m}}, \bar{n})}\ [u'v']$$

$$\vdash \overline{A_i(\mathfrak{P}, \bar{\mathfrak{m}}, \bar{n})} + B_i(\mathfrak{P}, \bar{\mathfrak{m}}, \bar{n})\ [u', v']$$

in $(\mathfrak{F}^k)$, where $u' < \varphi(k)$ and $v' < \psi(k)$, for all $i = 1, \dots, p$ and all $\mathfrak{m}$ and n. $\mathfrak{P}$ is a list of predicate variables, $\mathfrak{X}$ a corresponding list of bound variables, $\mathfrak{m}$ is a list of numbers, and A_i

and B_i are finitary Σ^1_1 af. But, by the analysis of Δ^1_1 CR above, this yields

$$\vdash \bigwedge \underline{X} \bigwedge_{\underline{m}} \bigvee Z \bigwedge_{n} (Z\bar{n} \leftrightarrow A_i(X,\bar{\underline{m}},\bar{n}))\ [x,y]$$

in $(\mathfrak{F}^{k+1})$, where $x < \max(2\cdot\omega^k + \varphi(k), 2(\omega^k + \omega))$ and $y < \max(\omega^k \oplus \psi(k), \omega^k + r + 1)$. I.e., $x < \omega^k + \varphi(k)$ and $y < \omega^k + \psi(k)$. So, by p cuts in $(\mathfrak{F}^{k+1})$,

$$\vdash \Delta\ [u_1, v_1]$$

with $u_1 < \varphi(k+1)$ and $v_1 < \psi(k+1)$.

So $\vdash \Delta$ in $\mathfrak{H}^k$ yields a normal derivation in $(\mathfrak{F}^k)$ of Δ with rank $\leq X^k(\cdots X^1(2^u_r)\cdots)$ for some $r < \omega$ and $u < \omega^k + \cdots + \omega^1 + \omega^2$. For example, $X^1(X^1(0)) = \varepsilon_{\varepsilon_0}$ is a bound on the provable ordinals of $\mathfrak{H}^1$ and $X^\omega(0)$ is a bound on the provable ordinals of $\mathfrak{H}$. I do not know whether or not these bounds are optimal.

In his analysis in [2] of predicative proof, Feferman considers the hierarchy $(\mathfrak{H}_d)$ of formal systems where d ranges over the primitive recursive ordinal notations (i.e., $\mathfrak{O}$ relativized to the primitive recursive functions). $\mathfrak{H}_1 = \mathfrak{H}$, $\mathfrak{H}_{3\cdot 5^e}$ is the union of the $\mathfrak{H}_{[e](n)}$ $(n \geq 0)$, where $[e]$ is the primitive recursive function with Gödel number e (or is the constant function $= 0$ if e is not such a Gödel number), and $\mathfrak{H}_{2^d}$ is obtained by adding to $\mathfrak{H}_d$ the <u>formalized</u> ω-<u>rule</u> for $\mathfrak{H}_d$:

$$\bigwedge x \bigvee y P_d(y, \ulcorner A(\bar{x})\urcorner) \rightarrow \bigwedge x A(x)\ .$$

P_d is the primitive recursive proof predicate for $\mathfrak{H}_d$, $\ulcorner A(\bar{n})\urcorner$ is the Gödel number of $A(\bar{n})$, and $\bigwedge x A(x)$ is a closed formula. If d is a primitive recursive ordinal notation, the relation $d \leq z$ (d is of rank $\leq z$ in the partial ordering of the notations) is defined as in Section 3.

<u>If</u> $d < z$, <u>then there is an</u> $\mathfrak{F}_z$ <u>of rank</u> $\leq \omega^{\omega\cdot z}$ <u>such that</u> $\vdash \Delta$ <u>in</u> $\mathfrak{H}_d$ <u>implies</u> $\vdash \Delta\ [u,v]$ <u>in</u> $(\mathfrak{F}_z)$ <u>for some</u> u <u>and</u> $v < \omega^{\omega\cdot z}$.

The proof is by induction on d. If $d = 1$, then $z \geq 1$, and we have already proved this case. If $d = 3\cdot 5^e$, then for each n there is a z_n with $[e](n) < z_n < z$. By the induction hypothesis, we have $\mathfrak{F}_{[e](n)} \leq \omega^{\omega\cdot z_n}$ which satisfies the assertion for $\mathfrak{H}_{[e](n)}$. Let $\mathfrak{F}_d$ enumerate the closure of the union of the $\mathfrak{J}_{[e](n)}$. This will clearly have rank $\leq \omega^{\omega\cdot z}$. Assume the assertion holds for $\aleph_d$. We will prove it for $\aleph_{2^d}$. $\aleph^{\rho}_{2^d}$ is $\aleph_{2^d}$ with no instances of (*). A derivation in $\aleph^{k+1}_{2^d}$ contains an instance of (*) only when the subderivations of the premises are derivations in $\aleph^{k}_{2^d}$. Every derivation in $\aleph^{\rho}_{2^d}$ of Δ transforms into a derivation of some $\Gamma = \Delta + \{\bigwedge x \bigvee y\, P_d(y,\ulcorner A_i(\overline{x})\urcorner) \wedge \overline{\bigvee x A_i(x)}$: $i = 1,\ldots,p\}$ in $\aleph_d$. So, $\vdash \Gamma\,[u,v]$ in $(\mathfrak{J}_x)$, where $d < x < z$ and u and v are $< \omega^{\omega\cdot x}$. When $P_d(\overline{m},\ulcorner A_i(\overline{n})\urcorner)$ is false, its complement is an axiom; and when it is true, we obtain from m a proof of $A_i(\overline{n})$ in $\aleph_d$. So, in any case, $\vdash \overline{P_d(\overline{m},\ulcorner A_i(\overline{n})\urcorner)} + A_i(\overline{n})$ in $\aleph_d$, for each m and n. So

$$\vdash \overline{\bigwedge x \bigvee y\, P_d(y,\ulcorner A_i(\overline{x})\urcorner)} \vee \bigwedge x A_i(x)[\omega^{\omega\cdot x} + 4, \omega^{\omega\cdot x}]$$

in $(\mathfrak{J}_x)$. So, by p cuts, $\vdash \Delta\,(u',v']$ in $(\mathfrak{J}_x)$ with u' and $v' < \omega^{\omega\cdot x} + \omega$. Set $\mathfrak{J}^0 = \mathfrak{J}_x$, and let $\mathfrak{J}^{k+1}$ enumerate the closure of $\mathfrak{J}^k$ and the predicates $A(\mathfrak{J}^k)$, where $A = A(b)$ is a finitary Σ^1_1 af. So, $\mathfrak{J}^k \leq \omega^{\omega\cdot x}\cdot\omega^k$. Set $\varphi(0) = \omega^{\omega\cdot x} + \omega$, and $\varphi(k+1) = \omega^{\omega\cdot x}\cdot\omega^k + \varphi(k)$. We have already shown that $\vdash \Delta$ in $\aleph^{\rho}_{2^d}$ implies $\vdash \Delta\,[u,v]$ in $(\mathfrak{J}^0)$ for some u and $v < \varphi(0)$. Assume that $\vdash \Delta$ in $\aleph^{k}_{2^d}$ implies $\vdash \Delta\,[u,v]$ in $(\mathfrak{J}^k)$ for some u and $v < \varphi(k)$. Exactly as in the case of $\aleph$ (in place of $\aleph_{2^d}$) above, it follows that $\vdash \Delta$ in $\aleph^{k+1}_{2^d}$ implies $\vdash \Delta\,[u,v]$ in $(\mathfrak{J}^{k+1})$ with u and $v < \omega^{\omega\cdot x}\cdot\omega^k + \varphi(k) = \varphi(k+1)$, since $\mathfrak{J}^k \leq \omega^{\omega\cdot x}\cdot\omega^k$. Let $\mathfrak{J}_{2^d}$ enumerate the union of the $\mathfrak{J}^k$. $\mathfrak{J}_{2^d}$ is closed and is of rank $\leq \omega^{\omega\cdot x}\cdot\omega^{\omega} \leq \omega^{\omega\cdot z}$. Also, each derivation of Δ in $\aleph_{2^d}$ is a derivation in some $\aleph^{k}_{2^d}$, and so yields $\vdash \Delta\,[u,v]$ in $(\mathfrak{J}_{2^d})$ with u and $v < \omega^{\omega\cdot z}$.

Thus, $\chi^{\omega\cdot z}(\chi^{\omega\cdot z}(\omega^{\omega\cdot z}\cdot 2))$ <u>is a bound on the provable ordinals of</u> $\aleph_d$ <u>when</u> $d < z$.

§8. <u>The</u> Σ^1_1 <u>axiom of choice</u>. Let $\langle a,b\rangle$ denote some standard computable bijection of $N \times N$ onto N whose inverse is $c \to ((c)_0,(c)_1)$. $P^n s$ is an abbreviation for $P\langle\overline{n},s\rangle$. The Σ^1_1 axiom of choice is

$$\bigwedge_n \bigvee Z A(\overline{n},Z) \to \bigvee Z \bigwedge_n A(\overline{n},Z^n),$$

where $A(b,P)$ is a predicate (i.e. containing no quantifiers). It is called Σ^1_1 because, using it, we can derive the corresponding implications for arbitrary Σ^1_1 af $A(b,P)$. An equivalent form of the Σ^1_1 axiom of choice is the rule

Σ_1^1 $\underset{\sim}{AC}$
$$\frac{\Gamma + \bigwedge_n \bigvee Z A(\overline{n},Z)}{\Gamma + \bigvee Z \bigwedge_n A(\overline{n},Z^n)}$$

where $A(b,P)$ is restricted to predicates. Using this rule, it is easy to derive the Δ_1^1 comprehension axiom

$$\bigwedge_n (A(\overline{n}) \leftrightarrow \overline{B(\overline{n})}) \to \bigvee Z \bigwedge_n (Z\overline{n} \leftrightarrow A(\overline{n})),$$

where $A(b)$ and $B(b)$ are Σ_1^1. In particular, Δ_1^1 $\underset{\sim}{CR}$ is derivable using Σ_1^1 $\underset{\sim}{AC}$.

$\Sigma(\mathfrak{F})$ will denote the system $(\mathfrak{F})$ of elementary analysis with the rule Σ_1^1 $\underset{\sim}{AC}$ added, but with the restriction that $A(b,P)$ be in the list $\mathfrak{F}$ of predicates. I will show that, if Δ is Σ_1^1 (i.e. contains only Σ_1^1 af) and $\vdash \Delta\ [u,v]$ in $\Sigma(\mathfrak{F})$, then $\vdash \Delta\ [u']$ in $(\mathfrak{F}')$ where $\mathfrak{F}'$ and u' depend on u and v. It is not known whether there is such a constructive elimination theorem for arbitrary sets Δ of af. (Of course, nonstructively, we obtain such an elimination theorem for the full analytic comprehension axiom by taking $\mathfrak{F}'$ to be a countable model of constructible sets.)

For the sake of simplicity, I will consider only $\Sigma(\mathfrak{F}_0)$, where $\mathfrak{F}_0$ is an enumeration of the finitary predicates; but the extension to arbitrary $\mathfrak{F}$ is straightforward.

A derivation in $\Sigma(\mathfrak{F}_0)$ is called strict if all its cut formulae are of the form $\bigvee ZA(Z)$ or $\bigwedge ZA(Z)$, where $A(P)$ is finitary. $\vdash' \Delta\ [u]$ will mean that there is a strict derivation of Δ in $\Sigma(\mathfrak{F}_0)$ of rank $\leq u$.

III. If $\omega^2 \leq u$, $v \neq 0$ and $\vdash \Delta\ [u,\omega^v]$ in $\Sigma(\mathfrak{F}_0)$, then $\vdash' \Delta\ [\chi^v(u)]$ in $\Sigma(\mathfrak{F}_0)$.

This follows from the proof of the Elimination Theorem for elementary analysis. For, the Elimination Lemma remains valid except for the case that the cut formula A or $\overline{A}$ is pt of an instance of Σ_1^1 $\underset{\sim}{AC}$; and in that case it is of the required form.

Let $P_0, P_1, \ldots$ be variables not occuring in the finitary predicate $A(b,P)$. Set

$$C_n(b) = A(b,P_n) \wedge \bigwedge_{k<n} \overline{A(b,P_k)},$$

$$H(b) = \bigvee_k (C_k((b)_0) \wedge P_k(b)_1).$$

Then it is easy to verify that, for each n and k, there are normal derivations of finite rank of

$$\overline{A(\bar{n},P_k)} + \bigvee_m C_m(\bar{n}),\quad \overline{C_m(\bar{n})} + A(\bar{n},H^n).$$

So,

$$\vdash \overline{\bigwedge_n \bigvee_k A(\bar{n},P_k)} + \bigwedge_n A(\bar{n},H^n)\ [\omega + 2].$$

Let $\mathfrak{F} \leq z$ and let K be the result of replacing each P_k in H by F_k. Then $K < z + w$ and

$$\vdash \overline{\bigwedge_n \bigvee_k A(\bar{n},F_k)} + \bigwedge_n A(\bar{n},K^n)\ [2 \cdot z + \omega + 2].$$

We call K the <u>choice predicate for</u> $A(b,P)$ <u>relative to</u> $\mathfrak{F}$.

Let $\mathfrak{F}_0$ be the finitary predicates; and for $z > 0$, let $\mathfrak{F}_z$ be the closure of the union of all the $\mathfrak{F}_x$ with $x < z$ and all choice predicates for finitary $A(b,P)$ relative to lists $\mathfrak{F}$ <u>derived</u> from some $\mathfrak{F}_x$ with $x < z$. $\mathfrak{F}$ is derived from $\mathfrak{F}_x$ if it is obtained by replacing all but a finite number of predicate variables in each predicate in $\mathfrak{F}_x$ by the constant Q. Here, Q is to be fixed, once and for all. Note that a choice predicate relative to such a derived list <u>is</u> a predicate: It contains only a finite number of variables. For each z, we clearly have $\mathfrak{F}_z \leq \omega^{1+z}$.

IV. <u>If</u> Δ <u>is</u> Σ_1^1 <u>and</u> $\vdash' \Delta\ [u]$, <u>then</u> $\vdash \Delta\ [\chi^{1+u}(0)]$ <u>in</u> $(\mathfrak{F}_u)$.

The proof is by induction on u. When the last step in the strict derivation of Δ is by $\underset{\sim}{\bigwedge}$, $\underset{\sim}{\bigvee}$, $\underset{\sim}{\triangle}$ or $\underset{\sim}{\exists}$, the result immediately follows from the induction hypothesis, since $\chi^u(0)$ is strictly increasing in u. The last step is not by $\underset{\sim}{\forall}$, since Δ is Σ_1^1. So, we need only consider the cases $\Sigma_1^1\ \underset{\sim}{AC}$ and $\underset{\sim}{C}$. Let the last step be by $\Sigma_1^1\ \underset{\sim}{AC}$, so that Δ is of the form $\Gamma + \bigvee Z \bigwedge_n A(\bar{n},Z^n)$ and $\vdash' \Gamma + \bigwedge_n \bigvee Z A(\bar{n},Z)\ [v]$ for some $v < u$. By the ind. hyp., $\vdash \Gamma + \bigwedge_n \bigvee Z A(\bar{n},Z)\ [\chi^{1+v}(0)]$ in $(\mathfrak{F}_v)$. Since there are no instances of $\underset{\sim}{\forall}$ in this latter derivation, it easily follows that $\vdash \Gamma + \bigwedge_n \bigvee_k A(\bar{n},F_k^v)\ [\chi^{1+v}(0)]$ in $(\mathfrak{F}_v)$, where $\mathfrak{F}_v = (F_0^v, F_1^v, \ldots)$. Let $\mathfrak{F} = (F_0, F_1, \ldots)$ be the list derived from $\mathfrak{F}_v$ by replacing each variable which does not occur in Δ by Q. Then, $\vdash \Gamma + \bigwedge_n \bigvee_k A(\bar{n},F_k)\ [\chi^{1+v}(0)]$ in $(\mathfrak{F}_v)$, by substitution throughtout the derivation. Let K be the choice predicate for $A(b,P)$ relative to $\mathfrak{F}$. Then $\vdash \overline{\bigwedge_n \bigvee_k A(\bar{n},F_k)} + \bigwedge_n A(\bar{n},K^n)\ [\omega^{1+v} + \omega + 2]$. So by

$\underset{\sim}{C}$, $\vdash \Gamma + \bigwedge_n A(\bar{n},K^n)\ [x,y]$, where $x < \chi^{1+u}(0)$ and $y = \omega^{1+v} + 2$, in $(\mathfrak{F}_v)$. So by the Elimination Theorem for $(\mathfrak{F}_v)$, $\vdash \Gamma + \bigwedge_n A(\bar{n},K^n)\ [z]$ in $(\mathfrak{F}_v)$, where $z \leq \chi^0(\chi^0(\chi^{1+v}(\chi^{1+v}(2\cdot\omega^{1+v} + x)))) < \chi^{1+u}(0)$. So by $\underset{\sim}{\exists}$, $\vdash \Delta\ [\chi^{1+u}(0)]$ in $(\mathfrak{F}_u)$ since K is in the list $\mathfrak{F}_u$. Finally, suppose that the last step is by $\underset{\sim}{C}$. Then, since the derivation is strict, the last step has the form

$$\frac{\Delta + \bigvee ZA(Z) \quad \Delta + \bigwedge \overline{ZA(Z)}}{\Delta}.$$

By reduction and the inductive hypothesis, $\vdash \Delta + \bigvee ZA(Z)\ [\chi^{1+v}(0)]$ and $\vdash \Delta + \overline{A(P)}\ [\chi^{1+v}(0)]$ in $(\mathfrak{F}_v)$ for some $v < u$. So by $\underset{\sim}{\forall}$ and $\underset{\sim}{C}$, $\vdash \Delta\ [x,k]$ in $(\mathfrak{F}_v)$, with $x = \chi^{1+v}(0) + 2$ and $k < \omega$. Hence, by the Elimination Theorem for $(\mathfrak{F}_v)$, $\vdash \Delta\ [z]$ in $(\mathfrak{F}_v)$ - and so, $(\mathfrak{F}_u)$ - where $z \leq 2_k^{\chi^{1+v}(\omega^{1+v}+x)} < \chi^{1+u}(0)$.

q.e.d.

EXAMPLE. Let $\mathcal{K}$ be pure second order Peano arithmetic with $\Sigma_1^1\ \underset{\sim}{AC}$ and the arithmetic comprehension axiom

$$\frac{\Gamma + A(F)}{\Gamma + \bigvee ZA(Z)}$$

added (where $A(P)$ contains no predicate quantifiers). Of course, in $\Sigma_1^1\ \underset{\sim}{AC}$, $\bigwedge_n$ should be interpreted as $\bigwedge x$ and $A(b,P)$ is a formula of (first order) Peano arithmetic. Suppose that all the predicate quantifiers in Δ are existential and $\vdash \Delta$ in $\mathcal{K}$. Then $\vdash \Delta\ [u,v]$ in $\Sigma(\mathfrak{F}_0)$, with $u < \omega^2$ and $v < \omega$. Then by III, $\vdash' \Delta\ [x]$ with $x < \varepsilon_0$. So, $\vdash \Delta\ [\chi^x(0)]$ in $\mathfrak{F}_x$, by IV. Since the principle of induction $\theta(n)$ is Σ_1^1:

$\chi^{\varepsilon_0}(0)$ <u>is a bound on the provable ordinals of</u> $\mathcal{K}$.

This result is proved by H. Friedman in his dissertation (<u>Subsystems of set theory and analysis</u>, M.I.T.) by a translation of $\mathcal{K}$ into ramified analysis. Theorem (6.19) of Feferman [2] seems to show that the bound $\chi^{\varepsilon_0}(0)$ is optimal.

REFERENCES

[1] Barwise, J., Infinitary logic and admissible sets, J. Symbolic Logic, to appear.

[2] Feferman, S., Systems of predicative analysis, Journal of Symbolic Logic, vol. 29 (1964), pp. 1-30.

[3] Gentzen, G., Untersuchungen über das logische Schlussen, Mathematische Zeitschrift, vol. 39 (1935), pp. 176-221.

[4] ________, Die Widerspruchspreiheit der reinen Zahlentheorie, Mathematische Annalen, vol. 112 (1936), pp. 493-565.

[5] ________, Beweisbarkeit und Unbeweisbarkeit von Anfangsfällen der transfiniten Induktion in der reinen Zahlentheorie, Mathematische Annalen, vol. 119 (1943), pp. 140-161.

[6] Karp, C. R., Languages with expressions of infinite length, Amsterdam, 1964.

[7] Lopez-Escobar, E. G. K., An interpolation theorem for denumerably long formulas, Fundamenta Mathematicae, LVII (1965).

[8] Lorenzen, P., Algebraische und logistische Untersuchungen über freie Verbände, Journal of Symbolic Logic, vol. 16 (1951), pp. 81-106.

[9] Schütte, K., Beweistheoretische Erfassung der unendlichen Induktion in der Zahlentheorie, Mathematische Annalen, vol. 122, (1955), pp. 369-389.

[10] ________, Kennzeichnung von Ordnungszahlen durch rekursiv erklärte Funktionen, Mathematische Annalen, vol. 127, (1954), pp. 15-32.

[11] ________, Beweistheorie, Berlin, Gottingen, Heidelberg, 1960.

[12] ________, Predicative well-orderings, Formal Systems and Recursive Functions, Amsterdam 1965, pp. 280-303.

[13] ________, Eine Grenze für die Beweisbarkeit der transfiniten Induktion in der verzweigten Typenlogic, Archiv fur Mathematische Logik und Grundlagenferschung, vol. 7, pp. 45-60.

[14] Tait, W. W., Cut elimination in infinite propositional logic (Abstract), Journal of Symbolic Logic, vol. 31, (1966), p. 151.

UNIVERSITY OF ILLINOIS, CHICAGO CIRCLE

A DETERMINATE LOGIC

GAISI TAKEUTI*

Let L be a fixed language and K be a set of structures related to L. A sentence ψ in L is said to be K-valid if and only if for every structure $\mathfrak{A}$ in K, $\mathfrak{A} \models \psi$. If there exists a logical system S such that for every sentence ψ in L $S \vdash \psi$ is equivalent to "ψ is K-valid," then S said to be a K-logic.

In this paper, we shall consider only the following particular kind of language. Let α_0 and β_0 be ordinals. A language L is said to be of type (α_0,β_0) if and only if the propositional connectives in L consist precisely of those of length less than α_0 and the quantifiers of L are precisely the homogeneous and inhomogeneous quantifiers of length less than β_0. A language of type (α_0,β_0) is sometimes expressed by $L(\alpha_0,\beta_0)$. The sequence $x_0 x_1 \cdots x_\beta \cdots (\beta < \alpha)$ is expressed by $x_0 \cdots \hat{x}_\alpha$. If f is a map from α into $\{\forall, \exists\}$, $Q^f x_0 \cdots \hat{x}_\alpha$ expresses a quantifier of length α. If all the values of f are constantly $\forall$ or $\exists$, then $Q^f x_0 \cdots \hat{x}_\alpha$ is also written $\forall x_0 \cdots \hat{x}_\alpha$ or $\exists x_0 \cdots \hat{x}_\alpha$ respectively and is said to be a homogeneous quantifier. If a quantifier is not homogeneous, it is said to be inhomogeneous. If $\alpha = \omega$ and $f(n) = \forall$ for each even number n and $f(n) = \exists$ for each odd number n, then $Q^f x_0 \cdots \hat{x}_\omega$ is written $\forall x_0 \exists x_1 \forall x_2 \exists x_3 \cdots$. The function $\bar{f}$ defined by the following conditions is said to be the dual of f.

1. The lengths of f and $\bar{f}$ are the same.
2. $\bar{f}(\beta) = \forall$ if and only if $f(\beta) = \exists$.
3. $\bar{f}(\beta) = \exists$ if and only if $f(\beta) = \forall$.

* Work partially supported by National Science Foundation grant GP-6132.

If f and g are dual, then Q^f and Q^g are also said to be <u>dual</u>. The dual quantifier of $\forall x_0 \exists x_1 \forall x_2 \cdots$ is written $\exists x_0 \forall x_1 \exists x_2 \cdots$. κ is always understood to be the least regular infinite cardinal not less than $\max(\alpha_0,\beta_0)$. We always assume that $L(\alpha_0,\beta_0)$ has κ individual variables.

Let $\mathfrak{A}$ be a structure related to $L(\alpha_0,\beta_0)$. $\mathfrak{A}$ is said to be <u>determinate</u> if and only if the following holds for every formula ψ in $L(\alpha_0,\beta_0)$.

Exactly one of the two formulas $Q^f x_0 \cdots \hat{x}_\alpha \psi(x_0,\ldots,\hat{x}_\alpha,a_0,\ldots,\hat{a}_\beta)$, or $Q^{\bar{f}} x_0 \cdots \hat{x}_\alpha \neg\psi(x_0,\ldots,\hat{x}_\alpha,a_0,\ldots,\hat{a}_\beta)$ is satisfied in $\mathfrak{A}$ for every sequence $a_0,\ldots,\hat{a}_\beta$ of members of the universe of $\mathfrak{A}$.

(This schema is called the <u>axiom of determinateness for the quantifier</u> Q^f.) It should be remarked that β may be greater than β_0. If K is the class of all determinate structures, then a K-logic is also said to be a <u>determinate logic</u>. The word "determinate" comes from the axiom of determinateness in [5] and [6]. Roughly speaking, a structure $\mathfrak{A}$ is determinate if there exists a winning strategy for every definable game in $\mathfrak{A}$.

This paper is a continuation of [3]. Only homogeneous quantifiers are considered in [3]. For the system of [3], we proved the completeness theorem, the cut-elimination theorem and the interpolation theorem. However Malitz [4] found a counterexample of our interpolation theorem and later we found an error in case 2 of the proof of our Theorem 5 [3]. In this paper, we shall generalize the system in [3] by introducing inhomogeneous quantifiers and prove that the system so generalized is a determinate logic. Then we shall prove that if a formula ψ is provable in our determinate logic and the inference for an inhomogeneous quantifier is used only once at the end of the proof of ψ, then ψ is valid. (See Theorem 3 in §3 for the precise form.) By using this property, the same method as in [3] proves the following interpolation theorem. Let A and B be formulas without inhomogeneous quantifiers. If $A \to B$ is valid, then there exists a formula C possibly with inhomogeneous quantifiers such that $A \to C$ and $C \to B$ are valid and every predicate constant or individual constant except $=$ in C is contained both in A and B.

1. <u>A logical system</u>.

We use sequents in our logical system. <u>Sequents</u> are of the form $\Gamma \to \Delta$, where Γ, Δ, and Greek capital letters in general are well-ordered sequences of formulas. The length of Γ or Δ may be very big. Let λ be the cardinal number of the set of all formulas in $L(\alpha_0,\beta_0)$. Then the length of Γ or Δ can be any ordinal less than λ^+.

The postulates of our system is the following.

1. Beginning sequents

$D \to D$, where D is an arbitrary formula.

$\to a = a$.

2. Rules of inference

2.1. Structural rule

$$\frac{\Gamma \to \Delta}{\Pi \to \Lambda}$$

where every formula occurring in Γ occurs in Π, and every formula occurring in Δ occurs in Λ.

2.2. Introduction of $\neg$ in the succedent.

$$\frac{\{A_\lambda\}_{\lambda<\gamma},\ \Gamma \to \Delta}{\Gamma \to \Delta,\ \{\neg\ A_\lambda\}_{\lambda<\gamma}},$$

where $\{A_\lambda\}_{\lambda<\gamma}$ stands for the sequence $A_0, A_1, \ldots, \hat{A}_\gamma$.

2.3. Introduction of $\neg$ in the antecedent.

$$\frac{\Gamma \to \Delta,\ \{A_\lambda\}_{\lambda<\gamma}}{\{\neg\ A_\lambda\}_{\lambda<\gamma},\ \Gamma \to \Delta}$$

2.4. Introduction of $\vee$ in the succedent.

$$\frac{\Gamma \to \Delta,\ \{A_{\lambda,\mu}\}_{\mu<\beta_\lambda,\lambda<\gamma}}{\Gamma \to \Delta,\ \{\bigvee_{\mu<\beta_\lambda} A_{\lambda,\mu}\}_{\lambda<\gamma}}$$

2.5. Introduction of $\vee$ in the antecedent.

$$\frac{\{A_{\lambda,\mu_\lambda}\}_{\lambda<\gamma},\ \Gamma \to \Delta \quad \text{for all } \{\mu_\lambda\}_{\lambda<\gamma} \text{ such that } \mu_\lambda < \beta_\lambda\ (\lambda<\gamma)}{\{\bigvee_{\mu<\beta_\lambda} A_{\lambda,\mu}\}_{\lambda<\gamma},\ \Gamma \to \Delta}$$

2.6. Introduction of $\wedge$ in the succedent.

$$\frac{\Gamma \to \Delta, \{A_{\lambda,\mu_\lambda}\}_{\lambda<\gamma} \quad \text{for all } \{\mu_\lambda\}_{\lambda<\gamma} \text{ such that } \mu_\lambda < \beta_\lambda(\lambda < \gamma)}{\Gamma \to \Delta, \{\bigwedge_{\mu<\beta_\lambda} A_{\lambda,\mu}\}_{\lambda<\gamma}}$$

2.7. Introduction of $\wedge$ in the antecedent.

$$\frac{\{A_{\lambda,\mu}\}_{\mu<\beta_\lambda,\lambda<\gamma}, \Gamma \to \Delta}{\{\bigwedge_{\mu<\beta_\lambda} A_{\lambda,\mu}\}_{\lambda<\gamma}, \Gamma \to \Delta}$$

2.8. Introduction of Q in the succedent.

$$\frac{\Gamma \to \Delta, \{A_\lambda(\vec{a}_\lambda)\}_{\lambda<\gamma}}{\Gamma \to \Delta, \{Q^{f_\lambda} \vec{x}_\lambda A_\lambda(\vec{x}_\lambda)\}_{\lambda<\gamma}}$$

Here $\vec{a}_\lambda$ means a sequence $a_{\lambda,0}, a_{\lambda,1}, \ldots, a_{\lambda,\mu}, \ldots$.

If $f_\lambda(\mu) = \forall$, then $a_{\lambda,\mu}$ is said to be an <u>eigenvariable</u> of this inference.

2.9. Introduction of Q in the antecedent.

$$\frac{\{A_\lambda(\vec{a}_\lambda)\}_{\lambda<\gamma}, \Gamma \to \Delta}{\{Q^{f_\lambda} \vec{x}_\lambda A_\lambda(\vec{x}_\lambda)\}_{\lambda<\gamma}, \Gamma \to \Delta}$$

If $f_\lambda(\mu) = \exists$, then $a_{\lambda,\mu}$ is said to be an <u>eigenvariable</u> of this inference.

In both 2.8 and 2.9 we use the following terminology. When an eigenvariable occurs in $\vec{a}_\lambda$, $Q^{f_\lambda} \vec{x}_\lambda A_\lambda(\vec{x}_\lambda)$ in the schema is said to be a <u>principal formula</u> of this eigenvariable and also a <u>principal formula</u> of the inference. $A_\lambda(\vec{a}_\lambda)$ is said to be the <u>side formula</u> of a principal formula $Q^{f_\lambda} \vec{x}_\lambda A_\lambda(\vec{x}_\lambda)$. μ is said to be the <u>order</u> of the variable $a_{\lambda,\mu}$ with respect to a principal formula $Q^{f_\lambda} \vec{x}_\lambda A_\lambda(\vec{x}_\lambda)$. If the orders of two variables a and b w.r.t. a same principal formula are μ and ν and $\mu < \nu$, then a is said to be <u>before</u> b w.r.t. this principal formula.

3. Cut

$$\frac{\Gamma \to \Delta, A_\lambda \quad \text{for all } \lambda < \gamma \quad \{A_\lambda\}_{\lambda<\gamma}, \Pi \to \Lambda}{\Gamma, \Pi \to \Delta, \Lambda}$$

4. Rules for equality

4.1. First rules for equality.

$$\frac{\Gamma^{(\vec{a})} \to \Delta^{(\vec{a})}}{\vec{a} = \vec{b},\ \Gamma^{(\vec{b})} \to \Delta^{(\vec{b})}} \; ; \quad \frac{\Gamma^{(\vec{a})} \to \Delta^{(\vec{a})}}{\vec{b} = \vec{a},\ \Gamma^{(\vec{b})} \to \Delta^{(\vec{b})}} \; ;$$

where $\vec{a} = \vec{b}$ means the sequence $\{a_\lambda = b_\lambda\}_{\lambda<\gamma}$ and $\Gamma^{(\vec{b})} \to \Delta^{(\vec{b})}$ means the result obtained from $\Gamma^{(\vec{a})} \to \Delta^{(\vec{a})}$ by replacing one or more occurrences of a_λ by b_λ for all $\lambda < \gamma$.

4.2. Second rule for equality.

Let Σ be an arbitrary set of variables and $\tilde{\Sigma}$ be a set consisting of all prime formulas of the form $a = b$ such that a and b belong to Σ. $(\Phi \mid \Psi)$ is said to be a <u>decomposition</u> of Σ if and only if $\Phi \cup \Psi = \tilde{\Sigma}$ and $\Phi \cap \Psi = 0$.

$$\frac{\Phi,\ \Gamma \to \Delta,\ \Psi \text{ for all decompositions } (\Phi \mid \Psi) \text{ of } \tilde{\Sigma}}{\Gamma \to \Delta} .$$

Every formal proof must satisfy the following eigenvariable conditions.

5.1. If a free variable occurs in two or more places as eigenvariables, the principal formulas of these eigenvariables are the same formula and the orders of this eigenvariable w.r.t. each principal are the same. If a occurs in two different side formulas $A(\vec{a}_1)$ and $A(\vec{a}_2)$ as an eigenvariable of a principal formula $Q^f \vec{x} A(\vec{x})$ and $a_{1,\mu}$ and $a_{2,\mu}$ are a, then $a_{1,\nu}$ and $a_{2,\nu}$ are the same for all $\nu < \mu$.

5.2. To each free variable a, an ordinal number named the height $h(a)$ of a must be assigned and satisfy the conditions.

5.2.1. The height of an eigenvariable is greater than the height $h(b)$ of every free variable b in the principal formula of the eigenvariable.

5.2.2. The <u>height</u> of an eigenvariable a is greater than the height of b if b is before a w.r.t. a principal formula of an eigenvariable a.

5.3. Each variable occurring in an inference as an eigenvariable must not occur in the end sequent.

REMARK. The following weaker modification of eigenvariable conditions is enough to get a determinate logic. Replace the last half of 5.1 by the following. If $A(\vec{a})$ is a side formula of a principal formula $Q^f \vec{x} A(\vec{x})$ and a_ν and a_μ are eigenvariables of $Q^f \vec{x} A(\vec{x})$ and $\nu \neq \mu$, then a_ν and a_μ are different. If a occurs in two different side formulas $A(\vec{a}_1)$ and $A(\vec{a}_2)$ as an eigenvariable of a principal formula $Q^f \vec{x} A(\vec{x})$ and $a_{1,\mu}$ and $a_{2,\mu}$ are a, then $a_{1,\nu}$ and $a_{2,\nu}$ are the same for any noneigenvariable $a_{1,\nu}$ of $Q^f \vec{x} A(\vec{x})$, for each $\nu < \mu$.

5.2.2 can be replaced by the following 5.2.2'.

5.2.2'. The height of an eigenvariable a is greater than the height of b if a is an eigenvariable of a principal formula and b is before a w.r.t. this principal formula but b is not an eigenvariable of this principal formula.

2. Examples of cut-free formal proofs.

(1) If $\Gamma \to \Delta$ has no inhomogeneous quantifiers and is valid, then there exists a cut-free proof of $\Gamma \to \Delta$ because a completeness theorem and a cut-elimination theorem can be proved for such a sequent. As one of the simplest cases of this kind, we shall show a cut-free proof of the axiom of dependent choice.

$$\dfrac{\dfrac{\dfrac{\dfrac{\dfrac{F(a_n,a_{n+1}) \to F(a_n,a_{n+1})}{\{F(a_m,a_{m+1})\}_{m<\omega} \to F(a_n,a_{n+1}) \quad \text{for every } n<\omega}}{\{F(a_m,a_{m+1})\}_{m<\omega} \to \bigwedge_{n<\omega} F(a_n,a_{n+1})}}{\{F(a_m,a_{m+1})\}_{m<\omega} \to \forall x_0 \exists x_1 x_2 \cdots \bigwedge_{n<\omega} F(x_n,x_{n+1})}}{\{\forall x \exists y F(x,y)\}_{m<\omega} \to \forall x_0 \exists x_1 x_2 \cdots \bigwedge_{n<\omega} F(x_n,x_{n+1})}}{\forall x \exists y F(x,y) \to \forall x_0 \exists x_1 x_2 \cdots \bigwedge_{n<\omega} F(x_n,x_{n+1})}$$

In this proof, $h(a_m)$ is defined to be m for each $m < \omega$.

(2) A proof of the axiom of determinateness

Let $\vec{a}$ be $\{a_\lambda\}_{\lambda<\alpha}$ and $\vec{b}$ be $\{b_\mu\}_{\mu<\beta}$.

$$\dfrac{\dfrac{A(\vec{a},\vec{b}) \to A(\vec{a},\vec{b})}{\to A(\vec{a},\vec{b}), \neg A(\vec{a},\vec{b})}}{\to Q^f \vec{x} A(\vec{x},\vec{b}), Q^{\bar{f}} \vec{x} \neg A(\vec{x},\vec{b})}$$

In this proof, $h(a_\lambda) = 1 + \lambda$ and $h(b_\mu) = 0$.

(3) Malitz's example.

Malitz found a counterexample to the interpolation theorem in homogeneous infinitary logic. His example is the following. Let A and B be two well-ordered sets with the same order type. If F and G are two order preserving one-to-one map from A onto B, then F and G are the same. Let $Ln(=, <)$ be a formula which expresses that $<$ together with $=$ is a linear ordering relation. Let Γ be a sequence of the following formulas.

$$Ln(\overset{1}{=}, \overset{1}{<}), Ln(\overset{2}{=}, \overset{2}{<}),$$

$$\forall x \forall y \forall u \forall v(x \overset{1}{=} y \wedge u \overset{2}{=} v \to (F(x,u) \leftrightarrow F(y,v))),$$

$$\forall x \forall y \forall u \forall v(x \overset{1}{=} y \wedge u \overset{2}{=} v \to (G(x,u) \leftrightarrow G(y,v))),$$

$$\forall x \forall y \forall u \forall v(F(x,u) \wedge F(y,v) \to (x \overset{1}{<} y \leftrightarrow u \overset{2}{<} v) \wedge (x \overset{1}{=} y \leftrightarrow u \overset{2}{=} v))$$

$$\forall x \forall y \forall u \forall v(G(x,u) \wedge G(y,v) \to (x \overset{1}{<} y \leftrightarrow u \overset{2}{<} v) \wedge (x \overset{1}{=} y \leftrightarrow u \overset{2}{=} v))$$

Note that all the quantifiers in Γ are universal at the beginning of a formula. The following sequent is easily proved to be valid.

$$\Gamma, \forall x \exists y F(x,y), \forall x \exists y G(x,y),$$
$$\forall x \exists y F(y,x), \forall x \exists y G(y,x), F(a,b) \to G(a,b), \exists x_0 x_1 \cdots \bigwedge_n (x_{n+1} \overset{1}{<} x_n).$$

We shall obtain a cut-free proof of this sequent. Let T be the set of all finite sequences of 1's and 2's. It is understood that the empty sequence is a member of T. We use τ as a variable ranging over members of T. The set D of free variables is defined as follows.

(1) $a \in D$. (a is a a^τ, where τ is the empty sequence.)

(2) $a^\tau \in D$, then $b^{\tau 1}$ and $b^{\tau 2}$ are members of D.

(3) $b^\tau \in D$, then $a^{\tau 1}$ and $a^{\tau 2}$ are members of D.

(4) The only members of D are those obtained by (1), (2) and (3).

The members of D are $a, b^1, b^2, a^{11}, a^{12}, a^{21}, a^{22}, b^{111}, b^{112}, \ldots$. Γ' is a sequence containing all formulas which are obtained from a formula in Γ by deleting all the universal quantifiers and replacing bound variables by the members of D. (From one formula, infinitely many formulas will be obtained. Of course, in one instance of substitution, the same member of D should be substituted for the same bound variable throughout a formula.) Δ' is a sequence containing all formulas of the forms

$$F(a^{\tau},b^{\tau 1}),\ F(a^{\tau 1},b^{\tau}),\ G(a^{\tau},b^{\tau 2}),\ G(a^{\tau 2},b^{\tau})\quad (\tau \in T)\ .$$

In the following lemmas, we state several sequents which are provable in the ordinary first order predicate calculus and hence cut-free provable in Gentzen's LK.

LEMMA 1. The following are LK-provable.

(1) $\Gamma'\ \Delta' \rightarrow b^{\tau 11} = b^{\tau}$, where $b^{\tau_1} = b^{\tau_2}$ is an abbreviation of $b^{\tau_1} \overset{2}{=} b^{\tau_2}$. In the same way, $a^{\tau_1} = a^{\tau_2}$ is an abbreviation of $a^{\tau_1} \overset{1}{=} a^{\tau_2}$.

(2) $\Gamma',\ \Delta' \rightarrow b^{\tau 22} = b^{\tau}$.

(3) $\Gamma',\ \Delta' \rightarrow a^{\tau 11} = a^{\tau}$.

(4) $\Gamma',\ \Delta' \rightarrow a^{\tau 22} = a^{\tau}$.

Proof. Obviously, $\Gamma',\ F(a^{\tau 1},b^{\tau 11}),\ F(a^{\tau 1},b^{\tau}) \rightarrow b^{\tau 11} = b^{\tau}$ whence follows (1) trivially. The The proof of (2), (3) and (4) are similar.

LEMMA 2. The following are LK-provable.

(1) $\Gamma',\ \Delta',\ b^{\tau} = b^{\tau 12} \rightarrow a^{\tau 1} = a^{\tau 2}$

(2) $\Gamma',\ \Delta',\ a^{\tau} = a^{\tau 12} \rightarrow b^{\tau 1} = b^{\tau 2}$.

Proof. Under the hypotheses of Γ' and Δ', $b^{\tau} = b^{\tau 12}$ implies $a^{\tau 2} = a^{\tau 122}$. Using the previous lemma, we have $a^{\tau 2} = a^{\tau 1}$. The proof of (2) is similar.

LEMMA 3. The following are provable in LK.

(1) $\Gamma',\ \Delta',\ b^{\tau i1} = b^{\tau i2} \rightarrow a^{\tau 1} = a^{\tau 2}\quad (i = 1,2)$.

(2) $\Gamma',\ \Delta',\ a^{\tau i1} = a^{\tau i2} \rightarrow b^{\tau 1} = b^{\tau 2}\quad (i = 1,2)$.

Proof. Under the hypotheses of Γ' and Δ', $b^{\tau 11} = b^{\tau 12} \rightarrow b^{\tau} = b^{\tau 12} \rightarrow a^{\tau 1} = a^{\tau 2}$ (Lemmas 1 and 2). The other cases are similarly proved.

LEMMA 4. The following is provable in LK.

$$\Gamma', \Delta', b^{\tau 1} = b^{\tau 2} \rightarrow b^{1} = b^{2} .$$

Proof. This is easily proved by induction on the length of τ, using Lemma 3.

LEMMA 5. The following are provable in LK.

(1) $\Gamma', \Delta', b^{1} = b^{2} \rightarrow G(a,b^{1})$

(2) $\Gamma', \Delta'\ b^{1} < b^{2} \rightarrow a^{12} < a$ where $b^{\tau_1} < b^{\tau_2}$ and $a^{\tau_1} < a^{\tau_2}$ are abbreviations of $b^{\tau_1} \overset{2}{<} b^{\tau_2}$ and $a^{\tau_1} \overset{1}{<} a^{\tau_2}$ respectively.

(3) $\Gamma', \Delta', b^{2} < b^{1} \rightarrow a^{21} < a$.

Proof. (1) $\Gamma', G(a,b^{2}), b^{1} = b^{2} \rightarrow G(a,b^{1})$

(2) $\Gamma', F(a,b^{1}), b^{1} < b^{2}, G(a,b^{2}), G(a^{12},b^{1}) \rightarrow a^{12} < a$

(3) $\Gamma', F(a,b^{1}), b^{2} < b^{1}, F(a^{21},b^{2}) \rightarrow a^{21} < a$.

LEMMA 6. The following are provable in LK.

(1) $\Gamma', \Delta', b^{\tau 1} = b^{\tau 2} \rightarrow G(a,b^{1})$

(2) $\Gamma', \Delta', b^{\tau 1} < b^{\tau 2} \rightarrow a^{\tau 12} < a^{\tau}$

(3) $\Gamma', \Delta', b^{\tau 2} < b^{\tau 1} \rightarrow a^{\tau 21} < a^{\tau}$.

Proof. The proofs of (2) and (3) are similar to the proof of Lemma 5. (1) follows from Lemma 4 and (1) of Lemma 5.

DEFINITION. $R^{i}(\tau)$ means $b^{\tau 1} = b^{\tau 2}$ if $i = 0$; $b^{\tau 1} < b^{\tau 2}$, if $i = 1$; and $b^{\tau 2} < b^{\tau 1}$, if $i = 2$. T_0 is a set of all members τ in T such that the length of τ is odd.

The following immediately follows from Lemma 6.

LEMMA 7. The following is cut-free provable for each sequence of i_{τ} ($= 0,1,2$) ($\tau \in T_0$).

$$\{R^{i_\tau}(\tau)\}_{\tau\in T_0},\ \Gamma',\ \Delta' \rightarrow \bigwedge_n t_{n+1} \overset{1}{<} t_n,\ G(a,b^1)\ ,$$

where t_n *is a member of* D *whose length is* $2n$.

LEMMA 8. *The following is cut-free probable.*

$$\Gamma,\ \Delta',\ \forall x_0 x_1 \cdots \neg \bigwedge_n (x_{n+1} \overset{1}{<} x_n) \rightarrow G(a,b^1)\ .$$

Proof. This follows from Lemma 7, since $\forall x \forall y(x \overset{2}{<} y \vee x \overset{2}{=} y \vee y \overset{2}{<} x)$ is contained in Γ.

THEOREM. *The following is cut-free provable*.

$$\Gamma,\ \Delta,\ \forall x_0 x_1 \cdots \neg \bigwedge_n (x_{n+1} \overset{1}{<} x_n),\ F(a,b) \rightarrow G(a,b)\ .$$

Proof. Take b to be b^1. Then define $h(a^\tau)$ and $h(b^{\tau'})$ to be the length of τ and the length of τ' respectively. Then the theorem follows from Lemma 8.

3. *Validity of probable formulas*.

First of all, we shall prove the following theorem.

THEOREM 1. *Let* $\mathfrak{A}$ *be a determinate structure and* $\Gamma \rightarrow \Delta$ *be provable in the system in* §1. *Then* $\Gamma \rightarrow \Delta$ *is satisfied in* $\mathfrak{A}$.

Proof. Take an arbitrary formula with a quantifier at the beginning, say

$$Q^f\, \vec{x}\, A(\vec{x},\ \vec{a})\ ,$$

where $\vec{a}$ is the sequence of all free variables in this formula and the length of $\vec{x}$ is α. For each $\gamma < \alpha$, we introduce a Skolem function

$$g_A^{f,\gamma}(x_{\xi_0},\ldots,x_{\xi_\mu},\ldots,\vec{a}) \quad \text{or} \quad \overline{g}_A^{f,\gamma}(x_{\eta_0},\ldots,x_{\eta_\mu},\ldots,\vec{a})$$

according as $f(\gamma) = \exists$ or $f(\gamma) = \forall$, where $\xi_0,\ldots,\xi_\mu,\ldots$ are all ordinals $\xi < \gamma$ satisfying $f(\xi) = \forall$ and $\eta_0,\ldots,\eta_\mu,\ldots$ are all ordinals $\eta < \gamma$ satisfying $f(\eta) = \exists$. We define the following interpretation of $g_A^{f,\gamma}$ and $g_A^{\bar{f},\gamma}$ w.r.t. $\mathfrak{A}$.

If $Q^f\,\vec{x}\,A(\vec{x}, \vec{a})$ is satisfied in $\mathfrak{A}$, then the $g_A^{f,\gamma}$'s are functions satisfying

1.1. $\forall x_{\xi_0} x_{\xi_1} \cdots A(\tilde{x}_0, \ldots, \vec{a})$,

where $\tilde{x}_\gamma$ is x_γ if $f(\gamma) = \forall$ and $\tilde{x}_\gamma$ is $g_A^{f,\gamma}(x_{\xi_0}, \ldots, \vec{a})$ if $f(\gamma) = \exists$. Let D be the universe of $\mathfrak{A}$ and 0 be a member of D. $\vec{a}$ is understood to be a sequence of members of D. If $Q^f\,\vec{x}\,A(\vec{x}, \vec{a})$ is not satisfied in $\mathfrak{A}$, then the $g_A^{f,\gamma}$'s are interpreted to be the constant function 0 in $\mathfrak{A}$.

If $Q^{\bar{f}}\,\vec{x}\,\neg A(\vec{x}, \vec{a})$ is satisfied in $\mathfrak{A}$, then the $g_A^{\bar{f},\gamma}$'s are functions satisfying

1.2. $\forall x_{\eta_0} x_{\eta_1} \cdots \neg A(\tilde{x}_0, \ldots, \vec{a})$,

where $\tilde{x}_\gamma$ is x_γ if $f(\gamma) = \exists$ and $\tilde{x}_\gamma$ is $g_A^{\bar{f},\gamma}(x_{\xi_0}, \ldots, \vec{a})$ if $f(\gamma) = \forall$. If $Q^{\bar{f}}\,\vec{x}\,\neg A(\vec{x}, \vec{a})$ is not satisfied in $\mathfrak{A}$, then the $g_A^{\bar{f},\gamma}$'s are interpreted to be the constant function 0 in $\mathfrak{A}$.

Now let P be a proof-figure in our logical system. Well-order all the eigenvariables in P in such a way that $h(a_\beta) \leq h(a_\gamma)$ if $\beta < \gamma$ and let the well-ordered sequence be $a_0, a_1, \ldots, a_\beta, \ldots$. We shall define $t_0, t_1, \ldots, t_\beta, \ldots$ by transfinite induction on β. Assume that $t_0, \ldots, \hat{t}_\beta$ have been defined: we shall define t_β. Let $Q^f\,\vec{x}\,A(\vec{x}, \vec{b})$ and $A(\vec{d}, \vec{b})$ be the principal formula and a side formula of a_β and the order of a_β w.r.t. this principal formula be γ i.e., let a_β be d_γ. Therefore $u_0, \ldots, \hat{u}_\gamma$ and $s_0, \ldots, \hat{s}_\delta$ have been defined for $d_0, \ldots, \hat{d}_\gamma$ and $b_0, \ldots, \hat{b}_\delta$, where δ is the length of b. (If d_μ or b_ν is not an eigenvariable, then define u_μ or s_ν to be d_μ or b_ν itself respectively.) Then t_β is defined to be $g_A^{f,\gamma}(u_{\xi_0}, \ldots; s_0, \ldots, \hat{s}_\delta)$ or $g_A^{\bar{f},\gamma}(u_{\eta_0}, \ldots; s_0, \ldots, \hat{s}_\delta)$ according as $f(\gamma)$ is $\exists$ or $\forall$. This definition does not depend on the choice of $A(\vec{d}, \vec{b})$ because of 5.1 of §1.

Now substitute $t_0, t_1, \ldots, t_\beta, \ldots$ for $a_0, a_1, \ldots a_\beta, \ldots$ respectively in P. Let P' be the proof-figure thus obtained from P. The end-sequents of P' and P are the same because the end-sequent of P has no eigenvariables. We shall show that every sequent of P' is satisfied in $\mathfrak{A}$. We have only to show that if the upper sequents of an inference in P' are satisfied in $\mathfrak{A}$, then the lower sequent of this inference is also satisfied in $\mathfrak{A}$. Since the other cases are obvious, we only consider the quantifier inferences. Introduction of Q in the antecedent in P' is of the following form

2.1. $$\frac{\ldots, A(\vec{u}, \vec{s}), \ldots, \Gamma \to \Delta}{\ldots, Q^f \vec{x} A(\vec{x}, \vec{s}), \ldots, \Gamma \to \Delta},$$

where u_γ is of the form $g_A^{f,\gamma}(u_{\xi_0}, \ldots, \vec{s})$ if $f(\gamma) = \exists$.

Introduction of Q in the succedent in P' is of the following form

2.2. $$\frac{\Gamma \to \Delta, \ldots, A(\vec{u}', \vec{s}), \ldots}{\Gamma \to \Delta, \ldots, Q^f \vec{x} A(\vec{x}, \vec{s}), \ldots},$$

where u'_γ is of the form $\bar{g}_A^{f,\gamma}(u'_{\eta_0}, \ldots, \vec{s})$, if $f(\gamma) = \forall$. Therefore, what we have to show are

3.1. $Q^f \vec{x} A(\vec{x}, \vec{s}) \to A(\vec{u}, \vec{s})$ for 2.1 and

3.2. $A(\vec{u}', \vec{s}) \to Q^f \vec{x} A(\vec{x}, \vec{s})$ for 2.2.

However 3.1 immediately follows from 1.1. Now we shall consider 3.2. Assume that $\neg Q^f \vec{x} A(\vec{x}, \vec{s})$ holds in $\mathfrak{A}$. Since $\mathfrak{A}$ is determinate, $Q^{\bar{f}} \vec{x} \neg A(\vec{x}, \vec{s})$ holds in $\mathfrak{A}$. Therefore what we have to show is

$$Q^{\bar{f}} \vec{x} \neg A(\vec{x}, \vec{s}) \to \neg A(\vec{u}', \vec{s}),$$

which follows from 1.2.

Since the determinateness of $\mathfrak{A}$ is used only for 3.2 and the axiom of determinateness always holds for a homogeneous quantifier, we have the following theorem.

THEOREM 2. *Let a proof-figure* P *in our determinate logic satisfy the following condition. Every quantifier introduced in the succedent in* P *is homogeneous. Then the end-sequent of* P *is valid.*

We can show a little more. First we shall define a logical system VSS which is a valid subsystem of our determinate logic.

DEFINITION. A proof-figure P in our determinate logic is said to be a proof-figure in VSS if every inference I in P on the introduction of Q in the succedent is homogeneous or satisfies the following condition 4.1.

4.1. I is of the form $\dfrac{\Gamma \to \Delta,\ A(\vec{d})}{\Gamma \to \Delta,\ Q^f\ \vec{x}\ A(\vec{x})}$

and no eigenvariable in P used prior to $\Gamma \to \Delta,\ Q^f\ \vec{x}\ A(\vec{x})$ occurs in $\Gamma \to \Delta,\ Q^f\ \vec{x}\ A(\vec{x})$.

THEOREM 3. <u>If a sequent</u> S <u>is provable in</u> VSS, <u>then</u> S <u>is valid</u>.

<u>Proof</u>. Define $\bar{g}_A^{f,\gamma}$ only for homogeneous f and $g_A^{f,\gamma}$ as in the proof of Theorem 1. Then define substitution also as in the proof of Theorem 1 except that all eigenvariables in the inference of 4.1 remain unsubstituted. Then P will be transformed into P'. What we have to show is that every sequent S' in P' is satisfied in $\mathfrak{A}$. This is shown by transfinite induction on the complexity of the proof ending with S. We can repeat the proof of Theorem 1 except in the following case. S is inferred by the inference I

$$\frac{\Gamma \to \Delta,\ A(\vec{d},\ \vec{b})}{\Gamma \to \Delta,\ Q^f\ \vec{x}\ A(\vec{x},\ \vec{b})},$$

where Q^f is not homogeneous. In order to illustrate the proof, we assume that $Q^f\ \vec{x}$ is $\forall x_0\ \exists x_1\ \forall x_2\ \exists x_3 \cdots$ and $\vec{d}$ is $d_0, d_1, d_2, \ldots$. Since I satisfies 4.1 and $h(d_0) < h(d_1) < h(d_2) < \cdots$, $(\Gamma \to \Delta,\ A(\vec{d},\ \vec{b}))'$ is of the form $\Gamma' \to \Delta',\ A(d_0,\ t_1(d_0,\ \vec{s}),\ d_2,\ t_3(d_0,\ d_2,\ \vec{s}), \ldots, \vec{s})$. It follows from the inductive hypothesis that $\Gamma' \to \Delta',\ A(d_0,\ t_1(d_0,\ \vec{s}),\ d_2,\ t_3(d_0,\ d_2,\ \vec{s}), \ldots, \vec{s})$ is satisfied in $\mathfrak{A}$ for every sequence $d_0, d_2, d_4, \ldots$ of members of $\mathfrak{A}$. Therefore $\Gamma' \to \Delta',\ Q^f\ \vec{x}\ A(\vec{x},\ \vec{s})$ is satisfied in $\mathfrak{A}$.

We shall consider another logical system.

DEFINITION. A figure P is said to be a <u>proof-figure in</u> RHS (restricted homogeneous system) if P satisfies the following conditions

5.1. All quantifiers in P are $\exists$.

5.2. P satisfies all conditions of proof-figure in §1 except 5.1 - 5.3.

5.3. Every inference in P of type 2.9 in §1 is of the following form

$$\frac{\{A_\lambda(\vec{a}_\lambda)\}_{\lambda<\gamma},\ \Gamma \to \Delta}{\{\exists \vec{x}_\lambda\, A_\lambda(\vec{x}_\lambda)\}_{\lambda<\gamma},\ \Gamma \to \Delta}$$

where no $a_{\lambda,\mu}$ occurs in

$$\{\exists \vec{x}_\lambda\, A_\lambda(\vec{x}_\lambda)\}_{\lambda<\gamma},\ \Gamma \to \Delta\ .$$

Then we have the following proposition.

PROPOSITION 1. If $\Gamma \to \Delta$ is provable in RHS and the heights h's are defined for all free variables in $\Gamma \to \Delta$, then there exists a proof-figure P' ending with $\Gamma \to \Delta$ in RHS such that the heights in P' of free variables in $\Gamma \to \Delta$ are the same as h.

Proof. We may assume that the same eigenvariable is never used in two different places. (Otherwise, we can rename some eigenvariables.) Then we define heights of free variables from the bottom up so that the proof-figure in RHS satisfies the conditions 5.1 - 5.3 in §1. Since our proof-figure satisfies 5.3 in the previous definition, this is easily done.

4. A completeness theorem.

First we shall prove the following theorem.

THEOREM 1. Let $\Gamma \to \Delta$ be a sequent. Then there exists a cut-free proof of $\Gamma \to \Delta$ in our determinate logic or else there exists a structure $\mathfrak{A}$ such that every formula in Γ is satisfied in $\mathfrak{A}$ and no formula in Δ is satisfied in $\mathfrak{A}$. (Remark that $\mathfrak{A}$ may not be determinate.)

Proof. Every free variable in Γ or Δ is understood to be an individual constant. Let D_0 be an arbitrary nonempty set containing all individual constants in Γ and Δ. Let D be the closure of D_0 w.r.t. all the functions $g_A^{f,\gamma}$ and $\bar{g}_A^{f,\gamma}$ for all formulas A in our language, i.e., let D be generated by all $g_A^{f,\gamma}$'s and $\bar{g}_A^{f,\gamma}$'s from D_0. (Actually it is enough for D to be closed w.r.t. all the functions $g_A^{f,\gamma}$ and $\bar{g}_A^{f,\gamma}$ for all subformulas A of a formula in Γ or Δ.) In this proof, a member of $D - D_0$ is said to be a free variable and a member of D_0 is said to be an individual constant. Let E be the set of all formulas of the form $s = t$, where s and t are members of D. Let $(\Phi \mid \Psi)$ be an arbitrary decomposition of E and consider the following sequent 1.1.

1.1. $\Phi, \Gamma \to \Delta, \Psi$.

If all the sequents of the form 1.1 are provable without cut, then $\Gamma \to \Delta$ is also provable without cut. If there exists a counterexample for a sequent of the form 1.1, then it is also a counterexample for $\Gamma \to \Delta$.

Let S be $\Gamma \to \Delta$. We shall define the figure P(S) in the following way.

(1) The lowest sequent is S.

(2) The immediate ancestors of S are all sequents of the form 1.1.

(3) When a sequent $\Pi \to \Lambda$ is

$$\{\neg C_\lambda\}_{\lambda<\gamma}, \Gamma' \to \Delta', \{\neg D_\mu\}_{\mu<\delta}$$

where Γ' and Δ' have no formulas whose outermost logical symbol is $\neg$, and $\Pi \to \Lambda$ is constructed by (2) or (8), the immediate ancestor of $\Pi \to \Lambda$ is

$$\{D_\mu\}_{\mu<\delta}, \Gamma' \to \Delta', \{C_\lambda\}_{\lambda<\gamma} .$$

(4) When a sequent $\Pi \to \Lambda$ is

$$\left\{\bigvee_{\lambda<\gamma_\mu} C_{\lambda,\mu}\right\}_{\mu<\gamma}, \Gamma' \to \Delta', \left\{\bigvee_{\rho<\delta_\sigma} D_{\rho,\sigma}\right\}_{\sigma<\delta},$$

where Γ' and Δ' have no formulas whose outermost logical symbol is $\vee$, and when $\Pi \to \Lambda$ is constructed by (3), then the immediate ancestors of $\Pi \to \Lambda$ are

$$\{C_{\lambda_\mu,\mu}\}_{\mu<\gamma}, \Gamma' \to \Delta', \{D_{\rho,\sigma}\}_{\rho<\delta_\sigma,\ \sigma<\delta}$$

for all sequences $\{\lambda_\mu\}_{\mu<\gamma}$ such that $\lambda_\mu < \gamma_\mu$.

(5) When a sequent $\Pi \to \Lambda$ is

$$\left\{\bigwedge_{\lambda<\gamma_\mu} C_{\lambda,\mu}\right\}_{\mu<\gamma}, \Gamma' \to \Delta', \left\{\bigwedge_{\rho<\delta_\sigma} D_{\rho,\sigma}\right\}_{\sigma<\delta},$$

where Γ' and Δ' have no formulas whose outermost logical symbol is $\wedge$, and when $\Pi \to \Lambda$ is constructed by (4), then the immediate ancestors of $\Pi \to \Lambda$ are

$$\{C_{\lambda,\mu}\}_{\lambda < \gamma_\mu, \mu < \gamma},\ \Gamma' \to \Delta',\ \{D_{\rho_\sigma,\sigma}\}_{\sigma < \delta}$$

for all sequences $\{\lambda_\mu\}_{\mu < \gamma}$ such that $\lambda_\mu < \gamma_\mu$.

(6) When a sequent $\Pi \to \Lambda$ is

$$\{Q^{f_\lambda} \vec{x}_\lambda\, A_\lambda(\vec{x}_\lambda, \vec{s}_\lambda)\}_{\lambda < \delta},\ \Gamma' \to \Delta'$$

where Γ' has no formulas whose outermost logical symbol is Q, and when $\Pi \to \Lambda$ is constructed by (5), then the immediate ancestors of $\Pi \to \Lambda$ are

$$\{A_\lambda(\vec{t}_{\lambda,\mu}, \vec{s}_\lambda)\}_{\mu,\lambda < \delta},\ \Gamma' \to \Delta'$$

for all $\vec{t}_{\lambda,\mu}$ satisfying the following.

$\vec{t}_{\lambda,\mu}$ is $\{t_{\lambda,\mu,0},\ldots,t_{\lambda,\mu,\nu},\ldots\}_{\nu < \gamma}$ where γ is the length of $\vec{x}_\lambda$. If $\xi_0, \xi_1, \ldots$ are all the ordinals $\xi < \gamma$ such that $f(\xi) = \forall$ and if $\eta_0, \eta_1, \ldots$ are all the ordinals $\eta < \gamma$ such that $f(\eta) = \exists$, then $t_{\lambda,\mu,\xi_0}, t_{\lambda,\mu,\xi_1}, \ldots$ is an arbitrary sequence of members of D and $t_{\lambda,\mu,\eta} = g^{f,\eta}_{A_\lambda}(t_{\lambda,\mu,\xi_0},\ldots,\vec{s}_\lambda)$. $\vec{t}_{\lambda,\mu}$ runs over all such sequences.

(7) When a sequent $\Pi \to \Lambda$ is

$$\Gamma' \to \Delta',\ \{Q^{f_\lambda} \vec{x}_\lambda\, A_\lambda\, (\vec{x}_\lambda, \vec{s}_\lambda)\}_{\lambda < \delta}$$

where Δ' has no formulas whose outermost logical symbol is Q, and when $\Pi \to \Lambda$ is constructed by (6), then the immediate ancestors of $\Pi \to \Lambda$ are

$$\Gamma' \to \Delta',\ \{A_\lambda(\vec{t}_{\lambda,\mu}, \vec{s}_\lambda)\}_{\mu,\lambda < \delta},$$

for all $\vec{t}_{\lambda,\mu}$ satisfying the following.

$\vec{t}_{\lambda,\mu}$ is $\{t_{\lambda,\mu,0},\ldots,t_{\lambda,\mu,\nu},\ldots\}_{\nu<\gamma}$ where γ is the length of $\vec{x}_\lambda$. If $\xi_0,\xi_1,\ldots$ are all the ordinals $\xi<\gamma$ such that $f(\xi) = \forall$ and if $\eta_0,\eta_1,\ldots$ are all the ordinals $\eta<\gamma$ such that $f(\eta) = \exists$, then t_{λ,μ,η_0}, $t_{\lambda,\mu,\eta_0},\ldots$ are arbitrary members of D and $t_{\lambda,\mu,\xi} = \bar{g}^{f,\xi}_{A_\lambda}(t_{\lambda,\mu,\eta_0},\ldots,\vec{s}_\lambda)$. $\vec{t}_{\lambda,\mu}$ runs over all such sequences.

(8) When a sequent $\Pi \to \Lambda$ is

$$\{s_\lambda = t_\lambda\}_{\lambda<\beta},\ \Gamma' \to \Delta',$$

where Γ' has no formulas of the form $s = t$ and when $\Pi \to \Lambda$ is constructed by (7), then the immediate ancestor of $\Pi \to \Lambda$ is the sequent $\Pi^1 \to \Lambda^1$, where Π^1 and Λ^1 are sequences of all the formulas obtained from a formula in Π and Λ respectively, by arbitrary interchanges of s_μ and $t_\mu (\mu<\beta)$. (So Π^1 and Λ^1 obviously include Π and Λ respectively.)

A <u>branch</u> of P(S) is an infinite sequence $S = S_0,S_1,S_2,\ldots$ such that S_{n+1} is an immediate ancestor of S_n. We have two cases.

<u>Case 1</u>. In every branch of P(S), there exists at least one sequent of the form

$$\Gamma_1,\ D,\ \Gamma_2 \to \Delta_1,\ D,\ \Delta_2 \quad \text{or} \quad \Gamma \to \Delta_1,\ s = s,\ \Delta_2\ .$$

<u>Case 2</u>. There exists at least one branch of P(S), in which there are no sequents of the form

$$\Gamma_1,\ D,\ \Gamma_2 \to \Delta_1,\ D,\ \Delta_2 \quad \text{or} \quad \Gamma \to \Delta_1,\ s = s,\ \Delta_2\ .$$

In Case 1, S is provable without cut. First we define the height of a free variable as follows.

2.1. If $a \in D_0$, then $h(a) = 0$.

2.2. If a is $g^{f,\gamma}_A(b_0,\ldots,b_\xi,\ldots)$ or $\bar{g}^{f,\gamma}_A(b_0,\ldots,b_\xi,\ldots)$, then $h(a)$ is the supremum of all $h(b_\xi)$'s.

It is easily shown that P(S) satisfies the conditions 5.1 and 5.3 in §1. In this proof, a figure P is said to be a semi-proof if and only if P satisfies all the conditions of a proof-figure except 5.1 - 5.3 in §1. P is said to be a quasi-proof if and only if P satisfies all the conditions of a proof-figure except 5.3 in §1. Now consider the following conditions on P.

3.1. P is a cut-free semi-proof.

3.2. Every individual constant or free variable in P occurs in P(S) and every quantifier inference in P occurs in P(S).

If P satisfies 3.1 and 3.2, then P obviously satisfies 5.1 - 5.2 in §1 and therefore P is a cut-free quasi-proof. Now consider the condition C on a sequent S' that S' have a quasi-proof P satisfying 3.1 and 3.2. Let S' be in P(S). It is easily seen that if every ancestor of S' satisfies C, then S' satisfies C. Suppose that S is not provable without cut. Then S does not satisfy C. Then some ancestor of S, say S_1, does not satisfy C. Continuing this argument, we get a sequence $S, S_1, S_2, \dots$, where S_{n+1} is an ancestor of S_n and does not satisfy C for each n. This contradicts the hypothesis of Case 1.

In Case 2, there exists a structure $\mathfrak{A}$ in which every formula in Γ is true and every formula in Δ is false. In the rest of this proof, we fix one branch, whose existence is assumed in the hypothesis of Case 2, and consider only the formulas and sequents in this branch, i.e. 'sequent' always means a sequent in this branch. We have only to define an interpretation which makes all the sequents in this branch false with respect to D.

LEMMA 1. If a formula $\neg A$ occurs in the antecedent (or succedent) of a sequent, then the formula A occurs in the succedent (or antecedent) of a sequent.

LEMMA 2. If a formula $\bigvee_{\lambda<\beta} A_\lambda$ occurs in the antecedent (or succedent) of a sequent, then a formula A_λ for some (or every) $\lambda < \beta$ occurs in the antecedent (or succedent) of a sequent.

LEMMA 3. If a formula $\bigwedge_{\lambda<\beta} A_\lambda$ occurs in the antecedent (or succedent) of a sequent, then a formula A_λ for every (or some) $\lambda < \beta$ occurs in the antecedent (or succedent) of a sequent.

LEMMA 4. If $Q^f \vec{x} A(\vec{x}, \vec{s})$ occurs in the antecedent of a sequent and $\xi_0, \xi_1, \dots$ are all the ordinals ξ such that $f(\xi) = \exists$ and $\eta_0, \eta_1, \dots$ are all the ordinals η such that $f(\eta) = \forall$, then for any arbitrary sequence $t_{\xi_0}, t_{\xi_1}, \dots$ of members of D, the formula $A(\vec{t})$ is in an antecedent of a sequent, where $t_\eta = g_A^{f,\eta}(t_{\xi_0},\ldots,\vec{s})$ for each $\eta = \eta_0, \eta_1, \ldots$.

LEMMA 5. If $Q^f \vec{x} A(\vec{x}, \vec{s})$ occurs in the succedent of a sequent and $\xi_0, \xi_1, \ldots$ are all the ordinals ξ such that $f(\xi) = \forall$ and $\eta_0, \eta_1, \ldots$ are all the ordinals η such that $f(\eta) = \exists$, then for an arbitrary sequence $t_{\eta_0}, t_{\eta_1}, \ldots$ of members of D, the formula $A(\vec{t})$ is in a succedent of a sequent, where $t_\xi = \bar{g}_A^{f,\xi}(t_{\eta_0},\ldots,\vec{s})$ for each $\xi = \xi_0, \xi_1, \ldots$.

These lemmas are obvious.

LEMMA 6. If a formula occurs in the antecedent of a sequent, then it does not occur in a succedent of any sequent.

Proof is by transfinite induction on the complexity of a formula using Lemmas 1 - 5.

LEMMA 7.

(1) For every member t of D, the formula $t = t$ occurs in the antecedent of a sequent.

(2) If s and t are members of D, and $s = t$ occurs in the antecedent of a sequent, then $t = s$ occurs in the antecedent of a sequent.

(3) If t_1, t_2 and t_3 are members of D, and $t_1 = t_2$ and $t_2 = t_3$ occur in the antecedent of a sequent, then the formula $t_1 = t_3$ occurs in an antecedent of a sequent.

(4) Let s_λ, $t_\lambda (\lambda < \beta)$ be members of D. If $A(s_0,\ldots,s_\lambda,\ldots)$ and $\{s_\lambda = t_\lambda\}_{\lambda<\beta}$ occur in the antecedent of a sequent, then $A(u_0,\ldots,u_\lambda,\ldots)$ occurs in the antecedent of a sequent for each sequence $u_0,\ldots,u_\lambda,\ldots$ such that u_λ is s_λ or t_λ.

Proof.

(1) $t = t$ must be contained in Φ or in Ψ in 1.1. Since $t = t$ cannot be contained in Ψ because of the hypothesis of Case 2, $t = t$ must be contained in Φ.

(2) Let $s = t$ occur in the antecedent of a sequent and let $t = s$ occur in the succedent of a sequent; then there is a sequent which contains $s = t$ in the antecedent and $t = s$ in the succedent. By the construction (8) of P(S), there must be a sequent of the form $\Gamma_1 \rightarrow \Delta_1$, $s = s$, Δ_2 which is a contradiction.

(3) and (4) can be proved in a similar way.

According to Lemma 7, D can be decomposed into equivalence-classes by $=$. Let D/= be the set of equivalence-classes so obtained; from now on we denote an element of $D/=$ by a representative of it. We define a structure $\mathfrak{A}$ over $D/=$ as follows. Let s be a variable in D. Then the value of s w.r.t. $\mathfrak{A}$ is defined to be the class represented by s. If P is a predicate constant, then $P(t_0,\ldots,t_\lambda,\ldots)$ is defined to be true w.r.t. $\mathfrak{A}$ if $P(t_0,\ldots,t_\lambda,\ldots)$ is in the antecedent of a sequent and is defined to be false w.r.t. $\mathfrak{A}$ otherwise. By transfinite induction on the complexity of A, we shall prove that A is true w.r.t. $\mathfrak{A}$ if A is in the antecedent of a sequent and A is false w.r.t. $\mathfrak{A}$ if A is in the succedent of a sequent. Since the other cases are easy, we only consider the cases where A is $Q^f\,\vec{x}\,A(\vec{x},\vec{s})$.

4.1. $Q^f\,\vec{x}\,A(\vec{x},\vec{s})$ is in the antecedent of a sequent. In this case, it follows from the inductive hypothesis and (6) of the construction of $P(S)$ that $A(\vec{t},\vec{s})$ is true w.r.t. $\mathfrak{A}$ for every $\vec{t}$ satisfying the following condition: If $\xi_0,\xi_1,\ldots$ are all the ordinals ξ such that $f(\xi)=\forall$ and if $\eta_0,\eta_1,\ldots$ are all the ordinals η such that $f(\eta)=\exists$, then $t_\eta = g_A^{f,\eta}(t_{\xi_0},\ldots,\vec{s})$ for every η. This implies that $Q^f\,\vec{x}\,A(\vec{x},\vec{s})$ is true w.r.t. $\mathfrak{A}$.

4.2. $Q^f\,\vec{x}\,A(\vec{x},\vec{s})$ is in the succedent of a sequent. In this case, it follows from the inductive hypothesis and (7) of the construction of $P(S)$ that $A(\vec{t},\vec{s})$ is false w.r.t. $\mathfrak{A}$ for every $\vec{t}$ satisfying the following condition. If $\xi_0,\xi_1,\ldots$ are all the ordinals ξ such that $f(\xi)=\forall$ and, if $\eta_0,\eta_1,\ldots$ are all the ordinals η such that $f(\eta)=\exists$, then $t_\xi = \bar{g}_A^{f,\xi}(t_{\eta_0},\ldots,\vec{s})$. This implies that $\neg A(\vec{t},\vec{s})$ is true w.r.t. $\mathfrak{A}$ for every such $\vec{t}$. Hence it follows that $Q^{\bar{f}}\,\vec{x}\,\neg A(\vec{x},\vec{s})$ is true w.r.t. $\mathfrak{A}$. Since $Q^{\bar{f}}\,\vec{x}\,\neg A(\vec{x},\vec{s}) \rightarrow \neg Q^f\,\vec{x}\,A(\vec{x},\vec{s})$ is satisfied in all structures, $Q^f\,\vec{x}\,A(\vec{x},\vec{s})$ is false w.r.t. $\mathfrak{A}$.

The following theorem is a completeness theorem for our determinate logic.

THEOREM 2. Let $\Gamma \rightarrow \Delta$ be a sequent. Then $\Gamma \rightarrow \Delta$ is provable in our determinate logic or else there exists a determinate structure $\mathfrak{A}$ such that every formula in Γ is satisfied in $\mathfrak{A}$ and no formula in Δ is satisfied in $\mathfrak{A}$.

Let D and D_0 be the same as in the proof of Theorem 1. Now Γ_0 is defined to be a sequence containing all formulas of the form

$$Q^f\,\vec{x}\,A(\vec{x},\vec{s}) \vee Q^{\bar{f}}\,\vec{x}\,\neg A(\vec{x},\vec{s}),$$

where $A(\vec{x}, \vec{s})$ is an arbitrary formula in our language, $\vec{x}$ and $\vec{s}$ are the only free variables in $A(\vec{x}, \vec{s})$, and $\vec{s}$ is an arbitrary sequence of members of D. $\tilde{\Gamma}$ is defined to be Γ_0, Γ. Without loss of generality, we may assume that a member of D_0 is never used as an eigenvariable in any quasi-proof. Then we have the following theorem.

THEOREM 3. $\tilde{\Gamma} \to \Delta$ <u>has a cut-free quasi-proof whose end-sequent is</u> $\tilde{\Gamma} \to \Delta$, <u>or else there exists a determinate structure</u> $\mathfrak{A}$ <u>such that every formula in</u> $\tilde{\Gamma}$ <u>is satisfied in</u> $\mathfrak{A}$ <u>and no formula in</u> Δ <u>is satisfied in</u> $\mathfrak{A}$.

Theorem 3 implies Theorem 2 as follows. Since every formula in Γ_0 is provable in our determinate logic as in §2, (2), $\Gamma \to \Delta$ is obtained from $\tilde{\Gamma} \to \Delta$ by a cut as follows

$$\frac{\to B_0 \quad \cdots \quad \to B_\beta \quad \cdots; \quad \overset{\vdots\, P}{B_0, \ldots, B_\beta, \ldots, \Gamma \to \Delta}}{\Gamma \to \Delta}$$

where $B_0, \ldots, B_\beta, \ldots$ is Γ_0. It is easily seen that the figure thus obtained satisfies all the properties of a proof-figure including 5.3 in §1.

The proof of Theorem 3 is obtained from the proof of Theorem 1 by replacing a proof-figure and Γ by a quasi-proof and $\tilde{\Gamma}$ respectively. Since $\tilde{\Gamma}$ includes Γ_0, it is easily shown that $\mathfrak{A}$ is determinate.

DEFINITION. HLS (homogeneous logic system) is obtained from our determinate logic by adding a restriction that every quantifier be homogeneous.

Then an argument similar to that of §3 and Theorem 1 in this section yields the following theorem.

THEOREM 4. <u>If</u> $\Gamma \to \Delta$ <u>is provable in</u> HLS, <u>then</u> $\Gamma \to \Delta$ <u>is valid. Conversely, if every quantifier in</u> $\Gamma \to \Delta$ <u>is homogeneous, then</u> $\Gamma \to \Delta$ <u>is provable without cuts in</u> HLS, <u>or else there exists a structure</u> $\mathfrak{A}$ <u>such that every formula in</u> Γ <u>is satisfied in</u> $\mathfrak{A}$ <u>and no formula in</u> Δ <u>is satisfied in</u> $\mathfrak{A}$.

REMARK. We cannot improve Theorem 2 by replacing "provable" by "provable without cuts." In order to see this, let α be the initial ordinal of the cardinality of ${}^{\omega}2$. Let $f \in {}^{\omega}2$. Then $\psi(f)$ is defined to be $a_0 = i_0 \wedge a_1 = i_1 \wedge \cdots$, where $i_k = 0$ or 1 according as $f(k) = 0$ or 1.

The formula $\psi(f)$ expresses the function f. Now let $A \subseteq {}^{\omega}2$. Then A is expressed by the formula $\bigvee_{f\in A} \psi(f)$, where $\bigvee_{f\in A}$ is defined in terms of $\bigvee_{\alpha}$. Now a theorem in [1] implies that there exists a set $A \subseteq {}^{\omega}2$ such that the axiom of determinateness fails for the game defined by A. If a formula ψ expresses A, then

$$\forall x(x = 0 \vee x = 1) \rightarrow 0 = 1,\ \neg\left(\forall x_0 \exists x_1 \forall x_2 \cdots \psi(x_0,x_1,\ldots) \vee \exists x_0 \forall x_1 \exists x_2 \cdots \neg \psi(x_0,x_1,\ldots)\right)$$

is provable in our determinate logic, where ψ is constructed from $0, 1, =, \bigwedge_{\omega}$, and $\bigvee_{\alpha}$. This means that $\forall x(x = 0 \vee x = 1) \rightarrow 0 = 1$ is provable in our determinate logic if our language has $\bigvee_{\alpha}$. However this is not provable without cuts even if our language has $\bigvee_{\alpha}$.

5. An interpolation theorem.

First, we shall define some proof-theoretic notions.

DEFINITION. Let P be a semi-proof without cut and I be an inference in P. Let A be a formula in an upper sequent of I and B be a formula in the lower sequent of I. B is said to be the immediate successor of A if and only if the following is satisfied.

Case 1. I is 2.1 in §1.

If A is a formula in Γ in 2.1, then B is the first formula in Π, which is the same as A. If A is a formula in Δ in 2.1, then B is the first formula in Λ, which is the same as A.

Case 2. I is one of 2.2 - 2.9 or 4.2, and A is a formula in Γ or Δ in the upper sequent of I.

If A is the α-th formula in Γ or Δ, then B is the α-th formula in Γ or Δ in the lower sequent respectively.

Case 3. If I is 2.2 or 2.3 and A is A_λ, then B is $\neg A_\lambda$. If I is 2.4 or 2.5 and A is $A_{\lambda,\mu}$ or A_{λ,μ_λ}, then B is $\bigvee_{\mu<\beta_\lambda} A_{\lambda,\mu}$. If I is 2.6 or 2.7 and A is A_{λ,μ_λ} or $A_{\lambda,\mu}$, then B is $\bigwedge_{\mu<\beta_\lambda} A_{\lambda,\mu}$. If I is 2.8 or 2.9 and A is $A_\lambda(\vec{a}_\lambda)$, then B is $Q^{f_\lambda} \vec{x}_\lambda A_\lambda(\vec{x}_\lambda)$. If I is 4.1 and A is the α-th formula in $\Gamma^{(\vec{a})}$ of $\Delta^{(\vec{a})}$, then B is the α-th formula in $\Gamma^{(\vec{b})}$ or $\Delta^{(\vec{b})}$ respectively.

B is said to be a <u>successor</u> of A, if there exists a sequence $A_0, A_1, \ldots, A_n$ such that $A = A_0$ and $B = A_n$ and A_{i+1} is the immediate successor of A_i for each $i < n$.

Now our interpolation theorem takes the following form.

THEOREM 1. *If a sequent* $\Gamma_1, \Gamma_2 \to \Delta_1, \Delta_2$ *is valid and does not have any inhomogeneous quantifier, then there exists a formula* C *such that both the sequents*

$$\Gamma_1 \to \Delta_1, C \quad \text{and} \quad C, \Gamma_2 \to \Delta_2$$

are valid and every free variable or predicate constant in C, *except* = , *occurs in both* Γ_1, Δ_1 *and* Γ_2, Δ_2. (*Remark that* C *may have inhomogeneous quantifiers and/or longer logical connective or quantifiers than those occurring in* $\Gamma_1, \Gamma_2, \Delta_1, \Delta_2$.)

Our proof follows the proof of Theorem 5 in [3]. First we shall prove the following lemma.

LEMMA 1. *Let* P *be a cut-free proof-figure to* $\Gamma_1, \Gamma_2 \to \Delta_1, \Delta_2$ *in* HLS *and let it satisfy the following conditions* 1.1 - 1.2.

1.1. *Every quantifier in* P *is* $\exists$.

1.2. *Every inference in* P *using quantifier introduction is an inference using introduction of* $\exists$ *in the succedent. Then there exist cut-free proof-figures* P_1 *and* P_2 *in* RHS, *and a formula* C *satisfying the following conditions*.

2.1. *The end-sequent of* P_1 *is* $C, \Gamma_1 \to \Delta_1$ *and the end-sequent of* P_2 *is* $\Gamma_2 \to \Delta_2, C$.

2.2. *Every free variable or predicate constant in* C, *except* = , *occurs in both* Γ_1, Δ_1 *and* Γ_2, Δ_2. (*Remark that* 1.1 *is not an essential restriction on* P *because* $\forall$ *can be expressed by* $\neg$ *and* $\exists$.)

Proof. The proof is done by transfinite induction on the complexity of P.

Case 1. P consists of a single beginning sequent. The theorem is obvious.

Case 2. The last inference of P is of the form

$$\frac{\Gamma_1, \Gamma_2 \to \Delta_1', \{A_\lambda(\vec{a}_\lambda)\}_{\lambda<\beta_1}, \Delta_2', \{B_\mu(\vec{b}_\mu)\}_{\mu<\beta_2}}{\Gamma_1, \Gamma_2 \to \Delta_1', \{\exists \vec{x}_\lambda A_\lambda(\vec{x}_\lambda)\}_{\lambda<\beta_1}, \Delta_2', \{\exists \vec{y}_\mu B_\mu(\vec{y}_\mu)\}_{\mu<\beta_2}}$$

where Δ_1 is $\Delta_1', \{\exists \vec{x}_\lambda A_\lambda(\vec{x}_\lambda)\}_{\lambda<\beta_1}$ and Δ_2 is $\Delta_2' \{\exists \vec{y}_\mu B_\mu(\vec{y}_\mu)\}_{\mu<\beta_2}$.

By the inductive hypothesis, there exists $C'(\vec{a}, \vec{b})$ satisfying the following conditions.

(1) $C'(\vec{a}, \vec{b}),\ \Gamma_1 \to \Delta_1', \{A_\lambda(\vec{a}_\lambda)\}_{\lambda<\beta_1}$ and $\Gamma_2 \to \Delta_2', \{B_\mu(\vec{b}_\mu)\}_{\mu<\beta_2}, C'(\vec{a}, \vec{b})$

are provable in RHS.

(2) Every free variable and predicate constant in $C'(\vec{a}, \vec{b})$ except $=$ is contained in both $\Gamma_1, \Delta_1', \{A_\lambda(\vec{a}_\lambda)\}_{\lambda<\beta_1}$ and $\Gamma_2, \Delta_2', \{B_\mu(\vec{b}_\mu)\}_{\mu<\beta_2}$. $\vec{a}$ is a sequence consisting of all variables in $C'(\vec{a}, \vec{b})$ which are not in Γ_1, Δ_1. $\vec{b}$ is a sequence consisting of all variables in $C'(\vec{a}, \vec{b})$ which are not in Γ_2, Δ_2. Then the required formula C is $\exists \vec{x} \forall \vec{y}\, C'(\vec{x}, \vec{y})$, where $\forall$ is an abbreviation of $\neg \exists \neg$.

Case 3. The last inference of P is of the form

$$\frac{\Gamma_1'^{(\vec{a})}, \Gamma_2'^{(\vec{a})} \to \Delta_1^{(\vec{a})}, \Delta_2^{(\vec{a})}}{\vec{a}_1 = \vec{b}_1, \vec{a}_2 = \vec{b}_2, \Gamma_1'^{(\vec{b})}, \Gamma_2'^{(\vec{b})} \to \Delta_1^{(\vec{b})}, \Delta_2^{(\vec{b})}}$$

where Γ_1 is $\vec{a}_1 = \vec{b}_1, \Gamma_1'^{(\vec{b})}$ and Γ_2 is $\vec{a}_2 = \vec{b}_2, \Gamma_2'^{(\vec{b})}$. This can be divided into two steps, namely: first, the substitution of $\vec{a}_1$ for $\vec{b}_1$; then, the substitution of $\vec{a}_2$ for $\vec{b}_2$. So we may assume that $\vec{a}_1 = \vec{b}_1$ is empty. By the inductive hypothesis, there exists a formula $C'(\vec{a}, \vec{b})$ which satisfies the theorem for $\Gamma_1'^{(\vec{a})}, \Gamma_2'^{(\vec{a})} \to \Delta_1^{(\vec{a})}, \Delta_2^{(\vec{a})}$. $\vec{a}$ is a sequence consisting of all variables in $C'(\vec{a}, \vec{b})$ which are not in Γ_1, Δ_1 and $\vec{b}$ is a sequence consisting of all variables in $C'(\vec{a}, \vec{b})$ which are not in Γ_2, Δ_2. Then take C to be $\exists \vec{x} \forall \vec{y} (\bigwedge_\mu \check{a}_{2,\mu} = \check{b}_{2,\mu} \wedge C'(\vec{x}, \vec{y}))$, where $\check{a}_{2,\mu}$ or $\check{b}_{2,\mu}$ means some x or some y if $a_{2,\mu}$ or $b_{2,\mu}$ are in $\vec{a}$ or $\vec{b}$ and $a_{2,\mu}$ or $b_{2,\mu}$ otherwise.

Case 4. The last inference of P is of the form

$$\frac{\Phi, \Gamma_1, \Gamma_2 \to \Delta_1, \Delta_2, \Psi}{\Gamma_1, \Gamma_2 \to \Delta_1, \Delta_2} \quad \text{for all } (\Phi \mid \Psi).$$

By the inductive hypothesis, there exist formulas $C_{(\Phi|\Psi)}$ such that $C_{(\Phi|\Psi)}, \Gamma_1 \to \Delta_1$

$\Phi, \Gamma_2 \to \Delta_2, \Psi, C_{(\Phi|\Psi)}$ are provable in RHS. So $\bigvee_{(\Phi|\Psi)} C_{(\Phi|\Psi)}, \Gamma_1 \to \Delta_1$ and $\Gamma_2 \to \Delta_2, \bigvee_{(\Phi|\Psi)} C_{(\Phi|\Psi)}$ are provable in RHS. Let $\vec{a}$ be a sequence consisting of all free variables in $\bigvee_{(\Phi|\Psi)} C_{(\Phi|\Psi)}$ which do not appear in Γ_2, Δ_2. We rewrite $\bigvee_{(\Phi|\Psi)} C_{(\Phi|\Psi)}$ as $C'(\vec{a})$. Then take C to be $\forall \vec{x}\, C'(\vec{x})$.

Other cases: The proof is similar.

Now we shall consider the proof of Theorem 1.

Proof of Theorem 1. Since $\Gamma_1, \Gamma_2 \to \Delta_1, \Delta_2$ is valid, there exists a cut-free proof-figure P ending with $\Gamma_1, \Gamma_2 \to \Delta_1, \Delta_2$ in HLS. From §4 it follows that P may be assumed to satisfy the following condition.

3.1. If a variable occurs in two different side formulas as an eigenvariable, then these two formulas are the same. Moreover we can assume the following on P without loss of generality.

3.2. Every quantifier in P is $\exists$.

3.3. The height of a free variable in $\Gamma_1, \Gamma_2 \to \Delta_1, \Delta_2$ is less than the height of any eigenvariable in P.

3.4. The heights of two different variables in P are different. Let $\Gamma_1' \to \Delta_1'$ be a sequent in P. Let $\Phi(\Gamma_1', \Delta_1)$ be the sequence $A_0, A_1, \ldots, A_\mu, \ldots$ of all A_μ's such that A_μ is of the form $\neg \exists \vec{x} A(\vec{x}) \vee A(\vec{a})$ where $\exists \vec{x} A(\vec{x})$ is a principal formula of an introduction of $\exists$ in the antecedent above $\Gamma_1' \to \Delta_1'$ and $A(\vec{a})$ is its side formula. Replacing $\Gamma_1' \to \Delta_1'$ by $\Phi(\Gamma_1', \Delta_1'), \Gamma_1' \to \Delta_1'$ and inserting some appropriate structural inferences, we get a new figure P' satisfying the following conditions.

4.1. P' satisfies 1.1 and 1.2 in Lemma 1.

4.2. The end-sequent of P' is of the following form

$$\{\neg \exists \vec{x}_\lambda A_\lambda(\vec{x}_\lambda, \vec{c}_\lambda) \vee A_\lambda(\vec{a}_\lambda, \vec{c}_\lambda)\}, \Gamma_1, \{\neg \exists \vec{y}_\mu B_\mu(\vec{y}_\mu, \vec{d}_\mu) \vee B_\mu(\vec{b}_\mu, \vec{d}_\mu)\}, \Gamma_2 \to \Delta_1, \Delta_2 .$$

4.3. The height of any $c_{\lambda,\alpha}$ is less than the height of any $a_{\lambda,\beta}$; the height of any $d_{\mu,\alpha}$ is less than the height of any $b_{\mu,\beta}$.

4.4. Every free variable or predicate constant except $=$ occurring in $\exists \vec{z}_\lambda \exists \vec{x}_\lambda A_\lambda(\vec{x}_\lambda, \vec{z}_\lambda)$ occurs in Γ_1 or in Δ_1 and every free variable or predicate constant except $=$ occuring in $\exists \vec{z}_\mu \exists \vec{y}_\mu B_\mu(\vec{y}_\mu, \vec{z}_\mu)$ occurs in Γ_2 or in Δ_2.

4.5. Any $a_{\lambda,\alpha}$ and $b_{\mu,\beta}$ are different. (Otherwise we can modify P' so that P' satisfies 4.5, because P satisfies 3.1.)

Applying Lemma 1, we have a formula $C(\vec{a})$ such that the following conditions are satisfied.

5.1. $C(\vec{a}), \{\neg \exists \vec{x}_\lambda A_\lambda(\vec{x}_\lambda, \vec{c}_\lambda) \vee A_\lambda(\vec{a}_\lambda, \vec{c}_\lambda)\}, \Gamma_1 \rightarrow \Delta_1$ and $\{\neg \exists \vec{y}_\mu B_\mu(\vec{y}_\mu, \vec{d}_\mu) \vee B_\mu(\vec{b}_\mu, \vec{d}_\mu)\}, \Gamma_2 \rightarrow \Delta_2, C(\vec{a})$ are provable in RHS and let Q_1 and Q_2 be proof-figures to these sequents in RHS.

5.2. Every free variable or predicate constant except $=$ occurring in $C(\vec{a})$ is in both $\{A_\lambda(\vec{a}_\lambda, \vec{c}_\lambda)\}, \Gamma_1, \Delta_1$ and $\{B_\mu(\vec{b}_\mu, \vec{d}_\mu)\}, \Gamma_2, \Delta_2$.

5.3. $\vec{a}$ is a sequence consisting of all variables in $C(\vec{a})$ which are not in both Γ_1, Δ_1 and Γ_2, Δ_2 and well-ordered according to heights.

Then consider the following figure

$$
\begin{array}{c}
\Big\downarrow Q_1 \\
C(\vec{a}), \{\neg \exists \vec{x}_\lambda A_\lambda(\vec{x}_\lambda, \vec{c}_\lambda) \vee A_\lambda(\vec{a}_\lambda, \vec{c}_\lambda)\}, \Gamma_1 \rightarrow \Delta_1 \\
\hline
C(\vec{a}), \{\exists \vec{x}'_\lambda(\neg \exists \vec{x}_\lambda A_\lambda(\vec{x}_\lambda, \vec{c}_\lambda) \vee A_\lambda(\vec{x}'_\lambda, \vec{c}_\lambda))\}, \Gamma_1 \rightarrow \Delta_1 \\
\hline
C(\vec{a}), \{\forall \vec{z}_\lambda \exists \vec{x}'_\lambda(\neg \exists \vec{x}_\lambda A_\lambda(\vec{x}_\lambda, \vec{z}_\lambda) \vee A_\lambda(\vec{x}'_\lambda, \vec{z}_\lambda))\}, \Gamma_1 \rightarrow \Delta_1 \\
\hline
Q^f \vec{x}\, C(\vec{x}), \{\forall \vec{z}_\lambda \exists \vec{x}'_\lambda(\neg \exists \vec{x}_\lambda A_\lambda(\vec{x}_\lambda, \vec{z}_\lambda) \vee A_\lambda(\vec{x}'_\lambda, \vec{z}_\lambda))\}, \Gamma_1 \rightarrow \Delta_1
\end{array}
$$

where f is defined as follows.

6.1. If a_α is not contained in Γ_1, Δ_1 or Γ_2, Δ_2 and a_α is one of $b_{\mu,\gamma}$, then $f(\alpha) = \exists$.

6.2. If a_α is not contained in Γ_1, Δ_1 or Γ_2, Δ_2 and a_α is one of $a_{\lambda,\gamma}$, then $f(\alpha) = \forall$.

6.3. If a_α is contained in Γ_1, Δ_1 but not in Γ_2, Δ_2, then $f(\alpha) = \forall$.

6.4. If a_α is contained in Γ_2, Δ_2, but not in Γ_1, Δ_1, then $f(\alpha) = \exists$.

6.5. Otherwise $f(\alpha) = \exists$.

The heights of variables in $\vec{a}_\lambda$, $\vec{c}_\lambda$, $C(\vec{a})$, Γ_1, Δ_1 are defined to be the heights in P. The heights of all other variables in Q_1 can be defined adequately according to Proposition 1 in §3 so that the whole proof-figure satisfies 5.1 - 5.3 in §1. This means that $Q^f \vec{x}\, C(\vec{x}), \Gamma_1 \to \Delta_1$ is valid. The validity of $\Gamma_2 \to \Delta_2, Q^f \vec{x}\, C(\vec{x})$ is also easily shown by observing the following proof-figure in VSS.

$$
\begin{array}{l}
\quad\quad\quad \downarrow Q_2 \\
\{\neg \exists \vec{y}_\mu\, B_\mu(\vec{y}_\mu, \vec{d}_\mu) \vee B_\mu(\vec{b}_\mu, \vec{d}_\mu)\}, \Gamma_2 \to \Delta_2, C(\vec{a}) \\
\hline
\{\exists \vec{y}'_\mu(\neg \exists \vec{y}_\mu\, B_\mu(\vec{y}_\mu, \vec{d}_\mu) \vee B_\mu(\vec{y}'_\mu, \vec{d}_\mu))\}, \Gamma_2 \to \Delta_2, C(\vec{a}) \\
\hline
\{\forall \vec{z}_\mu \exists \vec{y}'_\mu(\neg \exists \vec{y}_\mu\, B_\mu(\vec{y}_\mu, \vec{z}_\mu) \vee B_\mu(\vec{y}'_\mu, \vec{z}_\mu))\}, \Gamma_2 \to \Delta_2, C(\vec{a}) \\
\hline
\{\forall \vec{z}_\mu \exists \vec{y}'_\mu(\neg \exists \vec{y}_\mu\, B_\mu(\vec{y}_\mu, \vec{z}_\mu) \vee B_\mu(\vec{y}'_\mu, \vec{z}_\mu))\}, \Gamma_2 \to \Delta_2, Q^f \vec{x}\, C(\vec{x})
\end{array}
$$

Q.E.D.

In the same way, we can prove the following theorem.

THEOREM 2. <u>If every quantifier in</u> $\Gamma_1, \Gamma_2 \to \Delta_1, \Delta_2$ <u>is homogeneous and</u> $\Gamma_1, \Gamma_2 \to \Delta_1, \Delta_2$ <u>is valid and does not have</u> $=$, <u>and if</u> Γ_1, Δ_1 <u>and</u> Γ_2, Δ_2 <u>have at least one predicate constant in common, then there exists a formula</u> C <u>such that both the sequent</u>

$$C, \Gamma_1 \to \Delta_1 \quad \underline{\text{and}} \quad \Gamma_2 \to \Delta_2, C$$

<u>are valid and every free variable or predicate constant in</u> C <u>is contained in both</u> Γ_1, Δ_1 <u>and</u> Γ_2, Δ_2.

REMARK. In Theorems 1 and 2, we may add the condition that the inhomogeneous quantifier in C is only one at the beginning of C. As regards Malitz's Example in §2, we can construct an isomorphism between $\overset{1}{<}$ and $\overset{2}{<}$ by the following formula

$$\forall x_1 \exists y_1 \forall x_2 \exists y_2 \cdots \Big(\bigwedge_i x_i \overset{1}{<} a \to \bigwedge_i y_i \overset{2}{<} \wedge b \bigwedge_{i,j} ((x_i \overset{1}{<} x_j \leftrightarrow y_i \overset{2}{<} y_j) \wedge (x_i \overset{1}{=} x_j \leftrightarrow y_i \overset{2}{=} y_j))\Big)$$

$$\wedge \forall y_1 \exists x_1 \forall y_2 \exists x_2 \cdots \Big(\bigwedge_i y_i \overset{2}{<} b \to \bigwedge_i x_i \overset{1}{<} a \wedge \bigwedge_{ij} ((x_i \overset{1}{<} x_j \leftrightarrow y_i \overset{2}{<} y_j) \wedge (x_i \overset{1}{=} x_j \leftrightarrow y_i \overset{2}{=} y_j))\Big) .$$

The order type of a in $\left(\overset{1}{=}, \overset{1}{<}\right)$ is denoted by $|a|_1$ and the order type of b in $\left(\overset{2}{=}, \overset{2}{<}\right)$ is denoted by $|b|_2$. Then $|a|_1 \leq |b|_2$ is equivalent to

$$\forall x_1 \exists y_1 \forall x_2 \exists y_2 \cdots \Big(\bigwedge_i x_i \overset{1}{<} a \rightarrow \bigwedge_i y_i \overset{2}{<} b \wedge \bigwedge_{ij} (x_i \overset{1}{<} x_j \leftrightarrow y_i \overset{2}{<} y_j) \wedge (x_i \overset{1}{=} x_j \leftrightarrow y_i \overset{2}{=} y_j) \Big).$$

This is easily shown by transfinite induction on $|a|_1$.

REFERENCES

[1] Gale, D. and F. Stewart, Infinite games with perfect information, Annals of Mathematics Study No. 28, Princeton, (1953), 245-266.

[2] Henkin, L., Some remarks on infinitely long formulas, Infinitistic Methods Warszawa (1961), 167-183.

[3] Maehara, S. and G. Takeuti, A formal system of first order predicate calculus with infinitely long expressions, J. Math. Soc. Japan, 13 (1961), 357-370.

[4] Malitz, J., Problems in the Model Theory of Infinite Languages, Ph.D. Thesis, Berkeley, 1966.

[5] Mycielski, J., On the Axiom of Determinateness, Fundamenta Mathematicae, 53 (1964), 205-224.

[6] Mycielski, J. and H. Steinhaus, A mathematical axiom contradicting the axiom of choice, Bull. Pol. Acad., 10 (1962), p. 1.

THE INSTITUTE FOR ADVANCED STUDY

(ω_1,ω) PROPERTIES OF UNIONS OF MODELS

JOSEPH WEINSTEIN

Introduction.

Finitary model theory shows that the sentences preserved under extension are the Σ_1 sentences. Malitz has extended this result to infinitary (i.e. (ω_1,ω)) sentences. Can one similarly extend results on sentences preserved in unions of models? Here (in §2) we note some problems and results bearing on this question. §1 gives preliminaries and §3 gives some details of proofs.

1. Preliminaries. Let L be a language, possibly with equality, predicate letters, and individual constants, but with no function letters. A "formula" is any (ω_1,ω) formula of L having only finitely many free variables. Let F be any set of formulas. $\bigwedge F$ comprises all formulas equivalent with countable conjunctions of formulas of F; for each $n < \omega$ $\forall_n F$ comprises all formulas equivalent with formulas $\forall u\varphi$, where $\varphi \in F$ and u is a string of $< n$ variables; and $\forall F$ is the union of the $\forall_n F$, $n < \omega$. $\bigvee F$, $\exists F$ are defined similarly. ΠF is the least set G satisfying $F \subseteq G$, $\bigwedge G \subseteq G$, $\forall G \subseteq G$, $\bigvee G \subseteq G$. $\Pi' F$ is the least set G satisfying $F \subseteq G, \bigwedge G \subseteq G$, $\forall G \subseteq G$. (Observe that $\Pi' F = \bigwedge\forall F$.) Interchanging $\bigwedge$ with $\bigvee$ and $\forall$ with $\exists$, we define ΣF and $\Sigma' F$ like ΠF and $\Pi' F$. Put $\Sigma_0 = \Pi_0$ = open formulas; $\Sigma_{n+1} = \Sigma\Pi_n$ and $\Pi_{n+1} = \Pi\Sigma_n$ for each $n < \omega$; and $\Pi_2' = \Pi'\Sigma_1$.

An L-structure B is the union of a set S of L-structures if (1) the universe of B is the union of the universes of the members of S; (2) each predicate letter of L is interpreted in B by the union of its interpretations in the members of S; and (3) each constant of L has in each member of S the same interpretation as in B. S is a family if S has a union B with each member of S

The preparation of this paper was sponsored in part by NSF Grant GP-5600.

a subsystem of B.

Let k be a cardinal > 1, S a family with union B. S is a k-family if each subset of the universe of B with power < k is included in some member of S. S is a chain (in fact, a k-chain) if S is totally ordered by $\subseteq$ (with cofinality k).

For any property P of families, a P-sentence is a sentence φ which holds in the union B of a family S of models of φ whenever S has property P. A formula φ is a P-formula if replacement in φ of the free variables by new constants yields a P-sentence.

2. Problems and results.

Given a property P of families we seek to describe the P-formulas "syntactically." Further, given a pair (P,Q) of properties, we wish to learn which P-formulas are Q-formulas.

2.1. ω_1-formulas. Every (non-ω)-chain is an ω_1-family, so every ω_1-formula is a (non-ω)-chain formula, a fortiori is an ω_1-chain formula. Conversely, by a Löwenheim-Skolem argument (3.1), every ω_1-chain formula is an ω_1-formula. Thus, every chain formula is an ω_1-formula.

The set of ω_1-formulas clearly includes Σ_1 and is closed under the operation Π. Hence the set includes Π_2.

Open question. Does Π_2 = the ω_1-formulas?

2.2. Chain formulas and ω-formulas. By a Löwenheim-Skolem argument (3.2), every ω-chain formula is a chain formula.

One may readily verify the following inclusions (the middle inclusion holding because the set of ω-formulas includes Σ_1 and is closed under the operation Π'):

$$\Pi_2 \supseteq \Pi_2' \subseteq (\omega\text{-formulas}) \subseteq (\text{chain formulas}).$$

For finitary formulas, a well-known result of Chang et al. shows that the four displayed sets coincide. However, an infinitary Π_2 sentence need not be a chain sentence (e.g. $\bigvee_{n<\omega} \varphi_n$, where each φ_n is a Π_1 sentence asserting that the universe has < n members). Further, an interesting example (3.3) of Kueker shows that a chain sentence need not be an ω-sentence. Thus, the chain formulas do not coincide with either of Π_2, Π_2'.

Open "metaproblem." Propose a plausible syntactic characterization of the chain formulas.

Open question. Does $\Pi_2' =$ the ω-formulas?

2.3. n-formulas. One may readily verify that for each n, $1 < n < \omega$, the set of n-formulas includes $\forall_n\Sigma_1$. For finitary formulas Keisler has in fact shown that $\forall_n\Sigma_1$ coincides with the n-formulas.

Open question. Does $\forall_n\Sigma_1 =$ the n-formulas?

In partial "yes" to this and the preceding open question, we have:

THEOREM. For every sentence φ and countable structure A: In order that A be the union of an n-family (resp., ω-family) of models of φ it is sufficient (and clearly necessary) that A satisfy all $\forall_n\Sigma_1$ (resp. $\forall\Sigma_1$) consequences of φ.

The argument (3.4) uses two characteristic devices of (ω_1,ω) model theory: "Scott sentences" and the interpolation theorem.

3. Some proofs.

3.1. Let φ be an ω_1-chain sentence. Then φ is an ω_1-sentence. Let S, with union B, be an ω_1-family of models of φ: we must show that φ holds in B. If B is countable then B must belong to S and then φ holds in B, so we may suppose B uncountable. Then, by the hypothesis on S and by the downward Löwenheim-Skolem theorem, for every countable subsystem A of B there are countable subsystems A', A^* of B with $A \subsetneq A' \subseteq A^*$, $A' \in S$, and A^* a φ-elementary subsystem of B. Let A_0 be any countable subsystem of B, and define A_α, $0 < \alpha < \omega_1$, so that for each $\alpha < \omega_1$ $A_{\alpha+1}$ is A_α^* and for each limit $\alpha < \omega_1$ A_α is $\bigcup(A_\beta : \beta < \alpha)$. Let A be $\bigcup(A_\alpha : \alpha < \omega_1)$. Then A is the union of an ω_1-chain of models A_α' of φ, so φ holds in A. A is the union of the φ-elementary subsystems $(A_{\alpha+1} : \alpha < \omega_1)$ of B, hence is a φ-elementary subsystem of B. Therefore φ holds in B.

Q.E.D.

3.2. Suppose φ not a chain sentence. Then φ is not an ω-chain sentence. Let S be a chain of models of φ with union B, with $\neg\varphi$ true in B. Then there is a structure $(A,M,U,\ldots)$, where B is the structure $(U,\ldots)$ and $M = \{(s,x) : x \in s \in S\}$. For each $s \in S$ let $M_s = \{x : (s,x) \in M\}$. There is a sentence ψ which asserts that $\neg\varphi$ holds in $(U,\ldots)$, that φ holds in each subsystem of $(U,\ldots)$ determined by an M_s, and that the sets M_s form a chain with union U. By the downward

Löwenheim-Skolem theorem let $(A',M',U',\ldots')$ be a countable ψ-elementary subsystem of $(A,M,U,\ldots)$; and for each $s \in$ domain M' let $M'_s = \{x : (s,x) \in M'\}$. Then the subsystems of $(U',\ldots')$ determined by the M'_s form an ω-chain of models of φ whose union, $(U',\ldots')$, satisfies $\neg\varphi$. Thus φ is not an ω-chain sentence.

3.3. (Kueker). <u>A chain sentence</u> φ <u>which is not an</u> ω-<u>sentence</u>. One may construct pairwise incomparable subsets U_n, $n < \omega$, if ω such that each finite subset of ω is included in some U_n. Let φ_n be

$$\left(\bigwedge_{m\in U_n} \exists x\, P_m(x)\right) \wedge \left(\bigwedge_{m\notin U_n} \forall x \neg P_m(x)\right).$$

Let φ be $\bigvee_{n<\omega} \varphi_n$. φ is a chain sentence, for if φ holds in each member of a chain the incomparability of the U_n ensures that the same φ_n holds in all members of the chain, and in the union. On the other hand, let B be the structure with universe ω in which $\{m\}$ interprets P_m for each $m < \omega$, and let S comprise all subsystems with universe of the form U_n: then φ holds in all members of the ω-family S, but not in the union B.

3.4. <u>Let</u> φ <u>be a sentence of</u> L, A <u>a countable structure not the union of any</u> n-<u>family</u> (<u>resp</u>., ω-<u>family</u>) <u>of models of</u> φ. <u>Then some</u> $\forall_n\Sigma_1$ (<u>resp</u>., $\forall\Sigma_1$) <u>consequence of</u> φ <u>fails to hold in</u> A. For brevity of notation we illustrate the proof by taking $n = 2$. Then A has an element a such that no subsystem of the structure (A,a) satisfies φ. Obtain the language L' from L by adding a new constant c (to name a). By a theorem of Scott there is an L'-sentence σ' true in (A,a) for which all countable models are isomorphic. Then $\neg\varphi$ holds in all L'-subsystems of countable models of σ'. By a version (of Malitz) of the (ω_1,ω) interpolation theorem, there is a Σ_1 sentence ψ' of L' with $\sigma' \vdash \neg\psi' \vdash \neg\varphi$. Reindex bound variables so that v_0 does not occur in σ' or in ψ', and replace c by v_0 throughout σ' and ψ' to obtain L-formulas σ, ψ. Then $\varphi \vdash \forall v_0\psi$ and $\forall v_0\psi \in \forall_2\Sigma_1$. Now $\forall v_0\psi$ fails to hold in A (because $\forall v_0\psi \vdash \forall v_0 \neg\sigma$, whereas $\exists v_0\sigma$ holds in A).